Differential and Integral Equations

19⁵

DIFFERENTIAL AND INTEGRAL EQUATIONS

Peter J. Collins

Senior Research Fellow, St Edmund Hall, Oxford

OXFORD

UNIVERSITY PRESS

OXFORD
UNIVERSITY PRESS

Great Clarendon Street, Oxford OX2 6DP

Oxford University Press is a department of the University of Oxford.
It furthers the University's objective of excellence in research, scholarship,
and education by publishing worldwide in

Oxford New York

Auckland Cape Town Dar es Salaam Hong Kong Karachi
Kuala Lumpur Madrid Melbourne Mexico City Nairobi
New Delhi Shanghai Taipei Toronto

With offices in

Argentina Austria Brazil Chile Czech Republic France Greece
Guatemala Hungary Italy Japan Poland Portugal Singapore
South Korea Switzerland Thailand Turkey Ukraine Vietnam

Oxford is a registered trade mark of Oxford University Press
in the UK and in certain other countries

Published in the United States
by Oxford University Press Inc., New York

© Peter J. Collins, 2006

British Library Cataloguing in Publication Data
Data available

Library of Congress Cataloging in Publication Data
Data available

Typeset by the author and Dr J.D. Lotay using LaTeX
Printed in Great Britain
on acid-free paper by
Antony Rowe Ltd., Chippenham, Wiltshire

ISBN 0-19-853382-9   978-0-19-853382-5

0-19-929789-4   978-0-19-929789-4 (Pbk)

Preface

It is a truism that differential and integral equations lie at the heart of mathematics, being the inspiration of so many theoretical advances in analysis and applying to a wide range of situations in the natural and social sciences.

In a first-year course on differential equations, it is usual to give techniques for solving some very particular equations. In this 'second course', the theoretical basis underlying first-year work is explored and extended. Much of the discussion involves consideration of the complementary topic of integral equations, which form an important subject of study in their own right. Some basic results on integral equations of major significance are included in later chapters.

The approach here is thoroughly classical, depending only on some results from real variable theory (which are reviewed in Chapter 0) and from linear algebra (which are summarised where needed), and some knowledge of the solution of elementary ordinary differential equations (which is, in any case, treated in the Appendix). However, presentation and proofs have been chosen with a view to useful generalisations in important areas of functional analysis. I hope to satisfy, in both content and presentation, the (as I see it) uncontradictory demands of the 'pure' and the 'applied' mathematician.

Applications of the material included in these notes are myriad. With the limited space available here, a decision has been made to dwell on the understanding of basic ideas and therefore only to include necessarily short discussion of examples from other fields. The resulting gap may be filled by lectures and further reading.

The work here first appeared in the academic year 1986–87, as a course in the Methods of Applied Mathematics, designed for mathematicians, physicists and economists in their junior and senior undergraduate years, at the College of William and Mary in Virginia. It has subsequently appeared, in more extended incarnations, as Oxford lecture notes designed to cover the range of material most appropriate and most useful for modern mathematics courses in English-speaking universities. The author was pleased to discover that it has been found helpful not only to undergraduates who use mathematics, but also to other non-mathematicians (for example, research chemists) as a handy and accessible reference.

Thanks go to the many colleagues and students who, over the years, have taken the trouble to discuss points in the text with me; in particular, Dominic Welsh and Alan Day for their constructive criticism. Most recently, I have become indebted to Jason Lotay for his invaluable help, especially in updating the typesetting and presentation of the text, and to my patient editor at O.U.P.

Oxford, November 2005   P.J.C.

Contents

How to use this book

Differential and integral equations lie at the centre of mathematics, bringing together the best of the abstraction and power of pure mathematics to address some of the most important problems, both classical and modern, in applied mathematics. With this in mind, it is the author's view that *all* mathematicians should have come across the material in this book at some point of an undergraduate career.

That said, it has to be accepted that, in every institution of higher education, different syllabus balances are struck, and the book has been written to allow for many distinct interpretations. I refer not only to the material which can be retained or omitted (and as much to sections of chapters as to whole chapters) but also to the order in which the retained material is presented. For example, one may choose to read some or all of the material in Chapters 13–15 on series solutions, transform methods and phase-plane analysis before Chapters 8–10 on integral equations, or even (by starting with a course devoted to *ordinary* differential equations) before the partial differential equations of Chapters 5–7.

Further, the reader and lecturer have many choices in using individual chapters. The material in Chapter 13 on series solutions is a case in point, and different levels of appreciation of this topic can be obtained by choosing how far down the following list to proceed:

- just the use of power series in a neighbourhood of an ordinary point

- adding only an appreciation of the use of extended power series in a neighbourhood of a regular singular point

- then adding a rigorous treatment of how two independent solutions of a second-order equation may always be found in a neighbourhood of a regular singular point

- learning the elegant and powerful method of Frobenius

- extending to functions of a single complex variable

In more polemical vain, the author wonders how well-prepared the prospective student of, say, quantum mechanics would be, or what understanding of the aspects of the Legendre equation or Bessel's equation a student of applied analysis would have, if much of Chapter 13 is omitted.

Readers may find the following schematic presentation of the book's contents helpful in choosing their own route through the material.

> Basic material on ordinary
> differential equations:
>   Chapters 1–4

Essential material needed for the rest of the book, save section 4.2 on Green's functions, which may be delayed until Chapter 12

Chapters 13 (series solutions), 14 (transform methods) and 15 (phase-plane analysis) may be promoted here. The following blocks of material may also, to a great extent, be read independently.

> Basic material on partial
> differential equations:
>   Chapters 5–7

Those requiring only material on first-order equations may restrict attention to Chapter 5

> Basic material on integral
> equations:
>   Chapters 8–10

Crucial here are the Fredholm Alternative Theorem in Chapter 8 and the Expansion Theorem in Chapter 9

> The Calculus of Variations:
>   Chapter 11

This chapter should be of especial interest to students of geometry and mechanics

A full appreciation of some topics relies on results from a number of chapters. In particular, the Sturm–Liouville equation allows us to synthesize much of the more important material in earlier chapters: Chapter 12 relies, at various points, on sections 4.2, 8.3, 9.5 and 11.8. Discussion of Bessel's equation and the Legendre equation is spread over Chapters 7, 12, 13 and 14; indeed, such an observation gives good reason for studying material selected from a larger number of chapters of the book.

Prerequisites

Some familiarity with basic theory and technique in real analysis, such as is found in first-year university mathematics courses, will make the reading of this book much more straightforward. The same can be said, to a lesser extent, about elementary linear algebra, in particular the solution of simultaneous linear equations. A summary of much of the necessary analysis material involved may be found in Chapter 0, whereas that on linear algebra is introduced where it is required.

The author has assumed that the reader has come across a few standard elementary ordinary differential equations and learnt how simple examples may be solved. This material is, in any case, thoroughly reviewed in the Appendix.

Understanding of some parts of later chapters is aided by familiarity with other topics in analysis; in particular, the theory of complex functions in one variable is required in sections 13.8 and 14.5. Some chapters will mean more to readers who have taken elementary courses in applied mathematics; for example, on the wave motion of a taut string and the use of Fourier series in calculating solutions (see Chapter 7).

However, it is a main purpose of this text to take the reader with rather little knowledge of applied analysis to a reasonably complete understanding of a variety of important and much used mathematics, and to give confidence with techniques for solving a wide range of problems.

References to background material may be found in the Bibliography.

0 Some Preliminaries

A number of elementary results in real analysis will be used continually in the sequel. For ease of reference, we gather together here some that are more important or commonly found. Though we shall be careful, and encourage the reader to be careful, to give appropriate, sufficient conditions by means of justification, we shall not always be concerned with stating the sharpest possible results. In particular, we shall generally assume that a function is continuous when sometimes integrability over a bounded interval would suffice. Further, we shall not be concerned that differentiability is on occasion enough but normally, in these circumstances, assume that a function is continuously differentiable in the sense of the following definition.

Definition 1 If $A \subseteq \mathbb{R}$ and $B \subseteq \mathbb{R}^2$, the function $f : A \to \mathbb{R}$ is *continuously differentiable*, or *smooth*, *on* A if the derivative f' exists and is continuous at every point of A, and the function $g : B \to \mathbb{R}$ is *continuously differentiable on* B if the partial derivatives g_x, g_y exist and are continuous on B.

We now list some propositions, giving each a label in square brackets which we shall use subsequently for the purpose of reference. The notation (a, b) and $[a, b]$ will be used, respectively, for open and closed intervals in $\mathbb{R}$.

[A] If $f : [a, b] \to \mathbb{R}$ is continuous, then f is bounded. Further, if

$$m = \inf\{f(x) : a \leq x \leq b\}, \qquad M = \sup\{f(x) : a \leq x \leq b\},$$

then there exist x and X in $[a, b]$ such that $f(x) = m$ and $f(X) = M$. Similarly, a continuous real-valued function defined on a closed and bounded subset of $\mathbb{R}^2$ is bounded and attains its bounds.

[B] Mean Value Theorem If $f : [a, b] \to \mathbb{R}$ is continuous and f is differentiable on (a, b) then there exists x_0 in (a, b) for which

$$f(b) - f(a) \; = \; (b - a)f'(x_0).$$

[C] Chain Rule If $F : \mathbb{R}^3 \to \mathbb{R}$ and $F = F(x_1, x_2, x_3)$, where $x_i : \mathbb{R} \to \mathbb{R}$ $(i = 1, 2, 3)$ are continuously differentiable, then, for each x_0 in $\mathbb{R}$,

$$\frac{dF}{dx}(x_0) \; = \; \frac{\partial F}{\partial x_1} \cdot \frac{dx_1}{dx}(x_0) + \frac{\partial F}{\partial x_2} \cdot \frac{dx_2}{dx}(x_0) + \frac{\partial F}{\partial x_3} \cdot \frac{dx_3}{dx}(x_0),$$

where $\dfrac{\partial F}{\partial x_i}$ is evaluated at $(x_1(x_0), x_2(x_0), x_3(x_0))$ for $i = 1, 2, 3$.

Suitable modifications of [C], when either F or an x_i is defined only on a subset of the relevant $\mathbb{R}^n$, will be left to the reader.

An integration theory that integrates continuous functions, in the case of $\mathbb{R}$, over bounded intervals and, in the case of $\mathbb{R}^2$, over bounded rectangles will normally suffice, for the purposes of this book.

[D] If f is integrable on $[a, b]$, then $|f|$ is integrable on $[a, b]$ and

$$\left| \int_a^b f(x)\, dx \right| \; \leq \; \int_a^b |f(x)|\, dx$$

[E] Fundamental Theorem of Calculus If $f : [a, b] \to \mathbb{R}$ is continuously differentiable, then

$$\int_a^b f'(x)\, dx \; = \; f(b) - f(a).$$

In respect of double integration, we shall need the following weak form of Fubini's Theorem.

[F] Fubini's Theorem If $f : [a, b] \times [c, d] \to \mathbb{R}$ is continuous, then the repeated integrals

$$\int_c^d \int_a^b f(x, t)\, dx dt, \qquad \int_a^b \int_c^d f(x, t)\, dt dx$$

exist and are equal.

We shall often find the need to differentiate an integral with varying limits of integration. We shall lean on the following result which, if unknown to the reader, should be committed to memory. We again leave modifications made necessary by restriction of the domains of the functions involved to the reader.

[G] If $a : \mathbb{R} \to \mathbb{R}$, $b : \mathbb{R} \to \mathbb{R}$ are continuously differentiable and if $f : \mathbb{R}^2 \to \mathbb{R}$ and $\dfrac{\partial f}{\partial x}$ are continuous, then

$$\frac{d}{dx}\left(\int_{a(x)}^{b(x)} f(x,t)\, dt \right) = b'(x)f(x,b(x)) - a'(x)f(x,a(x)) + \int_{a(x)}^{b(x)} \frac{\partial f}{\partial x}(x,t)\, dt.$$

Exercise 1 Prove [G].
[HINT: Apply the chain rule to

$$F(x_1, x_2, x_3) = \int_{x_1}^{x_2} f(x_3, t)\, dt$$

for appropriate functions x_1, x_2, x_3.]

The other matter for preliminary discussion is uniform convergence of sequences and series of functions on a closed and bounded interval.

Definition 2 (a) A sequence (s_n) of functions $s_n : [a,b] \to \mathbb{R}$ is *uniformly convergent on* $[a,b]$ if there is a function $s : [a,b] \to \mathbb{R}$ for which, given $\varepsilon > 0$, there exists a positive integer N such that

$$|s_n(x) - s(x)| < \varepsilon \quad \text{for all } n \geq N \text{ and all } x \in [a,b].$$

Then, we say that (s_n) *converges uniformly to* s on $[a,b]$.
(b) A series $\sum_{n=1}^{\infty} u_n$ of functions $u_n : [a,b] \to \mathbb{R}$ is *uniformly convergent on* $[a,b]$ if the sequence (s_n) of partial sums $s_n : [a,b] \to \mathbb{R}$, defined by

$$s_n(x) = \sum_{i=1}^{n} u_i(x), \qquad (x \in [a,b])$$

is uniformly convergent on $[a,b]$. When (s_n) converges uniformly to $u : [a,b] \to \mathbb{R}$, we say that $\sum_{n=1}^{\infty} u_n$ *converges uniformly to* u on $[a,b]$.

Perhaps the most useful of tests for uniform convergence of series is the following.

[H] Weierstrass M-test   Suppose that, for each positive integer n, $u_n : [a, b] \to \mathbb{R}$ and M_n is a non-negative real constant for which

$$|u_n(x)| \leq M_n \quad \text{for all } x \in [a, b].$$

If the series $\sum_{n=1}^{\infty} M_n$ is convergent, then the series $\sum_{n=1}^{\infty} u_n$ is uniformly convergent on $[a, b]$.

The last proposition stated in this chapter will be used at crucial points in Chapters 1 and 2.

[I]   (a) If s_n is a sequence of continuous functions $s_n : [a, b] \to \mathbb{R}$, uniformly convergent to $s : [a, b] \to \mathbb{R}$ on $[a, b]$, then

  (i) s is also continuous on $[a, b]$,

  (ii) $\displaystyle\int_a^b s_n$ converges to $\displaystyle\int_a^b s$;

(b) If $\displaystyle\sum_{n=1}^{\infty} u_n$ is a series of continuous functions $u_n : [a, b] \to \mathbb{R}$, uniformly convergent to $u : [a, b] \to \mathbb{R}$ on $[a, b]$, then

  (i) u is also continuous on $[a, b]$,

  (ii) $\displaystyle\sum_{n=1}^{\infty} \int_a^b u_n = \int_a^b u.$

Note There will be a number of occasions in the text in which we shall wish, as in (b)(ii) above, to interchange an infinite sum with an integral. As here, this is usually justified by uniform convergence. However, the reader should note that other methods of justification are often open to the working mathematician.

1 Integral Equations and Picard's Method

1.1 Integral equations and their relationship to differential equations

Four main types of integral equations will appear in this book: their names occur in the table below. Suppose that $f : [a, b] \to \mathbb{R}$ and $K : [a, b]^2 \to \mathbb{R}$ are continuous, and that λ, a, b are constants.

Volterra non-homogeneous $\qquad y(x) \; = \; f(x) \; + \; \displaystyle\int_a^x K(x, t) y(t)\, dt$

Volterra homogeneous $\qquad y(x) \; = \; \displaystyle\int_a^x K(x, t) y(t)\, dt$

Fredholm non-homogeneous $\qquad y(x) \; = \; f(x) \; + \; \lambda \displaystyle\int_a^b K(x, t) y(t)\, dt$

Fredholm homogeneous $\qquad y(x) \; = \; \lambda \displaystyle\int_a^b K(x, t) y(t)\, dt$

where $x \in [a, b]$. Note that the Volterra equation can be considered as a special case of the Fredholm equation when $K(x, t) = 0$ for $t > x$ in $[a, b]$.

We will search for continuous solutions $y = y(x)$ to such equations. On occasion, x may range over a different domain from $[a, b]$; in which case, the domains of f and K will need appropriate modification.

The function $K = K(x, t)$ appearing in all four equations is called the *kernel* of the integral equation. Such a kernel is *symmetric* if $K(x, t) = K(t, x)$, for all $x, t \in [a, b]$.

A value of the constant λ, for which the homogeneous Fredholm equation has a solution $y = y(x)$ which is *not* identically zero on $[a, b]$, is called an *eigenvalue*, or *characteristic value*, of that equation, and such a non-zero solution $y = y(x)$ is called an *eigenfunction*, or *characteristic function*, 'corresponding to the eigenvalue λ'. The analogy with linear algebra is not accidental, as will be apparent in later chapters.

To investigate the relationship between integral and differential equations, we will need the following lemma which will allow us to replace a double integral by a single one.

Lemma 1 (Replacement Lemma) Suppose that $f : [a, b] \to \mathbb{R}$ is continuous. Then

$$\int_a^x \int_a^{x'} f(t) \, dt dx' = \int_a^x (x - t) f(t) \, dt, \qquad (x \in [a, b]).$$

Proof Define $F : [a, b] \to \mathbb{R}$ by

$$F(x) = \int_a^x (x - t) f(t) \, dt, \qquad (x \in [a, b]).$$

As $(x - t)f(t)$ and $\dfrac{\partial}{\partial x}[(x - t)f(t)]$ are continuous for all x and t in $[a, b]$, we can use [G] of Chapter 0 to differentiate F:

$$F'(x) = [(x - t)f(t)]\Big|_{t=x} \frac{d}{dx} x + \int_a^x \frac{\partial}{\partial x}[(x - t)f(t)] \, dt = \int_a^x f(t) \, dt.$$

Since, again by [G] of Chapter 0, $\displaystyle\int_a^x f(t) \, dt$, and hence $\dfrac{dF}{dx}$, are continuous functions of x on $[a, b]$, we may now apply the Fundamental Theorem of Calculus ([E] of Chapter 0) to deduce that

$$F(x') = F(x') - F(a) = \int_a^{x'} F'(x) \, dx = \int_a^{x'} \int_a^x f(t) \, dt dx.$$

Swapping the roles of x and x', we have the result as stated. $\square$

Alternatively, define, for $(t, x') \in [a, x]^2$,

$$
G(t, x') = \begin{cases} f(t) & \text{when } a \le t \le x' \le x, \\ 0 & \text{when } a \le x' \le t \le x. \end{cases}
$$

The function $G = G(t, x')$ is continuous, except on the line given by $t = x'$, and hence integrable. Using Fubini's Theorem ([F] of Chapter 0),

$$
\int_a^x \int_a^{x'} f(t) \, dt dx' = \int_a^x \left(\int_a^x G(t, x') \, dt \right) dx'
$$

$$
= \int_a^x \left(\int_a^x G(t, x') \, dx' \right) dt
$$

$$
= \int_a^x \left(\int_t^x f(t) \, dx' \right) dt
$$

$$
= \int_a^x (x - t) f(t) \, dt. \qquad \square
$$

We now give an example to show how Volterra and Fredholm integral equations can arise from a single differential equation (as we shall see, depending on which sort of conditions are applied at the boundary of the domain of its solution).

Example 1 Consider the differential equation

$$
y'' + \lambda y = g(x), \quad (x \in [0, L]),
$$

where λ is a positive constant and g is continuous on $[0, L]$. (Many readers will already be able to provide a method of solution. However, what we are considering here is equivalent formulations in terms of integral equations.) Integration from 0 to x ($x \in [0, L]$) gives

$$
y'(x) - y'(0) + \lambda \int_0^x y(t) \, dt = \int_0^x g(t) \, dt.
$$

(Note that, as y'' must exist for any solution y, both y and $y'' = g(x) - \lambda y$ are continuous, so that $\int_0^x y''(t)\,dt = y'(x) - y'(0)$ by [E] of Chapter 0.) As $y'(0)$ is a constant, a further integration from 0 to x and use of the Replacement Lemma twice now gives

$$(1) \qquad y(x) - y(0) - xy'(0) + \lambda \int_0^x (x-t)y(t)\,dt \;=\; \int_0^x (x-t)g(t)\,dt.$$

At this point comes the parting of the ways: we consider two ways in which conditions can be applied at the boundary of the domain of a solution.

(i) **Initial conditions** where y and y' are given at the 'initial' point. Suppose here that $y(0) = 0$ and $y'(0) = A$, a given real constant. Then

$$(2) \qquad y(x) \;=\; Ax + \int_0^x (x-t)g(t)\,dt - \lambda \int_0^x (x-t)y(t)\,dt.$$

Thus we have a Volterra non-homogeneous integral equation with, in the notation of the above table,

$$K(x,t) \;=\; \lambda(t-x),$$

$$f(x) \;=\; Ax + \int_0^x (x-t)g(t)\,dt,$$

which becomes homogeneous if and only if A and g satisfy

$$Ax + \int_0^x (x-t)g(t)\,dt \;=\; 0.$$

All equations are valid for x in $[0, L]$.

(ii) **Boundary conditions** where y is given at the end-points of an interval. Suppose here that $y(0) = 0$ and $y(L) = B$, another given constant. Then, putting $x = L$ in (1), we have

$$(3) \qquad y'(0) \;=\; \frac{1}{L}\left(\lambda \int_0^L (L-t)y(t)\,dt - \int_0^L (L-t)g(t)\,dt + B \right).$$

Substituting back into (1) and writing, for appropriate h,

$$\int_0^L h(t)\,dt \;=\; \int_0^x h(t)\,dt + \int_x^L h(t)\,dt,$$

one easily derives (and it is an exercise for the reader to check that)

(4)
$$y(x) = f(x) + \lambda \int_0^L K(x,t)y(t)\,dt \qquad (x \in [0,L])$$

where

$$K(x,t) = \begin{cases} \dfrac{t}{L}(L-x) & \text{when} \quad 0 \le t \le x \le L \\[2ex] \dfrac{x}{L}(L-t) & \text{when} \quad 0 \le x \le t \le L \end{cases}$$

and

$$f(x) = \frac{Bx}{L} - \int_0^L K(x,t)g(t)\,dt.$$

This time we have a non-homogeneous Fredholm equation (which becomes homogeneous when $f = 0$ on $[0,L]$). We will come across this type of kernel again in our discussion of Green's functions: note that the form of $K(x,t)$ 'changes' along the line $x = t$.

It is *important* to notice that, not only can the original differential equation be recovered from the integral equations (2), (4) by differentiation, but that, so can the initial and boundary conditions. Demonstration of these facts is left as exercises. □

Exercise 1 Recover $y'' + \lambda y = g(x)$, $y(0) = 0$, $y'(0) = A$ from (2), using differentiation and [G] of Chapter 0.

Exercise 2 Solve the integral equation

$$y(x) = e^x + 4 \int_0^x (x-t)y(t)\,dt$$

by first converting it to a differential equation with appropriate initial conditions.

Exercise 3 Suppose that p is a continuously differentiable function, nowhere zero on $[a,b]$, and define

$$P(x) = \int_a^x \frac{dt}{p(t)}, \qquad (x \in [a,b]).$$

Show that a solution of the differential equation

$$\frac{d}{dx}(p(x)y') = q(x)y + g(x),$$

(where q and g are continuous functions on $[a, b]$), with initial conditions $y(a) = A$, $y'(a) = B$, satisfies the Volterra integral equation

$$y(x) = f(x) + \int_a^x K(x, t)y(t)\, dt, \qquad (x \in [a, b]),$$

where

$$K(x, t) = q(t)(P(x) - P(t)),$$

and

$$f(x) = A + Bp(a)P(x) + \int_a^x (P(x) - P(t))g(t)\, dt.$$

Deduce that a solution of the equation

$$xy'' - y' - x^2 y = 8x^3,$$

with initial conditions $y(1) = 1$, $y'(1) = 4$, satisfies the equation

$$y(x) = x^4 + \tfrac{1}{2}\int_1^x (x^2 - t^2)y(t)\, dt, \qquad (x \geq 1).$$

Exercise 4 Find all the continuous eigenfunctions and the corresponding eigenvalues of the homogeneous Fredholm equation

$$y(x) = \lambda \int_0^1 K(x, t)y(t)\, dt,$$

where

$$K(x, t) = \begin{cases} x(1 - t) & \text{when} \quad 0 \leq x \leq t \leq 1 \\ t(1 - x) & \text{when} \quad 0 \leq t \leq x \leq 1 \end{cases}$$

by first converting it to a differential equation with appropriate boundary conditions.

Exercise 5 The thrice continuously differentiable real-valued function $y = y(x)$ satisfies the differential equation $y''' = f$ and is subject to the conditions $y(0) = y(1) = y(2) = 0$. By performing three integrations, show that a solution of the equation may be written,

$$y(x) = \int_0^2 L(x, t)f(t)\, dt,$$

for appropriate $L(x, t)$. You should determine such an $L(x, t)$.

1.2 Picard's method

In this section, we shall describe Picard's method, as used in the construction of a unique solution of an integral equation. This involves the construction of a sequence (y_n) of functions and, correspondingly, an infinite series Σu_n, where each u_n is defined by

$$u_n = y_n - y_{n-1}, \qquad (n = 1, 2, \ldots).$$

The Weierstrass M-test ([H] of Chapter 0) is used to show that the series is uniformly convergent to a function u. But notice that the N-th partial sum of the series is

$$\sum_{n=1}^{N} (y_n - y_{n-1}) = y_N - y_0.$$

So, the sequence (y_n) is uniformly convergent to $u + y_0$, which turns out (using the uniform convergence) to be the (unique) solution of the given integral equation.

We shall now put some clothes on this bare model by considering the Volterra integral equation,

$$y(x) = f(x) + \int_a^x K(x,t)y(t)\, dt, \qquad (x \in [a,b])$$

where f is continuous on $[a,b]$ and K and $\dfrac{\partial K}{\partial x}$ are continuous on $[a,b]^2$ (it is actually suffi-

cient for K and $\dfrac{\partial K}{\partial x}$ to be continuous on the triangular region $\{(x,t) : a \leq t \leq x \leq b\}$).

We show that the integral equation has a unique continuous solution.

Inductively, we first define the sequence (y_n) 'by iteration': put

(5) $y_0(x) = f(x), \qquad (x \in [a,b])$

which is continuous by hypothesis, and suppose that y_k $(1 \leq k \leq n-1)$ has been defined as a continuous function on $[a,b]$ by the formula

$$y_k(x) = f(x) + \int_a^x K(x,t)y_{k-1}(t)\, dt.$$

Then for each x in $[a,b]$, $K(x,t)y_{n-1}(t)$ is a continuous, and hence integrable, function of t on $[a,x]$. So, we may define

(6) $y_n(x) = f(x) + \int_a^x K(x,t)y_{n-1}(t)\, dt, \qquad (x \in [a,b]).$

By [G] of Chapter 0, the integral in (6) is a differentiable, and therefore continuous, function of x on $[a, b]$ and thus an inductive definition of the sequence (y_n) of functions continuous on $[a, b]$ via formulas (5) and (6) is complete. Second, we find non-negative constants M_n such that

$$|u_n(x)| \;=\; |y_n(x) - y_{n-1}(x)| \;\leq\; M_n,$$

for all $n \in \mathbb{N}$ and all $x \in [a, b]$, with

$$\sum_{n=1}^{\infty} M_n \text{ convergent.}$$

Again we proceed by induction. We start by noting that, as K and f are continuous functions defined on, respectively, $[a, b]^2$ and $[a, b]$, K and f are bounded (by [A] of Chapter 0). Suppose that

$$|K(x, t)| \;\leq\; L, \;\; |f(x)| \;\leq\; M \;\; \text{for all } x, t \in [a, b],$$

where L, M are non-negative constants. For the first step in the induction, we have

$$|y_1(x) - y_0(x)| \;=\; \left| \int_a^x K(x, t) y_0(t)\, dt \right|$$

$$\leq\; \int_a^x |K(x, t)||y_0(t)|\, dt \qquad \text{(by [D] of Chapter 0)}$$

$$\leq\; LM(x - a)$$

for all x in $[a, b]$. For an inductive hypothesis, we take

(7) $$\qquad |y_{n-1}(x) - y_{n-2}(x)| \;\leq\; L^{n-1} M \frac{(x - a)^{n-1}}{(n - 1)!}, \qquad \text{for all } x \in [a, b],$$

where $n \geq 2$. (The curious reader may wonder how one might *ab initio* strike on such a hypothesis: he is referred to Exercise 6 below.)

Then, again using [D] of Chapter 0,

$$|y_n(x) - y_{n-1}(x)| = \left| \int_a^x K(x,t)\{y_{n-1}(t) - y_{n-2}(t)\} \, dt \right|$$

$$\leq \int_a^x |K(x,t)| |y_{n-1}(t) - y_{n-2}(t)| \, dt$$

$$\leq \int_a^x L.L^{n-1} M \frac{(t-a)^{n-1}}{(n-1)!} \, dt$$

$$= L^n M \frac{(x-a)^n}{n!}$$

for all x in $[a,b]$. One should note that, in the middle of this argument, one substitutes a bound for $|y_{n-1} - y_{n-2}|$ as a function of t and not of x. This is what gives rise to the 'exponential term' $(x-a)^n/n!$ (As will be discovered below, the fixed limits of integration in the analogous Fredholm equation give rise to a term of a geometric series.) Having inductively found bounds for all the $|y_n - y_{n-1}|$ over $[a,b]$ we can now define the non-negative constants M_n as follows:

$$|y_n(x) - y_{n-1}(x)| \leq L^n M \frac{(x-a)^n}{n!} \leq L^n M \frac{(b-a)^n}{n!} \equiv M_n$$

for $n \geq 1$. Now,

$$\sum_{n=1}^{\infty} M_n = M \sum_{n=1}^{\infty} \frac{\{L(b-a)\}^n}{n!} = M(e^{L(b-a)} - 1),$$

the exponential series for e^x being convergent for all values of its argument x. So, all the hypotheses for the application of the Weierstrass M-test ([H] of Chapter 0) are satisfied and we can deduce that

$$\sum_{n=1}^{\infty} (y_n - y_{n-1})$$

is uniformly convergent on $[a,b]$, to $u : [a,b] \to \mathbb{R}$, say. Then, as we showed above in our general discussion, the sequence (y_n) converges uniformly to $y \equiv u + y_0$ on $[a,b]$, which must be *continuous* on $[a, b]$, since every y_n is (use [I](a) of Chapter 0). Hence, given $\varepsilon > 0$, there exists N such that

$$|y(x) - y_n(x)| \ < \ \varepsilon, \quad \text{for all } n \geq N \text{ and all } x \in [a, b].$$

So,

$$|K(x, t)y(t) - K(x, t)y_n(t)| \ \leq \ L\varepsilon, \quad \text{for all } n \geq N \text{ and all } x, t \in [a, b].$$

So, the sequence $(K(x, t)y_n(t))$ converges uniformly, as a function of t, to $K(x, t)y(t)$. Therefore, by [I](a) of Chapter 0,

$$\int_a^x K(x, t)y_n(t) \, dt \quad \text{converges to} \quad \int_a^x K(x, t)y(t) \, dt.$$

Letting n tend to infinity in (6), we have shown the existence of a continuous solution $y = y(x)$ of the given integral equation.

To proceed to a proof of uniqueness of the continuous solution, we suppose that there exists another such solution $Y = Y(x)$. The continuous function $y - Y$ is bounded on $[a, b]$ (by [A] of Chapter 0). Suppose that

$$|y(x) - Y(x)| \ \leq \ P, \quad \text{for all } x \in [a, b],$$

where P is a non-negative constant, Then, as both y and Y satisfy the integral equation,

$$|y(x) - Y(x)| \ = \ \left| \int_a^x K(x, t)(y(t) - Y(t)) \, dt \right|$$

$$\leq \ \int_a^x |K(x, t)||y(t) - Y(t)| \, dt$$

$$\leq \ LP(x - a), \qquad \text{for all } x \in [a, b].$$

Inductively, suppose that

$$|y(x) - Y(x)| \ \leq \ L^{n-1}P\frac{(x - a)^{n-1}}{(n - 1)!}, \qquad \text{for all } x \in [a, b] \text{ and } n \geq 2.$$

Then,

$$|y(x) - Y(x)| = \left| \int_a^x K(x,t)(y(t) - Y(t)) \, dt \right|$$

$$\leq \int_a^x L.L^{n-1} P \frac{(t-a)^{n-1}}{(n-1)!} \, dt$$

$$= L^n P \frac{(x-a)^n}{n!}, \qquad \text{for all } x \in [a,b].$$

So,

$$|y(x) - Y(x)| \leq L^n P \frac{(b-a)^n}{n!}, \qquad \text{for all } x \in [a,b] \text{ and } n = 1, 2, \dots$$

and, since the right hand side of the inequality tends to 0 as $n \to \infty$,

$$y(x) = Y(x), \qquad \text{for all } x \in [a,b]. \qquad \square$$

Digression on existence and uniqueness

The reader may wonder about the need or usefulness of such results as the above which establish the existence and uniqueness of a solution of, in this case, an integral equation. Why will a list of methods for solving particular kinds of equations, as a student is often encouraged to acquire, not suffice? The fact is that a number of equations, quite simple in form, do not possess solutions at all. An existence theorem can ensure that the seeker's search for a solution may not be fruitless. Sometimes one solution of a given equation is easy to find. Then a uniqueness theorem can ensure that the success so far achieved is complete, and no further search is needed.

One further word on the proof of existence theorems is in order here. There is no reason why such a proof should indicate any way in which one can actually find a solution to a given equation, and it often does not. However, many existence proofs do actually provide a recipe for obtaining solutions. The proof above does in fact provide a useful method which the reader should employ in completing Exercise 7 below.

Exercise 6 Calculate bounds for $|y_2(x) - y_1(x)|$ and $|y_3(x) - y_2(x)|$ to convince yourself of the reasonableness of the inductive hypothesis (7) in the above proof.

Exercise 7 Find the Volterra integral equation satisfied by the solution of the differential equation

$$y'' + xy = 1,$$

with initial conditions $y(0) = y'(0) = 0$. Use the above iterative method as applied to this integral equation to show that the first two terms in a convergent series expansion for this solution are

$$\frac{1}{2} x^2 - \frac{1}{40} x^5.$$

Be careful to prove that no other term in the expansion will contain a power of x less than or equal to 5.

Picard's method can also be applied to the solution of the Fredholm equation

$$(8) \qquad\qquad y(x) \;=\; f(x) + \lambda \int_a^b K(x, t) y(t) \, dt,$$

where f is continuous on $[a, b]$ and K continuous on $[a, b]^2$. On this occasion, the iterative procedure

$$(9) \qquad \begin{aligned} y_0(x) &\;=\; f(x) \\ y_n(x) &\;=\; f(x) + \lambda \int_a^b K(x, t) y_{n-1}(t) \, dt, \qquad (n \geq 1) \end{aligned}$$

for all x in $[a, b]$, gives rise (as the reader is asked to check) to the bound-inequalities

$$|y_n(x) - y_{n-1}(x)| \;\leq\; |\lambda|^n L^n M (b - a)^n \;\equiv\; M_n$$

for all $n \geq 1$ and all $x \in [a, b]$, where $|K| \leq L$, $|f| \leq M$, say. The series

$$\sum_{n=1}^{\infty} M_n$$

is geometric with common ratio $|\lambda| L (b - a)$, and so converges if $|\lambda L (b - a)| < 1$, that is, if

$$(10) \qquad\qquad |\lambda| \;<\; \frac{1}{L(b - a)},$$

assuming $L > 0$ and $b > a$, strictly. This additional sufficient condition ensures that a solution to the integral equation exists. The details of the remainder of the proof of existence and uniqueness here parallel those for the Volterra equation, and the reader is asked to supply them and note the differences.

No claim has been made above that the bound given for $|\lambda|$ is the best, that is, the largest, to ensure existence of a solution. We shall return to this point later during our discussion of the Fredholm Alternative.

Exercise 8 Suppose that

$$y(x) = 1 + \int_0^x e^{-t/x} y(t) \, dt$$

for $x > 0$ and that $y(0) = 1$. Show that the sequence $(y_n)_{n \geq 0}$ produced by Picard iteration is given by

$$y_0(x) = 1$$

$$y_n(x) = 1 + \sum_{j=1}^n a_j x^j, \qquad (n \geq 1)$$

for $x \geq 0$, where

$$a_1 = \int_0^1 e^{-s} \, ds$$

and

$$a_n = a_{n-1} \int_0^1 s^{n-1} e^{-s} \, ds \qquad (n \geq 2).$$

Exercise 9 For each of the following Fredholm equations, calculate the sequence $(y_n)_{n \geq 0}$ produced by Picard iteration and the bound on $|\lambda|$ for which the sequence converges to a solution (which should be determined) of the integral equation. Compare this bound with the bound given by the inequality (10) above.

(a) $\quad y(x) = x^2 + \lambda \int_0^1 x^2 t y(t) \, dt \qquad (x \in [0,1])$

(b) $\quad y(x) = \sin x + \lambda \int_0^{2\pi} \sin(x+t) y(t) \, dt \qquad (x \in [0, 2\pi])$

2 Existence and Uniqueness

We begin this chapter by asking the reader to review Picard's method introduced in Chapter 1, with particular reference to its application to Volterra equations. This done, we may fairly speedily reach the essential core of the theory of ordinary differential equations: existence and uniqueness theorems.

To see that this work is essential, we need go no further than consider some very simple problems. For example, the problem of finding a differentiable function $y = y(x)$ satisfying

$$y' = y^2 \quad \text{with} \quad y(0) = 1$$

has the solution $y = (1 - x)^{-1}$ which does not exist at $x = 1$; in fact, it tends to infinity ('blows up') as x tends to 1.

On the other hand, there are (Exercise 1) an infinite number of solutions of

$$y' = 3y^{\frac{2}{3}} \quad \text{with} \quad y(0) = 0$$

of which $y = 0$ identically and $y = x^3$ are the most obvious examples.

Further, most differential equations *cannot* be solved by performing a sequence of integrations, involving only 'elementary functions': polynomials, rational functions, trigonometric functions, exponentials and logarithms. The celebrated equation of Riccati,

$$y' = 1 + xy^2, \quad \text{with} \quad y(0) = 0,$$

is a case in point, amongst the most simple examples. In Exercise 2, the reader is asked to show that the method of proof of our main theorem provides a practical method of seeking a solution of this equation. In general, the theorem provides information about existence and uniqueness without the need for any attempt at integration whatsoever.

2.1 First-order differential equations in a single independent variable

We consider the existence and uniqueness of solutions $y = y(x)$ of the differential equation

$$(1) \qquad\qquad y' \;=\; f(x,y),$$

satisfying

$$(2) \qquad\qquad y(a) \;=\; c,$$

where a is a point in the domain of y and c is (another) constant. In order to achieve our aim, we must place restrictions on the function f:

(a) f is continuous in a region U of the (x,y)-plane which contains the rectangle

$$R \;=\; \{(x,y) : |x - a| \le h,\ |y - c| \le k\}$$

where h and k are positive constants,

(b) f satisfies the following 'Lipschitz condition' for all pairs of points $(x, y_1), (x, y_2)$ of U:

$$|f(x, y_1) - f(x, y_2)| \;\le\; A|y_1 - y_2|,$$

where A is a (fixed) positive constant.

Restriction (a) implies that f must be bounded on R (by [A] of Chapter 0). Letting

$$M \;=\; \sup\{|f(x,y)| : (x,y) \in R\},$$

we add just one further restriction, to ensure, as we shall see, that the functions to be introduced are well defined:

$$(c) \qquad\qquad Mh \;\le\; k.$$

We would also make a remark about the ubiquity of restriction (b). Such a Lipschitz condition must always occur when the partial derivative $\dfrac{\partial f}{\partial y}$ exists as a bounded function on U: if a bound on its modulus is $P > 0$, we can use the Mean Value Theorem of the differential calculus ([B] of Chapter 0), as applied to $f(x, y)$ considered as a function of y alone, to write, for some y_0 between y_1 and y_2,

$$|f(x, y_1) - f(x, y_2)| \;=\; \left|\frac{\partial f}{\partial y}(x, y_0)\right| |y_1 - y_2| \;\le\; P|y_1 - y_2|.$$

Theorem 1 (Cauchy–Picard) When the restrictions (a), (b), (c) are applied, there exists, for $|x - a| \leq h$, a solution to the problem consisting of the differential equation (1) together with the boundary condition (2). The solution is unique amongst functions with graphs lying in U.

Proof We apply Picard's method (see section 1.2) and define a sequence (y_n) of functions $y_n : [a - h, a + h] \to \mathbb{R}$ by the iteration:

$$y_0(x) = c,$$

$$y_n(x) = c + \int_a^x f(t, y_{n-1}(t))\, dt, \qquad (n \geq 1).$$

As f is continuous, $f(t, y_{n-1}(t))$ is a continuous function of t whenever $y_{n-1}(t)$ is. So, as in section 1.2, the iteration defines a sequence (y_n) of continuous functions on $[a-h, a+h]$, *provided that* $f(t, y_{n-1}(t))$ is defined on $[a - h, a + h]$; that is, provided that

$$|y_n(x) - c| \leq k, \quad \text{for all } x \in [a - h, a + h] \text{ and } n = 1, 2, \ldots$$

To see that this is true, we work by induction. Clearly,

$$|y_0(x) - c| \leq k, \quad \text{for each } x \in [a - h, a + h].$$

If $|y_{n-1}(x) - c| \leq k$ for all $x \in [a - h, a + h]$, where $n \geq 1$, then $f(t, y_{n-1}(t))$ is defined on $[a - h, a + h]$ and, for x in this interval,

$$|y_n(x) - c| = \left| \int_a^x f(t, y_{n-1}(t))\, dt \right| \leq M|x - a| \leq Mh \leq k, \quad (n \geq 1)$$

where we have used [D] of Chapter 0. The induction is complete.

What remains of the proof exactly parallels the procedure in section 1.2 and the reader is asked to fill in the details.

We provide next the inductive step of the proof of

$$|y_n(x) - y_{n-1}(x)| \leq \frac{A^{n-1}M}{n!}|x - a|^n, \quad \text{for all } x \in [a - h, a + h] \text{ and all } n \geq 1.$$

Suppose that

$$|y_{n-1}(x) - y_{n-2}(x)| \leq \frac{A^{n-2}M}{(n-1)!}|x - a|^{n-1} \quad \text{for all } x \in [a - h, a + h], \text{ where } n \geq 2.$$

Then, using the Lipschitz condition (b),

$$|y_n(x) - y_{n-1}(x)| = \left| \int_a^x (f(t, y_{n-1}(t)) - f(t, y_{n-2}(t))) \, dt \right|$$

(3)
$$\leq \left| \int_a^x |f(t, y_{n-1}(t)) - f(t, y_{n-2}(t))| \, dt \right|$$

$$\leq \left| \int_a^x A \, |y_{n-1}(t) - y_{n-2}(t)| \, dt \right|$$

$$\leq A \cdot \frac{A^{n-2} M}{(n-1)!} \left| \int_a^x |t - a|^{n-1} \, dt \right|$$

(4)
$$= \frac{A^{n-1} M}{n!} |x - a|^n, \qquad (n \geq 2)$$

for every x in $[a - h, a + h]$.

Note The reader may wonder why we have kept the outer modulus signs in (3) above, after an application of [D] of Chapter 0. The reason is that it is possible for x to be less than a, while remaining in the interval $[a - h, a + h]$. Putting

$$S = f(t, y_{n-1}(t)) - f(t, y_{n-2}(t)), \quad (n \geq 2)$$

[D] is actually being applied as follows when $x < a$:

$$\left| \int_a^x S \, dt \right| = \left| - \int_x^a S \, dt \right| = \left| \int_x^a S \, dt \right| \leq \int_x^a |S| \, dt = - \int_a^x |S| \, dt = \left| \int_a^x |S| \, dt \right|.$$

Similarly, for $x < a$,

$$\left| \int_a^x |t - a|^{n-1} \, dt \right| = \left| - \int_x^a (a - t)^{n-1} \, dt \right| = \frac{(a - x)^n}{n} = \frac{|x - a|^n}{n},$$

establishing (4).

Continuing with the proof and putting, for each $n \geq 1$,

$$M_n = \frac{A^{n-1} M h^n}{n!},$$

we see that we have shown that $|y_n(x) - y_{n-1}(x)| \leq M_n$ for all $n \geq 1$ and all x in $[a - h, a + h]$. However,

$$\sum_{n=1}^{\infty} M_n$$

is a series of constants, converging to

$$\frac{M}{A} \left(e^{Ah} - 1 \right).$$

So, the Weierstrass M-test ([H] of Chapter 0) may again be applied to deduce that the series

$$\sum_{n=1}^{\infty} (y_n - y_{n-1}),$$

converges uniformly on $[a - h, a + h]$. Hence, as in section 1.2, the sequence (y_n) converges uniformly to, say, y on $[a - h, a + h]$. As each y_n is continuous (see above) so is y by [I](a) of Chapter 0. Further, $y_n(t)$ belongs to the closed interval $[c - k, c + k]$ for each n and each $t \in [a - h, a + h]$. Hence, $y(t) \in [c - k, c + k]$ for each $t \in [a - h, a + h]$, and $f(t, y(t))$ is a well-defined continuous function on $[a - h, a + h]$. Using the Lipschitz condition, we see that

$$|f(t, y(t)) - f(t, y_n(t))| \leq A|y(t) - y_n(t)| \qquad (n \geq 0)$$

for each t in $[a - h, a + h]$; so, the sequence $(f(t, y_n(t)))$ converges uniformly to $f(t, y(t))$ on $[a - h, a + h]$. Applying [I](a) of Chapter 0,

$$\int_a^x f(t, y_{n-1}(t)) \, dt \to \int_a^x f(t, y(t)) \, dt$$

as $n \to \infty$. So, letting $n \to \infty$ in the equation

$$y_n(x) = c + \int_a^x f(t, y_{n-1}(t)) \, dt$$

defining our iteration, we obtain

(5) $$y(x) = c + \int_a^x f(t, y(t)) \, dt.$$

Note that $y(a) = c$. As the integrand in the right-hand side is continuous, we may (by [G] of Chapter 0) differentiate with respect to x to obtain

$$y'(x) = f(x, y(x)).$$

Thus $y = y(x)$ satisfies the differential equation (1) together with the condition (2). We have shown that there *exists* a solution to the problem.

The *uniqueness* of the solution again follows the pattern of our work in section 1.2. If $y = Y(x)$ is a second solution of (1) satisfying (2) with graph lying in U, then as $y(x)$ and $Y(x)$ are both continuous functions on the closed and bounded interval $[a-h, a+h]$, there must (by [A] of Chapter 0) be a constant N such that

$$|y(x) - Y(x)| \leq N \quad \text{for all } x \in [a-h, a+h].$$

Integrating $Y'(t) = f(t, Y(t))$ with respect to t, from a to x, we obtain

$$Y(x) = c + \int_a^x f(t, Y(t)) \, dt,$$

since $Y(a) = c$. So, using (5) and the Lipschitz condition made available to us by the graph of $y = Y(x)$ lying in U,

$$|y(x) - Y(x)| = \left| \int_a^x (f(t, y(t)) - f(t, Y(t))) \, dt \right|$$

(6)
$$\leq \left| \int_a^x A \, |y(t) - Y(t)| \, dt \right|$$

$$\leq AN \, |x - a|, \quad \text{for all } t \in [a-h, a+h].$$

We leave it to the reader to show by induction that, for every integer n and every $x \in [a-h, a+h]$,

$$|y(x) - Y(x)| \leq \frac{A^n N}{n!} \, |x - a|^n.$$

As the right-hand side of this inequality may be made arbitrarily small, $y(x) = Y(x)$ for each x in $[a-h, a+h]$. Our solution is thus unique. $\qquad \square$

Note (a) For continuous y, the differential equation (1) together with the condition (2) is equivalent to the integral equation (5).

(b) The analysis is simplified and condition (c) omitted if f is bounded and satisfies the Lipschitz condition in the strip

$$\{(x,y) : a - h \le x \le a + h\}.$$

(c) Notice that if the domain of f were sufficiently large and $\dfrac{\partial f}{\partial y}$ were to exist and be bounded there, then our work in the paragraph prior to the statement of the theorem would allow us to dispense with the graph condition for uniqueness.

Exercise 1 Find *all* the solutions of the differential equation

$$\frac{dy}{dx} = 3y^{2/3}$$

subject to the condition $y(0) = 0$. Which of the above restrictions does $f(x,y) = 3y^{2/3}$ not satisfy and why?

Exercise 2 By applying the method of proof of the above theorem, find the first three (non-zero) terms in the series expansion of the solution to the Riccati equation

$$y' = 1 + xy^2$$

satisfying $y(0) = 0$.

Exercise 3 Consider the initial value problem of finding a solution $y = y(x)$ in some neighbourhood of $x = 0$ to

$$\frac{dy}{dx} = f(x,y), \quad y(0) = c \quad (|x| < L)$$

where $f(x,y)$ is a continuous bounded real-valued function satisfying the Lipschitz condition

$$|f(x,y_1) - f(x,y_2)| \le A|y_1 - y_2| \quad (|x| < L, \text{ all } y_1, y_2)$$

for some positive constant A. For each of the following special cases

(i) $f = xy$, $c = 1$

(ii) $f = xy^2$, $c = 1$

(iii) $f = xy^{\frac{1}{2}}$, $c = 0$

determine if the Lipschitz condition is satisfied, find all the solutions of the problem and specify the region of validity of each solution.

Exercise 4 Show that the problem

$$y' = f(y), \qquad y(0) = 0$$

has an infinite number of solutions $y = y(x)$, for $x \in [0, a]$, if

(i) $$f(y) = \sqrt{1 + y} \quad \text{and} \quad a > 2,$$

or

(ii) $$f(y) = \sqrt{|y^2 - 1|} \quad \text{and} \quad a > \frac{\pi}{2}.$$

[Note that, in case (ii), the function $y = y(x)$ given by

$$y(x) = \begin{cases} \sin x & (0 \leq x < \tfrac{1}{2}\pi) \\ 1 & (\tfrac{1}{2}\pi \leq x \leq a) \end{cases}$$

is one solution of the problem.]

2.2   Two simultaneous equations in a single variable

It should be said at the outset that the methods of this section can be applied directly to the case of any finite number of simultaneous equations. The methods involve a straightforward extension of those employed in the last section and, for this reason, many of the details will be left for the reader to fill in.

We now seek solutions $y = y(x), z = z(x)$ of the simultaneous differential equations

(7) $$y' = f(x, y, z), \qquad z' = g(x, y, z)$$

which satisfy

(8) $$y(a) = c, \qquad z(a) = d,$$

where a is a point in the domains of y and z, c and d are also constants, and where

(d) f and g are continuous in a region V of (x, y, z)-space which contains the cuboid

$$S = \{(x, y, z) : |x - a| \leq h, \; \max(|y - c|, |z - d|) \leq k\}$$

where h, k are non-negative constants,

(e) f and g satisfy the following Lipschitz conditions at all points of V:

$$|f(x, y_1, z_1) - f(x, y_2, z_2)| \leq A \max(|y_1 - y_2|, |z_1 - z_2|)$$
$$|g(x, y_1, z_1) - g(x, y_2, z_2)| \leq B \max(|y_1 - y_2|, |z_1 - z_2|)$$

where A and B are positive constants,

(f) $$\max(M, N) . h \leq k,$$

where $M = \sup\{|f(x,y,z)| : (x,y,z) \in S\}$ and $N = \sup\{|g(x,y,z)| : (x,y,z) \in S\}$.

It is convenient (especially in the n-dimensional extension!) to employ the vector notation

$$\mathbf{y} = (y, z), \quad \mathbf{f} = (f, g), \quad \mathbf{c} = (c, d), \quad \mathbf{A} = (A, B), \quad \mathbf{M} = (M, N).$$

The reader can then easily check that, with use of the 'vector norm',

$$|\mathbf{y}| = \max(|y|, |z|),$$

where $\mathbf{y} = (y, z)$, the above problem reduces to

(7′) $$\mathbf{y}' = \mathbf{f}(x, \mathbf{y}),$$

satisfying

(8′) $$\mathbf{y}(a) = \mathbf{c}$$

where

(d′) $\mathbf{f}$ is continuous in a region V containing

$$S = \{(x, \mathbf{y}) : |x - a| \leq h, \ |\mathbf{y} - \mathbf{c}| \leq k\}$$

(e′) $\mathbf{f}$ satisfies the Lipschitz condition at all points of V:

$$|\mathbf{f}(x, \mathbf{y}_1) - \mathbf{f}(x, \mathbf{y}_2)| \leq |\mathbf{A}||\mathbf{y}_1 - \mathbf{y}_2|,$$

and

(f′) $$|\mathbf{M}|h \leq k.$$

The existence of a unique solution to (7′) subject to (8′) can now be demonstrated by employing the methods of section 2.1 to the iteration

(9)
$$\mathbf{y}_0(x) = \mathbf{c},$$
$$\mathbf{y}_n(x) = \mathbf{c} + \int_a^x \mathbf{f}(t, \mathbf{y}_{n-1}(t)) \, dt, \qquad (n \geq 1)$$

We thus have the following extension of Theorem 1.

Theorem 2 When the restrictions (d), (e), (f) are applied, there exists, for $|x-a| \leq h$, a solution to the problem consisting of the simultaneous differential equations (7) together with the boundary conditions (8). The solution is unique amongst functions with graphs lying in V.

Exercise 5 Consider the problem consisting of the simultaneous differential equations

$$y' = -yz, \quad z' = z^2$$

which satisfy

$$y(0) = 2, \quad z(0) = 3.$$

(i) Use Theorem 2 to prove that there is a unique solution to the problem on an interval containing 0.

(ii) Find the solution to the problem, specifying where this solution exists.

Exercise 6 Consider the problem

$$y' = 2 - yz, \quad z' = y^2 - xz, \quad y(0) = -1, \quad z(0) = 2.$$

Find the first three iterates $\mathbf{y}_0(x)$, $\mathbf{y}_1(x)$, $\mathbf{y}_2(x)$ in the vector iteration (9) corresponding to this problem.

Exercise 7 With the text's notation, prove that, if $f = f(x, y, z) = f(x, \mathbf{y})$ has continuous bounded partial derivatives in V, then f satisfies a Lipschitz condition on S of the form given in (e$'$).
[HINT: Write $f(x, y_1, z_1) - f(x, y_2, z_2) = f(x, y_1, z_1) - f(x, y_2, z_1) + f(x, y_2, z_1) - f(x, y_2, z_2)$ and use the Mean Value Theorem of the differential calculus ([B] of Chapter 0).]

Exercise 8 Compare the method employed in the text for solving (7) subject to (8) with that given by the simultaneous iterations

$$y_0(x) = c,$$

$$y_n(x) = c + \int_a^x f(t, y_{n-1}(t), z_{n-1}(t)) \, dt,$$

and

$$z_0(x) = d,$$

$$z_n(x) = d + \int_a^x g(t, y_{n-1}(t), z_{n-1}(t)) \, dt,$$

for $n \geq 1$ and $x \in [a - h, a + h]$. In particular, find bounds for $|y_1 - y_0|$, $|y_2 - y_1|$ and $|y_3 - y_2|$ on $[a - h, a + h]$ in terms of A, B, M, N and h.

2.3 A second-order equation

We now use Theorem 2 to find a solution $y = y(x)$ to the problem consisting of the differential equation

$$(10) \qquad \frac{d^2y}{dx^2} \equiv y'' = g(x, y, y')$$

together with the initial conditions

$$(11) \qquad y(a) = c, \quad y'(a) = d, \quad (c, d \text{ constants}).$$

(Note that y and y' are both given at the same point a.)

A problem of the type given by (10) taken together with 'initial conditions' (11), when a solution is only required for $x \geq a$, is called an **initial value problem** (IVP) – the variable x can be thought of as time (and customarily is then re-named t).

The trick is to convert (10) to the pair of simultaneous equations

$$(10') \qquad y' = z, \quad z' = g(x, y, z),$$

(the first equation defining the new function z). Corresponding to (11) we have

$$(11') \qquad y(a) = c, \quad z(a) = d.$$

We, of course, require certain restrictions on $g = g(x, y, z)$:

(d″) g is continuous on a region V of (x, y, z)-space which contains

$$S = \{(x, y, z) : |x - a| \leq h, \ \max(|y - c|, |z - d|) \leq k\}$$

where h, k are non-negative constants,

(e″) g satisfies the following Lipschitz condition at all points of V:

$$|g(x, y_1, z_1) - g(x, y_2, z_2)| \leq B \max(|y_1 - y_2|, |z_1 - z_2|),$$

where B is a constant.

Theorem 3 When the restrictions (d″), (e″) are imposed, then, for some $h > 0$, there exists, when $|x - a| \leq h$, a solution to the problem consisting of the second-order differential equation (10) together with the initial conditions (11). The solution is unique amongst functions with graphs lying in V.

The reader should deduce Theorem 3 from Theorem 2. It will be necessary, in particular, to check that $f(x, y, z) = z$ satisfies a Lipschitz condition on V and that an h can be found so that (f) of section 2.2 can be satisfied. We should note that the methods of this section can be extended so as to apply to the nth-order equation

$$y^{(n)} = f(x, y, y', \ldots, y^{(n-1)}),$$

when subject to initial conditions

$$y(a) = c_0, \quad y'(a) = c_1, \ldots, \quad y^{(n-1)}(a) = c_{n-1}.$$

We conclude our present discussion of the second-order equation by considering the special case of the *non-homogeneous linear equation*

$$(12) \qquad\qquad p_2(x)y'' + p_1(x)y' + p_0(x)y = f(x), \qquad (x \in [a, b])$$

where p_0, p_1, p_2, f are continuous on $[a, b]$, $p_2(x) > 0$ for each x in $[a, b]$, and the equation is subject to the initial conditions

$$(13) \qquad\qquad y(x_0) = c, \quad y'(x_0) = d, \qquad (x_0 \in [a, b]; \quad c, d \text{ constants}).$$

Theorem 4 There exists a unique solution to the problem consisting of the linear equation (12) together with the initial conditions (13).

As continuity of the function

$$g(x, y, z) = \frac{f(x)}{p_2(x)} - \frac{p_0(x)}{p_2(x)} y - \frac{p_1(x)}{p_2(x)} z$$

is clear, the reader need only check that this same function satisfies the relevant Lipschitz condition for all x in $[a, b]$. Note that the various continuous functions of x are bounded on $[a, b]$. (No such condition as (f) of section 2.2 is here necessary, nor is the graph condition of Theorem 3.)

By and large, the differential equations that appear in these notes are linear. It is of special note that the unique solution obtained for the equation of Theorem 4 is valid for the whole interval of definition of that linear equation. In our other theorems, although existence is only given 'locally' (for example, where $|x - a| \leq h$ *and* $Mh \leq k$ in Theorem 1), solutions are often valid in a larger domain. Often the argument used above to establish uniqueness in a limited domain can be used to extend this uniqueness to wherever a solution exists.

Exercise 9 Consider the problem

$$yy'' = -(y')^2, \quad y(0) = y'(0) = 1.$$

(i) Use Theorem 3 to show that the problem has a unique solution on an interval containing 0.

(ii) Find the solution and state where it exists.

3 The Homogeneous Linear Equation and Wronskians

The main aim of this chapter will be to use Theorem 4 of Chapter 2 – and specifically both existence and uniqueness of solutions – to develop a theory that will describe the solutions of the **homogeneous linear second-order equation**

$$(1) \qquad p_2(x)y'' + p_1(x)y' + p_0(x)y \;=\; 0, \qquad (x \in [a, b])$$

where p_0, p_1, p_2 are continuous real-valued functions on $[a, b]$ and $p_2(x) > 0$ for each x in $[a, b]$. ('Homogeneous' here reflects the zero on the right-hand side of the equation which allows λy to be a solution (for any real constant λ) whenever y is a given solution.) The language of elementary linear algebra will be used and the theory of simultaneous linear equations will be presumed.

Central to our discussion will be the Wronskian, or Wronskian determinant: if $y_1 : [a, b] \to \mathbb{R}$ and $y_2 : [a, b] \to \mathbb{R}$ are differentiable functions on the closed interval $[a, b]$, the *Wronskian of y_1 and y_2*, $W(y_1, y_2) : [a, b] \to \mathbb{R}$, is defined, for $x \in [a, b]$, by

$$(2) \qquad W(y_1, y_2)(x) \;=\; \begin{vmatrix} y_1(x) & y_2(x) \\ y_1'(x) & y_2'(x) \end{vmatrix} \;=\; y_1(x)y_2'(x) - y_1'(x)y_2(x).$$

If y_1 and y_2 are solutions of (1), it will turn out that *either* $W(y_1, y_2)$ is identically zero *or* never zero in $[a, b]$.

3.1   Some linear algebra

Suppose that $y_i : [a, b] \to \mathbb{R}$ and that c_i is a real constant, for $i = 1, \ldots, n$. By

$$c_1 y_1 + \ldots + c_n y_n = 0$$

is meant

$$c_1 y_1(x) + \ldots + c_n y_n(x) = 0 \quad \text{for each } x \text{ in } [a, b].$$

We may describe this situation by saying that

$$c_1 y_1 + \ldots + c_n y_n \quad \text{or} \quad \sum_{i=1}^{n} c_i y_i \quad \textit{is identically zero on } [a, b].$$

If $y_i : [a, b] \to \mathbb{R}$ for $i = 1, \ldots, n$, the set $\{y_1, \ldots, y_n\}$ is *linearly dependent on* $[a, b]$ if and only if there are real constants $c_1, \ldots, c_n$, not all zero, such that

$$c_1 y_1 + \ldots + c_n y_n = 0.$$

Otherwise, the set is *linearly independent on* $[a, b]$. So, $\{y_1, \ldots, y_n\}$ is linearly independent on $[a, b]$ if and only if

$$c_1 y_1 + \ldots + c_n y_n = 0,$$

with $c_1, \ldots, c_n$ real constants, necessitates

$$c_1 = \ldots = c_n = 0.$$

It is a common abuse of language to say that $y_1, \ldots, y_n$ *are* linearly dependent (or independent) when one means that the *set* $\{y_1, \ldots, y_n\}$ *is* linearly dependent (independent). We shall find ourselves abusing language in this manner.

We now turn to stating some elementary results relating to solutions $\{x_1, \ldots, x_n\}$ of the system of simultaneous linear equations

$$
\begin{aligned}
a_{11} x_1 \; + \; & \ldots \; + \; a_{1n} x_n \; = \; 0 \\
\vdots \quad\quad\quad & \quad\quad \vdots \quad\quad \vdots \\
a_{m1} x_1 \; + \; & \ldots \; + \; a_{mn} x_n \; = \; 0
\end{aligned}
$$

where a_{ij} is a real constant for $i = 1, \ldots, m$ and $j = 1, \ldots, n$.

(a) When $m = n$, the system has a solution *other than* the 'zero solution'

$$x_1 = \ldots = x_n = 0$$

if and only if the 'determinant of the coefficients' is zero, that is,

$$\begin{vmatrix} a_{11} & \cdots & a_{1n} \\ \vdots & & \vdots \\ a_{n1} & \cdots & a_{nn} \end{vmatrix} = 0.$$

(b) When $m < n$, the system has a solution other than the zero solution.

We conclude this section with an application of result (a), which will give us our first connection between linear dependence and Wronskians.

Proposition 1 If y_1, y_2 are differentiable real-valued functions, linearly dependent on $[a, b]$, then $W(y_1, y_2)$ is identically zero on $[a, b]$.

Proof As y_1, y_2 are linearly dependent, there are real constants c_1, c_2, not both zero, such that

$$(3) \qquad c_1 y_1(x) + c_2 y_2(x) = 0, \quad \text{for each } x \text{ in } [a, b].$$

Differentiating with respect to x we have

$$(4) \qquad c_1 y_1'(x) + c_2 y_2'(x) = 0, \quad \text{for each } x \text{ in } [a, b].$$

Treat the system consisting of (3) and (4) as equations in c_1 and c_2 (in the above notation, take $x_1 = c_1$, $x_2 = c_2$, $a_{11} = y_1(x)$, $a_{12} = y_2(x)$, $a_{21} = y_1'(x)$, $a_{22} = y_2'(x)$). Since c_1 and c_2 are not both zero, we may use result (a) to deduce that the determinant of the coefficients of c_1 and c_2 is zero. However, this determinant is precisely $W(y_1, y_2)(x)$, the Wronskian evaluated at x. We have thus shown that $W(y_1, y_2)(x) = 0$, for each x in $[a, b]$; that is, that $W(y_1, y_2)$ is identically zero on $[a, b]$. $\qquad \square$

The converse of Proposition 1 does not hold: the reader is asked to demonstrate this by providing a solution to the second exercise below.

Exercise 1 If $y_1(x) = \cos x$ and $y_2(x) = \sin x$ for $x \in [0, \pi/2]$, show that $\{y_1, y_2\}$ is linearly independent on $[0, \pi/2]$.

Exercise 2 Define $y_1 : [-1, 1] \to \mathbb{R}$, $y_2 : [-1, 1] \to \mathbb{R}$ by

$$y_1(x) = x^3, \qquad y_2(x) = 0 \qquad \text{for } x \in [0, 1],$$
$$y_1(x) = 0, \qquad y_2(x) = x^3 \qquad \text{for } x \in [-1, 0].$$

Show that y_1 and y_2 are twice continuously differentiable functions on $[-1, 1]$, that $W(y_1, y_2)$ is identically zero on $[-1, 1]$, but that $\{y_1, y_2\}$ is linearly independent on $[-1, 1]$.

3.2 Wronskians and the linear independence of solutions of the second-order homogeneous linear equation

We commence this section with an elementary result which is useful in any discussion of homogeneous linear equations. The proof is left to the reader.

Proposition 2 If $y_1, \ldots, y_n$ are solutions of (1) and $c_1, \ldots, c_n$ are real constants, then

$$c_1 y_1 + \ldots + c_n y_n$$

is also a solution of (1).

We now show that the converse of Proposition 1 holds when y_1 and y_2 are solutions of (1). The result we prove looks at first sight (and misleadingly) stronger.

Proposition 3 If y_1, y_2 are solutions of the linear equation (1) and if $W(y_1, y_2)(x_0) = 0$ for some x_0 in $[a, b]$, then y_1 and y_2 are linearly dependent on $[a, b]$ (and hence $W(y_1, y_2)$ is identically zero on $[a, b]$).

Proof Consider the following system as a pair of equations in c_1 and c_2:

(5)
$$c_1 y_1(x_0) + c_2 y_2(x_0) = 0,$$
$$c_1 y_1'(x_0) + c_2 y_2'(x_0) = 0.$$

(Note that $y_1'(x_0), y_2'(x_0)$ exist, as y_1, y_2 both solve (1) and hence even have second derivatives.) The determinant of the coefficients, here $W(y_1, y_2)(x_0)$, is zero. So, using result (a) of section 3.1, there is a solution

(6)
$$c_1 = C_1, \quad c_2 = C_2$$

with C_1, C_2 not both zero.

Proposition 2 allows us to conclude that the function $y : [a, b] \to \mathbb{R}$ defined by

$$y(x) \;=\; C_1 y_1(x) + C_2 y_2(x), \qquad (x \in [a, b])$$

is a solution of (1). But notice that equations (5), taken with the solution (6), state that

(7) $$y(x_0) \;=\; y'(x_0) \;=\; 0.$$

One solution of (1) together with the initial conditions (7) is clearly $y = 0$ identically on $[a, b]$. By Theorem 4 of Chapter 2, there can be no others; and so, necessarily,

$$C_1 y_1 + C_2 y_2 \;=\; 0$$

identically on $[a, b]$. Recalling that not both of C_1 and C_2 are zero, we have shown that y_1 and y_2 are linearly dependent on $[a, b]$. □

Propositions 1 and 3 are so important that we re-state them together as the following proposition.

Proposition 4 Suppose that y_1 and y_2 are solutions of the linear equation (1). Then

(a) $W(y_1, y_2)$ is identically zero on $[a, b]$ or never zero on $[a, b]$;
(b) $W(y_1, y_2)$ is identically zero on $[a, b]$ if and only if y_1 and y_2 are linearly dependent on $[a, b]$.

A proof similar to that of Proposition 3 establishes the following result which shows that we can never have more than two solutions to (1) which are linearly independent on $[a, b]$.

Proposition 5 If $n > 2$ and $y_1, \ldots, y_n$ are solutions of (1), then $\{y_1, \ldots, y_n\}$ is a linearly dependent set on $[a, b]$.

Proof Pick x_0 in $[a, b]$ and consider the pair of equations

$$c_1 y_1(x_0) + \ldots + c_n y_n(x_0) \;=\; 0$$

$$c_1 y_1'(x_0) + \ldots + c_n y_n'(x_0) \;=\; 0$$

in $c_1, \ldots, c_n$. As $n > 2$, result (b) of the last section implies that there is a solution

$$c_1 \;=\; C_1, \;\ldots, \; c_n \;=\; C_n$$

with $C_1, \ldots, C_n$ not all zero.

Using Proposition 2 above and Theorem 4 of Chapter 2, we deduce, as in the proof of Proposition 3, that

$$C_1 y_1 + \ldots + C_n y_n \;=\; 0$$

identically on $[a, b]$ and hence that $y_1, \ldots, y_n$ are linearly dependent on $[a, b]$. $\qquad \square$

We conclude our discussion of the solution of (1) by using Wronskians to show that this linear equation actually possesses two linearly independent solutions.

Proposition 6 There exist two solutions y_1, y_2 of (1) which are linearly independent on $[a, b]$. Further, any solution y of (1) may be written in terms of y_1 and y_2 in the form

$$(8) \qquad\qquad\qquad y \;=\; c_1 y_1 + c_2 y_2,$$

where c_1 and c_2 are constants.

Proof Pick x_0 in $[a, b]$. The existence part of Theorem 4 of Chapter 2 produces a solution y_1 of (1) satisfying the initial conditions

$$y_1(x_0) \;=\; 1, \qquad y_1'(x_0) \;=\; 0.$$

Similarly, there is a solution y_2 of (1) satisfying

$$y_2(x_0) \;=\; 0, \qquad y_2'(x_0) \;=\; 1.$$

As $W(y_1, y_2)(x_0) = 1$, we may use Proposition 1 to deduce that y_1 and y_2 are linearly independent on $[a, b]$.

If y is any solution of (1) then $\{y, y_1, y_2\}$ is linearly dependent on $[a, b]$ by Proposition 5. So, there are constants c, c_1', c_2' not all zero, such that

$$cy + c_1' y_1 + c_2' y_2 \;=\; 0.$$

The constant c must be non-zero; for otherwise,

$$c_1' y_1 + c_2' y_2 \;=\; 0,$$

which would necessitate, as $\{y_1, y_2\}$ is linearly independent, $c_1' = c_2' = 0$, contradicting the fact that not all of c, c_1', c_2' are zero. We may therefore define $c_1 = -c_1'/c$ and $c_2 = -c_2'/c$ to give

$$y \;=\; c_1 y_1 + c_2 y_2. \qquad\qquad \square$$

Note There are other ways of proving the propositions of this chapter. The proofs here have been chosen as they can be extended directly to cover the case of the n-th order homogeneous linear equation.

Exercise 3 Find the Wronskian $W(y_1, y_2)$ corresponding to linearly independent solutions y_1, y_2 of the following differential equations in $y = y(x)$, satisfying the given conditions. Methods for solving the equations may be found in the Appendix.

(a) $\qquad y'' = 0,$ $\qquad\qquad\qquad\qquad y_1(-1) = y_2(1) = 0$

(b) $\qquad y'' - y = 0,$ $\qquad\qquad\qquad\quad y_1(0) = y_2'(0) = 0$

(c) $\qquad y'' + 2y' + (1 + k^2)y = 0,$ $\qquad y_1(0) = y_2(\pi) = 0,$ $\qquad$ for $k > 0,$

(d) $\qquad x^2 y'' + xy' - k^2 y = 0,$ $\qquad\quad y_1(1) = y_2(2) = 0,$ $\qquad$ for $x, k > 0.$

(Note that the value of the Wronskian may still depend on constants of integration.)

Exercise 4 By considering (separately) the differential equations

$$y'' + y^2 = 0, \qquad y'' = 1,$$

show that linearity and homogeneity of (1) are necessary hypotheses in Proposition 2.

Exercise 5 Suppose that y_1, y_2 are solutions of (1). Show that the Wronskian $W = W(y_1, y_2)$ is a solution of the differential equation

(9) $\qquad\qquad\qquad p_2(x)W' + p_1(x)W = 0, \qquad (x \in [a, b]).$

Exercise 6 By first solving (9) of Exercise 5, give an alternative proof of Proposition 4(a).

Exercise 7 Show that at least one of the components of one of the solutions (y, z) of the simultaneous equations

$$\frac{dz}{dx} = \alpha \frac{df}{dx} \cdot \frac{dy}{dx} - xy, \qquad f(x)\frac{dy}{dx} = z,$$

where $f(x)$ is continuous, everywhere positive, and unbounded as $x \to \infty$, is unbounded as $x \to \infty$ if $\alpha > 0.$

[HINT: Form a second order equation for y and then solve the corresponding equation (9) for the Wronskian of its solutions.]

Exercise 8 Describe how equation (9) of Exercise 5 may be used to find the general solution of (1) when one (nowhere zero) solution $y = u$ is known. Compare this method with the one given in section (5) of the Appendix by showing that, if $y = u$ and $y = uv$ are solutions of (1), then (9) for $W = W(u, uv)$ gives rise to

$$p_2 u^2 v'' + (2uu'p_2 + p_1 u^2)v' \;\; = \;\; 0.$$

4 The Non-Homogeneous Linear Equation

In this chapter we shall consider the *non-homogeneous second-order linear equation*

$$(1) \qquad p_2(x)y'' + p_1(x)y' + p_0(x)y \; = \; f(x), \qquad (x \in [a, b])$$

where $f : [a, b] \to \mathbb{R}$ and each $p_i : [a, b] \to \mathbb{R}$ are continuous, and $p_2(x) > 0$, for each x in $[a, b]$. Any particular solution of (1) (or, indeed, of any differential equation) is called a *particular integral* of the equation, whereas the general solution $c_1 y_1 + c_2 y_2$ of the corresponding homogeneous equation

$$(2) \qquad p_2(x)y'' + p_1(x)y' + p_0(x)y \; = \; 0, \qquad (x \in [a, b])$$

given by Proposition 6 of Chapter 3 (where y_1, y_2 are linearly independent solutions of (2) and c_1, c_2 are arbitrary real constants) is called the *complementary function* of (1). If y_P is any particular integral and y_C denotes the complementary function of (1), $y_C + y_P$ is called the *general solution* of (1). We justify this terminology by the following proposition, which shows that, once two linearly independent solutions of (2) are found, all that remains to be done in solving (1) is to find one particular integral.

Proposition 1 Suppose that y_P is any particular integral of the non-homogeneous linear equation (1) and that y_1, y_2 are linearly independent solutions of the corresponding homogeneous equation (2). Then

(a) $c_1y_1 + c_2y_2 + y_P$ is a solution of (1) for any choice of the constants c_1, c_2,

(b) if y is any solution (that is, any particular integral) of (1), there exist (particular) real constants C_1, C_2 such that

$$y = C_1y_1 + C_2y_2 + y_P.$$

Proof (a) As $c_1y_1 + c_2y_2$ is a solution of (2) for any real choice of c_1, c_2 by Proposition 2 of Chapter 3, all we need to show is that, if Y is a solution of (2) and y_P a solution of (1), then $Y + y_P$ is a solution of (1). In these circumstances, we have

$$p_2Y'' + p_1Y' + p_0Y = 0$$

and

(3) $$p_2y_P'' + p_1y_P' + p_0y_P = f.$$

By adding,

$$p_2(Y + y_P)'' + p_1(Y + y_P)' + p_0(Y + y_P) = f;$$

and thus, $Y + y_P$ is a solution of (1) as required.

(b) As well as (3) above, we are given that

(4) $$p_2y'' + p_1y' + p_0y = f.$$

Subtracting,

$$p_2(y - y_P)'' + p_1(y - y_P)' + p_0(y - y_P) = 0;$$

so that, $y - y_P$ solves (2). Proposition 6 of Chapter 3 then finds real constants C_1, C_2 such that

$$y - y_P = C_1y_1 + C_2y_2.$$

The proposition is established.   □

4.1 The method of variation of parameters

The reader will probably already have encountered methods for finding particular integrals for certain (benign!) functions $f(x)$ occurring on the right-hand side of equation (1). This section produces a systematic method for finding particular integrals once one has determined two linearly independent solutions of (2), that is, once one has already determined the complementary function of (1).

All there really is to the method is to note that, if y_1, y_2 are linearly independent solutions of (2), then a particular integral of (1) is $k_1 y_1 + k_2 y_2$, where $k_i : [a, b] \to \mathbb{R}$ ($i = 1, 2$) are the continuously differentiable functions on $[a, b]$ defined, for any real constants α, β in $[a, b]$, by

$$(5) \qquad k_1(x) = -\int_\alpha^x \frac{y_2(t)f(t)}{p_2(t)W(t)}\, dt, \quad k_2(x) = \int_\beta^x \frac{y_1(t)f(t)}{p_2(t)W(t)}\, dt, \quad (x \in [a, b])$$

where

$$W = W(y_1, y_2) \equiv y_1 y_2' - y_2 y_1'$$

is the Wronskian of y_1 and y_2. (Notice that $W(t) \neq 0$ for each t in $[a, b]$ because y_1, y_2 are linearly independent.) To see that this is true, we differentiate k_1 and k_2 as given by (5):

$$k_1'(x) = -\frac{y_2(x)f(x)}{p_2(x)W(x)}, \qquad k_2'(x) = \frac{y_1(x)f(x)}{p_2(x)W(x)}, \qquad (x \in [a, b])$$

and notice that k_1', k_2' must therefore satisfy the simultaneous linear equations

$$(6) \qquad \begin{aligned} k_1'(x)y_1(x) + k_2'(x)y_2(x) &= 0, \\ k_1'(x)y_1'(x) + k_2'(x)y_2'(x) &= f(x)/p_2(x). \end{aligned} \qquad (x \in [a, b])$$

(The reader should check this.) Hence, putting

$$y = k_1 y_1 + k_2 y_2,$$

we have

$$y' = k_1 y_1' + k_2 y_2'$$

and

$$y'' = k_1 y_1'' + k_2 y_2'' + f/p_2.$$

So,

$$p_2 y'' + p_1 y' + p_0 y \;=\; k_1 (p_2 y_1'' + p_1 y_1' + p_0 y_1) + k_2 (p_2 y_2'' + p_1 y_2' + p_0 y_2) + f \;=\; f,$$

since y_1, y_2 solve (2). Thus, $y = k_1 y_1 + k_2 y_2$ solves (1) and the general solution of (1) is

$$(7) \qquad y(x) \;=\; c_1 y_1(x) + c_2 y_2(x) - \int_\alpha^x \frac{y_1(x) y_2(t)}{p_2(t) W(t)} f(t)\, dt + \int_\beta^x \frac{y_1(t) y_2(x)}{p_2(t) W(t)} f(t)\, dt$$

where c_1 and c_2 are real constants and $x \in [a, b]$.

The constants α, β should not be regarded as arbitrary in the sense that c_1, c_2 are. Rather, they should be considered as part of the definition of k_1 and k_2. Notice that, if $y = y(x)$ is the particular integral given by formula (7) with $c_1 = C_1$ and $c_2 = C_2$, and if α', β' are in $[a, b]$ then

$$y(x) \;=\; C_1' y_1(x) + C_2' y_2(x) - \int_{\alpha'}^x \frac{y_1(x) y_2(t)}{p_2(t) W(t)} f(t)\, dt + \int_{\beta'}^x \frac{y_1(t) y_2(x)}{p_2(t) W(t)} f(t)\, dt$$

for each x in $[a, b]$, where C_1', C_2' are the *constants* given by

$$(8) \qquad C_1' \;=\; C_1 - \int_\alpha^{\alpha'} \frac{y_2(t) f(t)}{p_2(t) W(t)}\, dt, \qquad C_2' \;=\; C_2 + \int_\beta^{\beta'} \frac{y_1(t) f(t)}{p_2(t) W(t)}\, dt.$$

Thus, changes in α, β just make corresponding changes in C_1, C_2.

An appropriate choice of α, β can often depend on conditions applied to the solution of (1).

(a) **Initial conditions** Suppose we are given values for $y(a)$ and $y'(a)$. Then it can be most convenient to choose $\alpha = \beta = a$. The particular integral in (7) is then

$$(9) \qquad \int_a^x \frac{(y_1(t) y_2(x) - y_1(x) y_2(t))}{p_2(t) W(t)} f(t)\, dt \;=\; \int_a^x \frac{(y_1(t) y_2(x) - y_1(x) y_2(t)) f(t)}{(y_1(t) y_2'(t) - y_2(t) y_1'(t)) p_2(t)}\, dt.$$

The reader should check that this particular integral and its derivative with respect to x are both zero at $x = a$. This is technically useful when given the above initial conditions, making it easier to calculate the constants c_1, c_2 in this case.

(b) **Boundary conditions** Suppose now that we are given values for $y(a)$ and $y(b)$. Convenient choices for α, β are often $\alpha = b$, $\beta = a$. The particular integral then becomes

$$(10) \qquad \int_x^b \frac{y_1(x)y_2(t)}{p_2(t)W(t)} f(t)\, dt + \int_a^x \frac{y_1(t)y_2(x)}{p_2(t)W(t)} f(t)\, dt \;=\; \int_a^b G(x,t)f(t)\, dt,$$

where

$$G(x,t) \;=\; \begin{cases} \dfrac{y_1(x)y_2(t)}{p_2(t)W(t)}, & \text{for } a \le x \le t \le b, \\[3mm] \dfrac{y_1(t)y_2(x)}{p_2(t)W(t)}, & \text{for } a \le t \le x \le b. \end{cases}$$

Notice that, at $x = a$, the first integral on the left-hand side of (10) is a multiple of $y_1(a)$ and the second integral vanishes. At $x = b$, it is the first integral that vanishes and the second is a multiple of $y_2(b)$. This can be especially useful if the linearly independent functions y_1, y_2 can be chosen so that $y_1(a) = y(a)$ and $y_2(b) = y(b)$. We shall return to these matters and the function $G = G(x,t)$ in our discussion of Green's functions in the next section.

An alternative view of the method

A commonly found presentation of the method of variation of parameters is the following.

Suppose that y_1 and y_2 are linearly independent solutions of (2). As $y = c_1 y_1 + c_2 y_2$ is a solution (the general solution) of the homogeneous equation (2), 'it is natural' to seek a solution to the non-homogeneous equation (1) by 'varying the parameters' c_1 and c_2. So, we seek functions k_1, k_2 such that $y = k_1 y_1 + k_2 y_2$ is a particular solution of (1) (where k_1 and k_2 are continuously differentiable). Then

$$(11) \qquad\qquad y' \;=\; k_1 y_1' + k_2 y_2' + k_1' y_1 + k_2' y_2.$$

The next step in the argument is to stipulate that k_1 and k_2 are to be chosen in order that

$$(12) \qquad\qquad k_1' y_1 + k_2' y_2 \;=\; 0.$$

(Yes, this is possible! We are often told that there is sufficient freedom in our choice of k_1 and k_2 to allow this and hence, of course, to simplify the subsequent 'working'.) Differentiating again and using (12),

$$(13) \qquad\qquad y'' \;=\; k_1 y_1'' + k_2 y_2'' + k_1' y_1' + k_2' y_2'.$$

In order to derive a further condition to permit $y = k_1 y_1 + k_2 y_2$ to be a solution to (1), we must substitute for this y, for y' given by (11) subject to (12), and for y'' given by (13) in equation (1). The reader should check that, since y_1 and y_2 solve (2), it is necessary that

$$(14) \qquad\qquad\qquad k_1' y_1' + k_2' y_2' \;=\; f/p_2.$$

Equations (12) and (14) are of course just equations (6). These may be solved for k_1' and k_2' as the determinant of the coefficients is the Wronskian $W(y_1, y_2)$ of the linearly independent solutions y_1, y_2 and hence is non-zero everywhere in $[a, b]$. Reversing steps in our earlier discussion of the method,

$$k_1(x) - k_1(\alpha) \;=\; - \int_\alpha^x \frac{y_2(t) f(t)}{p_2(t) W(t)}\, dt, \qquad k_2(x) - k_2(\beta) \;=\; \int_\beta^x \frac{y_1(t) f(t)}{p_2(t) W(t)}\, dt$$

for each x in $[a, b]$. Thus, the general solution of (1) is given by

$$y(x) \;=\; c_1 y_1(x) + c_2 y_2(x) + k_1(x) y_1(x) + k_2(x) y_2(x)$$

$$= \; c_1' y_1(x) + c_2' y_2(x) - \int_\alpha^x \frac{y_1(x) y_2(t)}{p_2(t) W(t)} f(t)\, dt + \int_\beta^x \frac{y_1(t) y_2(x)}{p_2(t) W(t)} f(t)\, dt$$

where $c_1' = c_1 + k_1(\alpha)$ and $c_2' = c_2 + k_2(\beta)$ are constants and $x \in [a, b]$.

This alternative view of the variation of parameters method does not present a purely deductive argument (in particular, it is necessary to stipulate (12)), but the reader may care to use it to aid his recall of k_1 and k_2 as defined in (5). The condition (12) will then need to be remembered!

It is our experience that computational errors in applying the method to practical problems are more easily avoided by deducing the general formula (7) before introducing particular functions y_1, y_2 and calculating $W(y_1, y_2)$.

It is also worth recalling that the variation of parameters particular integral is only one of many possible particular integrals. Practically, one should see if a particular integral can be more easily found otherwise. The Green's function method described in the next section, when applicable, is remarkably efficient.

Exercise 1 Check the following details:

(a) that equations (6) are satisfied by $k_1'(x)$ and $k_2'(x)$,

(b) that the values given for C_1' and C_2' in (8) are correct,

(c) that the integral (9) and its derivative with respect to x are both zero at $x = a$.

Use the method of variation of parameters to solve the following three problems.

Exercise 2 Find the general solution of $y'' + y = \tan x, \quad 0 < x < \pi/2$.

Exercise 3 Show that the solution of the equation

$$y'' + 2y' + 2y = f(x),$$

with f continuous and initial conditions $y(0) = y'(0) = 1$, can be written in the form

$$y(x) = e^{-x}(\cos x + 2\sin x) + \int_0^x e^{-(x-t)} \sin(x - t) f(t)\, dt.$$

Exercise 4 Find the necessary condition on the continuous function g for there to exist a solution of the equation

$$y'' + y = g(x),$$

satisfying $y(0) = y(\pi) = 0$.

Exercise 5 The function $y = y(x)$ satisfies the homogeneous differential equation

$$y'' + (1 - h(x))y = 0, \qquad (0 \le x \le K)$$

where h is a continuous function and K is a positive constant, together with the initial conditions $y(0) = 0$, $y'(0) = 1$. Using the variation of parameters method, show that y also satisfies the integral equation

$$y(x) = \sin x + \int_0^x y(t)h(t)\sin(x - t)\, dt \qquad (0 \le x \le K).$$

If $|h(x)| \le H$ for $0 \le x \le K$ and some positive constant H, show that

$$|y(x)| \le e^{Hx} \qquad (0 \le x \le K).$$

[HINTS: Re-write the differential equation in the form

$$y'' + y = h(x)y.$$

For the last part, use Picard's iterative method described in Chapter 1.]

4.2   Green's functions

Up to this point, in order to find the solution of a problem consisting of a differential equation together with initial or boundary conditions, we have first found a general solution of the equation and later fitted the other conditions (by finding values for constants occurring in the general solution in order that the conditions be met). The method produced by the theorem of this section builds the boundary condition into finding the solution from the start.

Theorem 1 Suppose that the operator L is defined by

$$Ly \;\equiv\; \frac{d}{dx}\left(p(x)\frac{dy}{dx}\right) + q(x)y \;\equiv\; (p(x)y')' + q(x)y,$$

where y is twice continuously differentiable, p is continuously differentiable, q is continuous and $p(x) > 0$ for all x in $[a,b]$. Suppose further that the homogeneous equation

(H) $Ly \;=\; 0 \qquad (x \in [a,b])$

has only the trivial solution (that is, the solution $y = 0$ identically on $[a,b]$) when subject to both boundary conditions

(α) $A_1 y(a) + B_1 y'(a) \;=\; 0$

(β) $A_2 y(b) + B_2 y'(b) \;=\; 0$

where A_1, A_2, B_1, B_2 are constants (A_1, B_1 not both zero and A_2, B_2 not both zero). If f is continuous on $[a,b]$, then the non-homogeneous equation

(N) $Ly \;=\; f(x), \qquad (x \in [a,b])$

taken together with both conditions (α) and (β), has a *unique* solution which may be written in the form

$$y(x) \;=\; \int_a^b G(x,t) f(t)\, dt, \qquad (x \in [a,b])$$

where G is continuous on $[a,b]^2$, is twice continuously differentiable on $\{(x,y) \in [a,b]^2 : x \neq y\}$, and satisfies

$$\frac{\partial G}{\partial x}(t+0,t) - \frac{\partial G}{\partial x}(t-0,t) \;=\; \frac{1}{p(t)}, \qquad (t \in [a,b]).$$

[We have used the notation: when $f = f(x, t)$,

$$f(t+0, t) = \lim_{x \downarrow t} f(x, t), \qquad f(t-0, t) = \lim_{x \uparrow t} f(x, t).]$$

The linear equations (H) and (N) here, with their left-hand sides written in the form $(py')' + qy$, are said to be in *self-adjoint form*.

Note The problem consisting of trying to find solutions of (N) together with (α) and (β), conditions applying at the two distinct points a and b, is called a **two-point boundary value problem (2-point BVP)**. It is to be distinguished from the **1-point BVP** or **initial value problem (IVP)** considered in Chapter 2, where y and y' are only given at the single point a and where, unlike here, a unique solution can, under very general conditions, always be found. The example in Exercise 4 of the previous section underlines the need to extend our theory.

The conditions (α), (β) are referred to as **homogeneous boundary conditions**, as it is only the ratios $A_1 : B_1$ and $A_2 : B_2$ that matter, not the actual values of A_1, A_2, B_1, B_2.

The function $G = G(x, t)$ occurring in the theorem is called the **Green's function** for the problem consisting of (N) together with the boundary conditions (α), (β). Such Green's functions appear throughout the theory of ordinary and partial differential equations. They are always defined (as below) in terms of solutions of a homogeneous equation and allow a solution of the corresponding non-homogeneous equation to be given (as in Theorem 1 here) in integral form.

Aside On the face of it, the equation (N) would seem to be a special case of the non-homogeneous equation (1) specified at the start of this chapter. In fact, any equation of the form (1) may be written in the form (N) by first 'multiplying it through' by

$$\frac{1}{p_2(x)} \exp\left(\int^x \frac{p_1(t)}{p_2(t)} \, dt\right).$$

The reader should check that the resulting equation is

$$\frac{d}{dx}\left(\exp\left(\int^x \frac{p_1}{p_2}\right) \frac{dy}{dx}\right) + \frac{p_0(x)}{p_2(x)} \exp\left(\int^x \frac{p_1}{p_2}\right) y = \frac{f(x)}{p_2(x)} \exp\left(\int^x \frac{p_1}{p_2}\right),$$

where each $\int^x$ is the indefinite integral. The equation is now in self-adjoint form and the coefficient functions satisfy the continuity, differentiability and positivity conditions of Theorem 1.

Before proceeding to the proof of the theorem, we shall establish two lemmas. The conditions of the theorem will continue to apply.

Lemma 1 Suppose that y_1 and y_2 are solutions of (H) and that W denotes the Wronskian $W(y_1, y_2)$. Then

$$p(x)W(x) \;=\; A, \quad \text{for each } x \in [a, b],$$

where A is a real constant. If y_1 and y_2 are linearly independent, then A is non-zero.

Proof Since y_1, y_2 solve (H), we have

(15)
$$Ly_1 \;=\; (py_1')' + qy_1 \;=\; 0,$$
$$Ly_2 \;=\; (py_2')' + qy_2 \;=\; 0.$$

Hence, on $[a, b]$ we have

$$\begin{aligned}
(pW)' &\;=\; (p(y_1 y_2' - y_2 y_1'))' \\
&\;=\; (y_1(py_2') - y_2(py_1'))' \\
&\;=\; y_1(py_2')' - y_2(py_1')' \\
&\;=\; y_1(-qy_2) - y_2(-qy_1), \quad \text{using (15),} \\
&\;=\; 0.
\end{aligned}$$

Thus pW is constant on $[a, b]$. If y_1 and y_2 are linearly independent on $[a, b]$, then Proposition 4 of Chapter 3 tells us that W is never zero in $[a, b]$, and (since we have insisted that p_2 is never zero in $[a, b]$) the proof is complete. $\square$

Note An alternative proof can be based on Exercise 5 of Chapter 3, with

$$p(x) \;=\; \exp\left(\int^x \frac{p_1}{p_2}\right)$$

as in the Aside above.

Lemma 2 There is a solution $y = u$ of $Ly = 0$ satisfying (α) and a solution $y = v$ of $Ly = 0$ satisfying (β) such that u and v are linearly independent over $[a, b]$.

Proof Noting that $Ly = 0$ is the linear equation

$$p(x)y'' + p'(x)y' + q(x)y = 0, \qquad (x \in [a, b])$$

where p, p' and q are continuous on $[a, b]$ and $p(x) > 0$ for each x in $[a, b]$, we see that we may apply the existence part of Theorem 4 of Chapter 2 to find functions u, v solving

$$Lu = 0, \quad u(a) = B_1, \quad u'(a) = -A_1,$$

$$Lv = 0, \quad v(b) = B_2, \quad v'(b) = -A_2.$$

Then, certainly $y = u$ solves $Ly = 0$ taken with (α) and $y = v$ solves $Ly = 0$ taken with (β). Further, u, v must both be not identically zero in $[a, b]$, because of the conditions placed on A_1, B_1, A_2, B_2.

Suppose that C, D are constants such that

(16) $$Cu(x) + Dv(x) = 0, \quad \text{for each } x \text{ in } [a, b].$$

As u, v solve $Ly = 0$, they must be differentiable and hence we may deduce

(17) $$Cu'(x) + Dv'(x) = 0, \quad \text{for each } x \text{ in } [a, b].$$

Multiplying (16) by A_2 and (17) by B_2, adding and evaluating at $x = b$ gives

$$C(A_2 u(b) + B_2 u'(b)) = 0.$$

If $C \neq 0$, u satisfies (β), as well as $Ly = 0$ and (α), and must be the trivial solution $u \equiv 0$ by one of the hypotheses of the theorem. This contradicts the fact that u is not identically zero in $[a, b]$. Similarly, we can show that

$$D(A_1 v(a) + B_1 v'(a)) = 0$$

and hence, if $D \neq 0$, that $v \equiv 0$ in $[a, b]$, also giving a contradiction. So, C and D must both be zero and u, v must be linearly independent over $[a, b]$.   □

Proof of Theorem 1 We first establish uniqueness of the solution. Suppose that y_1, y_2 solve

$$Ly = f(x) \quad \text{together with } (\alpha), (\beta).$$

Then, it is easy to see that $y = y_1 - y_2$ solves

$$Ly \;=\; 0 \quad \text{together with } (\alpha), \ (\beta).$$

By hypothesis, y must be the trivial solution $y \equiv 0$. So, $y_1 = y_2$ and, if a solution to (N) together with (α), (β) exists, it must be unique.

Letting u, v be the functions given by Lemma 2, we can deduce from Lemma 1 that

$$p(x)W(x) \;=\; A, \quad \text{for each } x \text{ in } [a,b],$$

where A is a *non-zero* constant and W is the Wronskian $W(u,v)$. We may therefore define $G : [a,b]^2 \to \mathbb{R}$ by

$$G(x,t) \;=\; \begin{cases} \dfrac{u(x)v(t)}{A}, & \text{for } a \le x \le t \le b, \\[3mm] \dfrac{u(t)v(x)}{A}, & \text{for } a \le t \le x \le b. \end{cases}$$

Then G is continuous on $[a,b]^2$, as can be seen by letting t tend to x both from above and below. Clearly, G is also twice continuously differentiable when $x < t$ and when $x > t$, and

$$\frac{\partial G}{\partial x}(t+0,t) - \frac{\partial G}{\partial x}(t-0,t) \;=\; \lim_{x \to t}\left(\frac{u(t)v'(x)}{A} - \frac{u'(x)v(t)}{A} \right)$$

$$= \;\frac{W(t)}{A}$$

$$= \;\frac{1}{p(t)}, \qquad \text{using Lemma 1,}$$

for each t in $[a,b]$. It thus remains to prove that, for x in $[a,b]$,

$$y(x) \;=\; \int_a^b G(x,t)f(t)\,dt;$$

that is,

(18) $$y(x) \;=\; \frac{v(x)}{A}\int_a^x u(t)f(t)\,dt + \frac{u(x)}{A}\int_x^b v(t)f(t)\,dt$$

solves (N) and satisfies the boundary conditions (α), (β). First, note that y is well defined (as G is continuous) and that

(19) $$y(a) \;=\; \frac{u(a)}{A} \int_a^b vf, \qquad y(b) \;=\; \frac{v(b)}{A} \int_a^b uf.$$

Using [G] of Chapter 0 to differentiate (18),

$$y'(x) \;=\; \frac{v'(x)}{A} \int_a^x uf \;+\; \frac{v(x)u(x)f(x)}{A} \;+\; \frac{u'(x)}{A} \int_x^b vf \;-\; \frac{u(x)v(x)f(x)}{A}$$

$$=\; \frac{v'(x)}{A} \int_a^x uf \;+\; \frac{u'(x)}{A} \int_x^b vf,$$

for each x in $[a, b]$. So,

(20) $$y'(a) \;=\; \frac{u'(a)}{A} \int_a^b vf, \qquad y'(b) \;=\; \frac{v'(b)}{A} \int_a^b uf.$$

From (19), (20),

$$A_1 y(a) + B_1 y'(a) \;=\; \frac{1}{A} \left(A_1 u(a) + B_1 u'(a) \right) \int_a^b vf \;=\; 0,$$

since u satisfies (α). Thus, y satisfies (α). Similarly, y satisfies (β), using the fact that v satisfies this latter condition. Now,

$$p(x)y'(x) \;=\; \frac{p(x)v'(x)}{A} \int_a^x uf \;+\; \frac{p(x)u'(x)}{A} \int_x^b vf,$$

and so, differentiating again,

$$\frac{d}{dx}\left(p(x)y'(x) \right) \;=\; \frac{1}{A}\frac{d}{dx}\left(p(x)v'(x) \right) \int_a^x uf \;+\; \frac{1}{A} p(x)v'(x)u(x)f(x)$$

$$+\; \frac{1}{A}\frac{d}{dx}\left(p(x)u'(x) \right) \int_x^b vf \;-\; \frac{1}{A} p(x)u'(x)v(x)f(x).$$

Hence, as $pW = p(uv' - u'v) = A$ and $Lu = Lv = 0$,

$$Ly = \frac{Lv}{A} \int_a^x uf + \frac{Lu}{A} \int_x^b vf + f = f,$$

and our proof is complete. □

Note The Green's function, $G = G(x, t)$, as constructed in the above proof, is independent of the function f on the right-hand side of (N). Put another way: the same G 'works' for every continuous f.

　　The triviality condition We should like to make three points concerning the theorem's hypotheses that (H), together with (α), (β), only has the identically zero solution.

(i) The condition is used in two places in the proof, in the deduction of Lemma 2 and the proof of uniqueness.

(ii) The condition is by no means always met. The reader should check that

$$y'' + y = 0, \qquad y(0) = y(\pi) = 0$$

does not satisfy it. Variation of parameters is a tool in this case, as a solution of Exercise 4 above will have discovered.

(iii) The condition may be recovered from the linear independence of functions u satisfying $Lu = 0$ together with (α) and v satisfying $Lv = 0$ together with (β); that is, the converse of Lemma 2 is also true, as we shall now show.

　　Suppose that u satisfies (H) and (α), that v satisfies (H) and (β), and that u, v are linearly independent, so that the Wronskian $W = W(u, v)$ is never zero in $[a, b]$. Further, suppose y satisfies both (α) and (β) as well as (H). Then, by Proposition 6 of Chapter 3, there are constants c_1, c_2 such that

$$y = c_1 u + c_2 v$$

and hence

$$y' = c_1 u' + c_2 v'$$

on $[a, b]$. Therefore,

$$A_1 y(a) + B_1 y'(a) = c_1(A_1 u(a) + B_1 u'(a)) + c_2(A_1 v(a) + B_1 v'(a))$$

and since both u and y satisfy (α),

$$c_2(A_1 v(a) + B_1 v'(a)) = 0,$$

as well as
$$c_2(A_1 u(a) + B_1 u'(a)) = 0.$$

The reader should check that both
$$c_2 A_1 (u(a)v'(a) - v(a)u'(a)) = c_2 A_1 W(a) = 0$$

and
$$c_2 B_1 (u(a)v'(a) - v(a)u'(a)) = c_2 B_1 W(a) = 0.$$

However, $W(a) \neq 0$ (as already noted) and A_1, B_1 are assumed not both zero. So, $c_2 = 0$ and similarly $c_1 = 0$. Thus, y is identically zero on $[a, b]$ and the triviality condition is met. □

The method of proof of Theorem 1 actually gives a technique for solving equations of the form (N), with boundary conditions (α), (β), which we now outline and exemplify.

(i)   Find u satisfying (H), (α) and v satisfying (H), (β).

(ii)   Calculate $W = W(u, v)$. *If it is non-zero, the method applies* and $A = pW$ is also non-zero.

(iii)   Write down the unique solution

$$y(x) = \int_a^b G(x, t) f(t) \, dt, \qquad (x \in [a, b])$$

where

$$G(x, t) = \begin{cases} \dfrac{u(x)v(t)}{A}, & \text{for } a \leq x \leq t \leq b, \\[2mm] \dfrac{u(t)v(x)}{A}, & \text{for } a \leq t \leq x \leq b. \end{cases}$$

(Note that G is well-defined if and only if A is non-zero.)

Example 1 Find a Green's function $G(x, t)$ which allows the unique solution of the problem
$$y'' + y = f(x), \qquad y(0) = y(\pi/2) = 0, \qquad (x \in [0, \pi/2])$$

where f is continuous on $[0, \pi/2]$, to be expressed in the form

$$y(x) = \int_0^{\frac{\pi}{2}} G(x, t) f(t) \, dt \qquad (0 \leq x \leq \pi/2).$$

We proceed as above.

(i) The function $u = \sin x$ satisfies

$$y'' + y = 0, \qquad y(0) = 0$$

and $v = \cos x$ satisfies

$$y'' + y = 0, \qquad y(\pi/2) = 0.$$

(ii) The Wronskian $W = W(u, v)$ is given by

$$W(u, v)(x) = \begin{vmatrix} \sin x & \cos x \\ \cos x & -\sin x \end{vmatrix} = -1,$$

so that $A = pW = -1$ (as $p \equiv 1$).

(iii) The required Green's function is

$$G(x, t) = \begin{cases} -\sin x \cos t, & \text{for } 0 \le x \le t \le \pi/2, \\ -\sin t \cos x, & \text{for } 0 \le t \le x \le \pi/2. \end{cases} \qquad \square$$

The reader will notice that the method does not apply if the second boundary condition in the above example is changed to $y(\pi) = 0$ (and a solution is sought in $[0, \pi]$). For then u and v would both need to be multiples of $\sin x$ and their Wronskian would be zero.

In Exercises 6 and 9 below, the relevant Wronskians will have been determined in attempts at Exercise 3 of Chapter 3.

Exercise 6 Given that $f : [-1, 1] \to \mathbb{R}$ is continuous, determine a Green's function $G(x, t)$ such that the equation
$$y'' = f(x), \qquad (-1 \le x \le 1)$$

together with the boundary conditions

$$y(-1) = y(1) = 0$$

has the solution

$$y(x) = \int_{-1}^{1} G(x, t) f(t) \, dt.$$

Exercise 7 By constructing a suitable Green's function $G(x,t)$, show that the solution $y = y(x)$ of the boundary value problem

$$(xy')' - \frac{y}{x} = f(x), \qquad y(1) = y(2) = 0,$$

for $1 \leq x \leq 2$ and f continuous on $[1,2]$, can be written in the form

$$y(x) = \int_1^2 G(x,t)f(t)\,dt.$$

Find the solution when $f(x) = 3/x$, for $x \in [1,2]$.

Exercise 8 Suppose that

$$Ly = x^2 y'' - x(x+2)y' + (x+2)y.$$

Given that $y = x$ is a solution of $Ly = 0$, solve the equation $Ly = x^4$ with boundary conditions $y(1) = y(2) = 0$ by first constructing an appropriate Green's function.

Exercise 9 If $f : [0,\pi] \to \mathbb{R}$ is continuous, find a Green's function for the problem

$$y'' + 2y' + (1+k^2)y = f(x), \qquad y(0) = y(\pi) = 0,$$

whenever the positive constant k is *not* an integer. When k is an integer, use the method of variation of parameters to determine a condition which permits a solution, and determine all solutions in this case.

Exercise 10 By constructing a Green's function, show that the solution to the problem consisting of the equation

$$y'' - 2y' + 2y = f(x), \qquad (0 \leq x \leq \pi)$$

together with the boundary conditions $y(0) = 0$ and $y'(\pi) = 0$, can be expressed in the form

$$y(x) = \int_0^\pi K(x,t)f(t)\,dt, \qquad (0 \leq x \leq \pi).$$

The second boundary condition is now replaced by $y(\pi) = 0$. Find a condition that $f(x)$ must satisfy in order that the modified problem can have a solution. Determine all the solutions in this case.

[Comment: the solution of the unmodified problem can be used as a particular integral for the modified problem, obviating the need to use the method of variation of parameters.]

Exercise 11 Let $K(x,t)$ be the function defined by

$$K(x,t) = \begin{cases} xt^{-1}\sin x \cos t & (0 < x \leq t), \\[2mm] xt^{-1}\sin t \cos x & (0 < t < x). \end{cases}$$

Show that the function $y = y(x)$ defined by

$$y(x) \;=\; \int_0^\pi K(x,t)f(t)\,dt$$

is a solution of the inhomogeneous differential equation

(IDE) $$y'' - \frac{2}{x}y' + \left(\frac{2}{x^2} + 1\right)y \;=\; -f(x) \qquad (0 < x < \pi)$$

which satisfies the boundary condition

(BC1) $$\lim_{x \downarrow 0} y'(x) \;=\; 0.$$

Show that y satisfies the second boundary condition

(BC2) $$y(\pi) \;=\; 0$$

provided that

$$\int_0^\pi f(x)\frac{\sin x}{x}\,dx \;=\; 0.$$

Is y the only solution of (IDE) that satisfies (BC1) and (BC2) in these circumstances? Give reasons.

[Note that $y(x) = x\sin x$ and $y(x) = x\cos x$ are linearly independent solutions of the homogeneous differential equation

$$y'' - \frac{2}{x}y' + \left(\frac{2}{x^2} + 1\right)y \;=\; 0 \qquad (x > 0).]$$

Exercise 12 A twice continuously differentiable function u, defined on the real interval $[a,b]$, satisfies $u(a) = 0, u'(b) \neq 0$ and the differential equation

$$Lu \;\equiv\; (pu')' + qu \;=\; 0,$$

where p' and q are continuous functions and $p(x) > 0$ for all x in $[a,b]$. Another such function v satisfies $v'(b) = 0$ and $Lv = 0$. Neither u nor v is identically zero on $[a,b]$.

(i) Prove that $v(a) \neq 0$ and u, v are linearly independent on $[a,b]$.

(ii) Use the functions u, v to show that the problem

$$Ly \;=\; f(x), \qquad y(a) \;=\; 0, \qquad y'(b) \;=\; 0, \qquad (x \in [a,b])$$

where f is a given continuous function, has a unique twice continuously differentiable solution expressible in the form

$$y(x) = \int_a^b G(x,t)f(t)\,dt.$$

[HINTS: Consider pW at a and b. Also, use the uniqueness given by Theorem 4 of Chapter 2.]

5 First-Order Partial Differential Equations

This chapter considers the equation

$$(1) \qquad\qquad Pp + Qq \;=\; R,$$

where p, q denote the partial derivatives

$$p \;=\; z_x \;=\; \frac{\partial z}{\partial x}, \qquad q \;=\; z_y \;=\; \frac{\partial z}{\partial y}$$

of the real-valued function $z = z(x,y)$ and P, Q, R are continuous real-valued functions of $\mathbf{r} = (x,y,z)$ in an appropriate domain in $\mathbb{R}^3$. The equation is linear in p, q and *quasi-linear* because, in addition, P, Q, R are functions of z as well as of x and y.

We search for real-valued solutions $z = z(x,y)$ of (1). It is often useful to visualise our methods geometrically, and we shall refer to such solutions, because of the objects they represent, as *solution surfaces*. There is one important picture to hold in one's mind, and this is illustrated in section 5.5, where it forms an integral part of our discussion. Otherwise, our aim is to encourage the reader's *analytic* intuition.

The main problem, the *Cauchy problem*, is to discover whether there is a unique solution surface which contains ('passes through') a given 'boundary' curve. This is in general not possible and when it is, there is usually a restriction to be placed on the domain of the independent variables x, y in order that such a unique solution exists. The restricted domain is called the *domain of definition* of the solution.

In this respect, the theory is not as tidy as with ordinary differential equations. The justification of all the claims we shall make are beyond the scope of this book. Not only will we omit a complete set of proofs, but we shall often argue by example. Nevertheless, it should be possible for the reader to gain a feel for the subject and the technique necessary to solve elementary examples which may arise, say, in probability theory, together with an awareness as to when solutions are going to break down.

Throughout our theoretical considerations, the functions P, Q, R will be assumed to satisfy Lipschitz conditions of the form (e) of section 2.2; that is, P, Q, R will satisfy such inequalities as

$$|P(x_1, y_1, z_1) - P(x_2, y_2, z_2)| \leq A \max(|x_1 - x_2|, |y_1 - y_2|, |z_1 - z_2|)$$

(for some positive constant A) in an appropriate domain in $\mathbb{R}^3$.

5.1   Characteristics and some geometrical considerations

Our discussion of solutions of equation (1) will involve an associated family of continuously differentiable curves, with equation $\mathbf{r} = \mathbf{r}(s) = (x(s), y(s), z(s))$, called *characteristic curves*, or just *characteristics*. These curves are defined as those satisfying the *characteristic equations*

$$(2) \qquad \frac{dx}{ds} = P(x(s), y(s), z(s)), \quad \frac{dy}{ds} = Q(x(s), y(s), z(s)), \quad \frac{dz}{ds} = R(x(s), y(s), z(s)).$$

(A solution of (2) is guaranteed by the obvious three-dimensional extension of Theorem 2 of section 2.2; more about this later.) Sometimes, to be concise, we shall re-write the characteristic equations as

$$(2') \qquad \frac{dx}{P} = \frac{dy}{Q} = \frac{dz}{R},$$

which some authors call *auxiliary equations*. It should be emphasised that $(2')$ will be considered only as an abbreviation of (2); there will be no need to discuss differentials dx, dy, dz as such.

Note One interprets a zero denominator in $(2')$ as implying a zero numerator: for example, $P = 0$ implies $\dfrac{dx}{ds} = 0$ (see (2)).

We now remind the reader of an elementary result from the theory of surfaces.

Lemma 1 Suppose that the surface defined by $f(x, y, z) = \lambda$, where f is continuously differentiable and λ is constant, has a well-defined tangent plane at $\mathbf{r}_0 = (x_0, y_0, z_0)$, so that $(\text{grad } f)(\mathbf{r}_0) = (f_x(x_0, y_0, z_0), f_y(x_0, y_0, z_0), f_z(x_0, y_0, z_0))$ is non-zero. Then the normal to the surface at $\mathbf{r}_0$ is in the direction $(\text{grad } f)(\mathbf{r}_0)$.

Corollary 1 If the continuously differentiable surface $z = z(x, y)$ has a well-defined tangent plane at $\mathbf{r}_0 = (x_0, y_0, z_0)$, then the normal to the surface at $\mathbf{r}_0$ is in the direction $(p(x_0, y_0), q(x_0, y_0), -1)$.

Exercise 1 Show that the problem of finding a surface $z = z(x, y)$ which cuts orthogonally each member of the one-parameter family of surfaces $f(x, y, z) = \lambda$ (λ varying) can be reduced, for continuously differentiable f, to finding a (continuously differentiable) solution $z = z(x, y)$ of the partial differential equation

$$f_x p + f_y q = f_z.$$

Equation (1) displays the fact that the two vectors (P, Q, R) and $(p, q, -1)$ are orthogonal, and by (2) we thus have that the tangent $d\mathbf{r}/ds$ to the characteristic is perpendicular to the surface normal $(p, q, -1)$. We certainly know therefore that, where a characteristic meets a solution surface, its tangent lies in the surface's tangent plane. We can in fact do rather better, as the next result demonstrates.

Proposition 1 Suppose that S is a solution surface of (1), with equation $z = z(x, y)$ and containing $\mathbf{r}_0 = (x_0, y_0, z_0)$, and that C is a characteristic curve for the equation through $\mathbf{r}_0$. Assuming P, Q, R are continuous and satisfy appropriate Lipschitz conditions, the curve C lies on S.

Proof Using the existence part of Cauchy–Picard's two-dimensional theorem (Chapter 2, Theorem 2), find a solution $(x, y) = (X(s), Y(s))$ of

$$\frac{dx}{ds} = P(x(s), y(s), z(x(s), y(s))), \qquad x(0) = x_0,$$

$$\frac{dy}{ds} = Q(x(s), y(s), z(x(s), y(s))), \qquad y(0) = y_0.$$

Putting $Z(s) = z(X(s), Y(s))$, we have

$$
\frac{dX}{ds} = P(X(s), Y(s), Z(s)), \qquad X(0) = x_0,
$$

$$
\frac{dY}{ds} = Q(X(s), Y(s), Z(s)), \qquad Y(0) = y_0
$$

and

$$
\frac{dZ}{ds} = z_x(X(s), Y(s)) \cdot \frac{dX}{ds} + z_y(X(s), Y(s)) \cdot \frac{dY}{ds}
$$

$$
= z_x(X(s), Y(s)) \cdot P(X(s), Y(s), Z(s)) + z_y(X(s), Y(s)) \cdot Q(X(s), Y(s), Z(s)).
$$

Hence,

$$
\frac{dZ}{ds} = R(X(s), Y(s), Z(s)), \qquad Z(0) = z_0.
$$

So, the curve C', with equation $\mathbf{r} = \mathbf{R}(s) = (X(s), Y(s), Z(s))$, is a characteristic curve through $\mathbf{r}_0$. Using the uniqueness part of Cauchy–Picard's theorem, C' must be C. But, by construction, C' lies on S; so, C must also. ☐

All characteristics meeting a solution surface thus lie on it. Since every point in space lies on a characteristic, a solution surface must be made up of characteristic curves – a matter to which we shall return later in the chapter.

5.2 Solving characteristic equations

We shall give two methods of finding a solution surface of the partial differential equation (1), containing a given boundary curve. Both depend on first solving the corresponding characteristic equations. They relate directly to two methods for finding characteristic curves:

(A) in parametric form,

(B) as the intersection of two families of surfaces.

Example 1 Find the characteristics of the partial differential equation $xp + yq = z$.

Method (A): The characteristic equations are

$$\frac{dx}{ds} = x, \qquad \frac{dy}{ds} = y, \qquad \frac{dz}{ds} = z.$$

These may be integrated directly to give a solution in parametric form, *viz.*:

$$x = Ae^s, \qquad y = Be^s, \qquad z = Ce^s,$$

where A, B, C are constants.

These are half-lines and may be shown to lie on the intersections of pairs of planes by eliminating s:

$$\frac{x}{A} = \frac{y}{B} = \frac{z}{C}$$

(with appropriate modification if any of A, B, C are zero). Note that only two of the constants are independent. □

Method (B): One may re-write the characteristic equations 'in auxiliary form':

$$\frac{\frac{dx}{ds}}{x} = \frac{\frac{dy}{ds}}{y} = \frac{\frac{dz}{ds}}{z} \qquad \text{or} \qquad \frac{dx}{x} = \frac{dy}{y} = \frac{dz}{z}$$

(the latter being a useful short-hand). Considering

$$\frac{dx}{x} = \frac{dy}{y}, \qquad \frac{dx}{x} = \frac{dz}{z},$$

say (any two different equations will do), we find, on integrating *with respect to* s, that

$$y = c_1 x, \qquad z = c_2 x,$$

where c_1, c_2 are constants. Both these equations must be satisfied, and we again arrive at straight lines, here given by

$$\frac{x}{1} = \frac{y}{c_1} = \frac{z}{c_2},$$

with two arbitrary constants. □

Which of the methods (A) or (B) is chosen may depend on

(i) the functions P, Q, R,

(ii) the boundary conditions to be fitted (which we shall come to below),

but it is also often just a question of personal preference.

We shall conclude this section by developing some techniques for solving characteristic equations in auxiliary form. The abbreviated notation gives, with practice, the chance of spotting solutions quickly. This often arises with the use of the following lemma.

Lemma 2 ('Componendo dividendo') If

$$E \equiv \frac{a}{l} = \frac{b}{m} = \frac{c}{n},$$

then

$$E = \frac{\lambda a + \mu b + \nu c}{\lambda l + \mu m + \nu n},$$

for all λ, μ, ν.

The proof of this result is, of course, trivial (in any appropriate field). The elements λ, μ, ν can be functions (not just constants), and we can even allow zero denominators which give zero numerators by means of the equality

$$\lambda a + \mu b + \nu c = E(\lambda l + \mu m + \nu n).$$

We can thus extend our methods beyond the 'separable', or 'separated', case

$$\frac{dx}{f(x)} = \frac{dy}{g(y)} = \frac{dz}{h(z)}$$

as the following example shows. The use of 'componendo dividendo', of course, corresponds precisely to linear manipulation of the characteristic equations (2), and zero denominators again imply zero numerators (see Note after equations (2')).

Example 2 For each of the following equations, find two families of surfaces that generate the characteristics of

(a) $(3y - 2z)p + (z - 3x)q = 2x - y$,

(b) $\dfrac{3}{x - y}(p - q) = 2,$ $(x \neq y)$

(c) $(3x - z)p + (3y - z)q = x + y$.

Ad (a): The auxiliary equations

$$\frac{dx}{3y - 2z} = \frac{dy}{z - 3x} = \frac{dz}{2x - y}$$

give rise to

$$dx + 2dy + 3dz = 0, \qquad (\lambda = 1,\ \mu = 2,\ \nu = 3)$$

$$xdx + ydy + zdz = 0, \qquad (\lambda = x,\ \mu = y,\ \nu = z)$$

which integrate to the two families of surfaces

$$x + 2y + 3z = c_1,$$

$$x^2 + y^2 + z^2 = c_2,$$

where c_1, c_2 are constants.

Ad (b): The auxiliary equations

$$\frac{(x - y)dx}{3} = \frac{(x - y)dy}{-3} = \frac{dz}{2}$$

give rise to

$$dx = -dy, \qquad \text{(separated already)}$$

$$\frac{(x - y)(dx - dy)}{6} = \frac{dz}{2}, \qquad (\lambda = 1,\ \mu = -1,\ \nu = 0 \text{ on left})$$

which integrate to

$$x + y = c_1,$$

$$(x - y)^2 - 6z = c_2.$$

Ad (c): The auxiliary equations

$$\frac{dx}{3x - z} = \frac{dy}{3y - z} = \frac{dz}{x + y}$$

give rise to

$$\frac{dx - dy}{3(x - y)} = \frac{dx + dy - dz}{2(x + y - z)} = \frac{dx + dy - 2dz}{x + y - 2z},$$

when one chooses, in turn,

$$
\begin{aligned}
\lambda &= 1, & \mu &= -1, & \nu &= 0; \\
\lambda &= 1, & \mu &= 1, & \nu &= -1; \\
\lambda &= 1, & \mu &= 1, & \nu &= -2.
\end{aligned}
$$

The new equations are in separable form (in the variables $x - y$, $x + y - z$, $x + y - 2z$) and can be integrated immediately (using logarithms) to give

$$\frac{(x + y - z)^3}{(x - y)^2} = c_1,$$

$$\frac{(x + y - 2z)^3}{x - y} = c_2,$$

whenever $x \neq y$. □

If one can only spot one integrable equation, the following 'trick' can sometimes be used to derive a second family of surfaces.

Ad (b) – alternative method: suppose one has found only

$$dx = -dy$$

giving the one-parameter family

$$x + y = c_1.$$

Substitute for y, so that

$$\frac{(2x - c_1)dx}{3} = \frac{dz}{2}.$$

Integrating:

$$2(x^2 - c_1 x) - 3z = -c_2',$$

and substitute for c_1 to give a second one-parameter family

$$3z + 2xy = c_2'.$$

The curves generated by

$$x + y = c_1, \qquad (x - y)^2 - 6z = c_2$$

are, of course, the same as those generated by

$$x + y = c_1, \qquad 3z + 2xy = c_2'.$$

(In the second pair of equations, substract twice the second from the square of the first, etc.) □

Exercise 2 Use *both* methods (A) and (B) to find the characteristics of the equation

$$zp + q + \lambda z = 0, \qquad (0 < \lambda < 1).$$

(Be wary of the position of the second z.)

Exercise 3 Find the characteristics of the equations

(a) $\quad (y + z)p - (z + x)q + (x - y) = 0,$

(b) $\quad (1 + z)p + 2(1 - z)q = z^2,$

(c) $\quad x(y + z^2)p + y(x^2 - z^2)q + z(x^2 + y) = 0.$

5.3 General solutions

With ordinary differential equations, an n-th order equation can be integrated to a 'general solution' containing n arbitrary constants. Conversely, n constants in an equation can often be eliminated by forming an appropriate n-th order ordinary differential equation. For example, the differential equation

$$x^2 y'' - 2xy' + 2y = 0$$

can be solved (by the methods of (7) of the Appendix) to give

$$y = Ax^2 + Bx,$$

where A, B are arbitrary constants. The second-order equation can be recovered (as the reader may verify directly) by eliminating these constants.

In the case of partial differential equations, the arbitrary constants are replaced by arbitrary (suitably differentiable) functions. For example, by successive integrations, the second-order partial differential equation

$$\frac{\partial^2 z}{\partial x \partial y} = 0$$

gives rise to

$$\frac{\partial z}{\partial x} = h(x),$$

and then

$$z = f(x) + g(y),$$

where f, g are arbitrary functions with $f'(x) = h(x)$. Clearly, partial differentiation, first with respect to x and then with respect to y (or in the other order), eliminates the two arbitrary functions f, g and recovers the second-order equation.

Definition 1 A *general solution* of an n-th order partial differential equation is a solution containing n arbitrary (suitably differentiable) functions.

So, in looking for solutions of the first-order equations studied in this chapter, we shall be looking for solutions containing one arbitrary continuously differentiable function. The following theorem gives a mechanism for finding such a solution.

Theorem 1 Suppose that the characteristics of the partial differential equation

(1) $$Pp + Qq = R$$

are represented as the intersection-curves of the two families of surfaces with equations

$$u(x, y, z) = c_1, \qquad v(x, y, z) = c_2,$$

where $u = u(x, y, z)$ and $v = v(x, y, z)$ are continuously differentiable functions. Then a general solution of (1) can be written in the form

$$F(u, v) = 0,$$

where F is an arbitrary continuously differentiable function of two variables, with partial derivatives F_u, F_v not both zero at any given point of their domain.

Proof We have to show that if $z = z(x, y)$ is continuously differentiable, then the elimination of the arbitrary continuously differentiable function F gives rise to z satisfying (1).

Using the chain rule ([C] of Chapter 0) to differentiate

$$F(u(x, y, z(x, y)), v(x, y, z(x, y))) = 0,$$

with respect to x and with respect to y, we have

$$F_u.(u_x + u_z p) + F_v.(v_x + v_z p) = 0,$$

$$F_u.(u_y + u_z q) + F_v.(v_y + v_z q) = 0.$$

Elimination of $F_u : F_v$ gives (and the reader should check that)

$$(3) \qquad \frac{\partial(u, v)}{\partial(y, z)} p + \frac{\partial(u, v)}{\partial(z, x)} q = \frac{\partial(u, v)}{\partial(x, y)},$$

where $\partial(u, v)/\partial(y, z)$ is the Jacobian determinant

$$\frac{\partial(u, v)}{\partial(y, z)} \equiv u_y v_z - u_z v_y,$$

etc. (Notice that in (3) the Jacobian with y, z in its denominator is taken with the partial derivative p of z with respect to x, with appropriate cyclic interchange of the variables in the other determinants.)

We now use the fact that the characteristics $\mathbf{r} = \mathbf{r}(s) = (x(s), y(s), z(s))$ of the original partial differential equation must lie on a member of each of the given families of surfaces; so that

$$u(x(s), y(s), z(s)) = c_1, \qquad v(x(s), y(s), z(s)) = c_2$$

and hence, by the chain rule,

$$u_x \frac{dx}{ds} + u_y \frac{dy}{ds} + u_z \frac{dz}{ds} = 0 = v_x \frac{dx}{ds} + v_y \frac{dy}{ds} + v_z \frac{dz}{ds}.$$

But the characteristics also satisfy the characteristic equation (2), and hence

$$u_x P + u_y Q + u_z R = 0 = v_x P + v_y Q + v_z R.$$

These last two equations may be re-written

(4)
$$\frac{P}{\dfrac{\partial(u,v)}{\partial(y,z)}} = \frac{Q}{\dfrac{\partial(u,v)}{\partial(z,x)}} = \frac{R}{\dfrac{\partial(u,v)}{\partial(x,y)}}.$$

Combining (3) and (4), we see that $z = z(x,y)$ must satisfy (1), and the proof is complete. □

Note The equations (4) follow from the elementary result:

$$\left.\begin{array}{l} a_1 x + a_2 y + a_3 z = 0 \\[2mm] b_1 x + b_2 y + b_3 z = 0 \end{array}\right\} \quad \text{if and only if} \quad \frac{x}{\begin{vmatrix} a_2 & a_3 \\ b_2 & b_3 \end{vmatrix}} = \frac{y}{\begin{vmatrix} a_3 & a_1 \\ b_3 & b_1 \end{vmatrix}} = \frac{z}{\begin{vmatrix} a_1 & a_2 \\ b_1 & b_2 \end{vmatrix}}$$

The reader might note the cyclic permutations of the a_i's and b_i's in the determinants.

Example 3 Find general solutions of each of the equations of Example 2.

By the theorem, general solutions are

(a) $F(x + 2y + 3z, x^2 + y^2 + z^2) = 0,$

(b) $G(x + y, (x - y)^2 - 6z) = 0 \quad$ or $\quad H(x + y, 3z + 2xy) = 0,$

(c) $K\left(\dfrac{(x + y - z)^3}{(x - y)^2}, \dfrac{(x + y - 2z)^3}{x - y}\right) = 0, \quad (x \neq y)$

where F, G, H, K are arbitrary continuously differentiable functions. □

Exercise 4 Find a general solution of each of the equations of Exercise 3.

5.4   Fitting boundary conditions to general solutions

As we noted at the start of this chapter, the main problem in the theory of the partial differential equation (1) is to discover whether there is a (unique) solution surface which contains a given 'boundary' curve. In this section, we concentrate attention on finding the surface when the curve permits such a solution to exist. We shall turn to theoretical considerations, as to when this is possible, in the next section.

The techniques we wish to impart here will be clear from the following worked examples.

Example 4 Find the solution to

$$(3y - 2z)p + (z - 3x)q \;=\; 2x - y$$

which contains the line of intersection of the two planes $x = y, z = 0$.

By Example 3(a), a general solution of the partial differential equation is

$$F(x + 2y + 3z, x^2 + y^2 + z^2) \;=\; 0,$$

where F is an arbitrary continuously differentiable function. For this to contain the curve $x = y, z = 0$, we must have

$$F(3x, 2x^2) \;=\; 0,$$

for every x. This suggests

$$F(u, v) \;\equiv\; 2u^2 - 9v$$

as a possible solution. This in turn gives

$$2(x + 2y + 3z)^2 - 9(x^2 + y^2 + z^2) \;=\; 0. \qquad\qquad \square$$

Alternatively, try to find a solution in the form

(5) $$x^2 + y^2 + z^2 \;=\; f(x + 2y + 3z).$$

This equation represents a cone with axis $(1, 2, 3)$. It can be solved for z in terms of x and y to give either the 'upper' or the 'lower' points of the surface of the cone. Only one contains the line $x = y, z = 0$ for some continuously differentiable function f. Putting $x = y, z = 0$,

$$f(3x) \;=\; 2x^2;$$

and so, with $3x = t$, the function f is given by

$$f(t) = \frac{2}{9} t^2.$$

Hence, substituting $x + 2y + 3z$ for t, we find the required solution

(6) $$x^2 + y^2 + z^2 = \frac{2}{9}(x + 2y + 3z)^2,$$

as above. $\qquad\qquad \square$

Note Though the choice of method above is very much a matter of personal preference (ours is for the second), the reader will find that having both methods available can be of great use. In the second method, we have sought a solution $F(c_1, c_2) = 0$, where

$$F(c_1, c_2) \;=\; f(c_1) - c_2.$$

The trial solution (6) involves expressing the more 'complicated' $v \equiv x^2 + y^2 + z^2$ in terms of the 'simpler' $u \equiv x + 2y + 3z$. Such organisation can simplify the working. The reader will notice that we have not bothered to 'reduce' the solution to $z = z(x, y)$ form. Whilst doing this, consider what kind of quadratic surface is represented by (7), containing, as it must, the line $x = y, z = 0$.

Example 5 Find a surface cutting each member of the family

$$3 \log(x - y) + 2z \;=\; \lambda \qquad (y < x, \; -\infty < \lambda < \infty)$$

orthogonally and containing the y-axis.

    By virtue of Exercise 1, we need to solve

$$\frac{3}{x - y}\,(p - q) \;=\; 2.$$

Example 2(b) provides the characteristics as the intersection curves of the families

$$x + y \;=\; c_1, \qquad (x - y)^2 - 6z \;=\; c_2.$$

So, we search for a solution in the form

$$(x - y)^2 - 6z \;=\; g(x + y).$$

On the y-axis, $x = z = 0$, and so

$$g(y) \;=\; y^2.$$

The solution is therefore

$$(x - y)^2 - 6z \;=\; (x + y)^2$$

or

(7)
$$z \;=\; -\frac{2xy}{3}. \qquad\qquad \square$$

Note If we had worked with the alternative general solution given in Example 3(b) and tried

$$3z + 2xy = h(x + y),$$

we should have discovered, on putting $x = z = 0$, that h is identically 0. But this gives the same solution (8), which should comfort the reader.

Example 6 The same as Example 5, save that the boundary curve is the line-segment given by $x = z = 0$, $0 < y < 1$.

   The analysis is the same as far as

$$g(y) = y^2$$

which now holds only for $0 < y < 1$. The solution now is therefore

$$z = -\frac{2xy}{3}$$

which is certainly valid for $0 < x + y < 1$ (as it is the substitution of $x + y$ for y which allows the derivation of the solution from the functional form for g). In fact, the solution still satisfies the equation everywhere.   □

Exercise 5 Solve the equations of Exercise 3 when subject to the (respective) boundary conditions:

(a) $x = z = 0$;

(b) $x = y$, $z = 1$;

(c) $x + z = 0$, $y = 1$.

Answers may be left in implicit form. Note any restrictions on the validity of the solutions you give.

Exercise 6 Find a surface cutting each member of the family

$$(x + y)z = \lambda$$

orthogonally and containing the line-segment given by

$$y + z = x = 1, \qquad 0 < y < 1.$$

Exercise 7 Find the solution surface in the half-plane $y \geq 0$ of the equation of Example 2(c), which contains the half-line given by $y = z = 0$, $x < 0$. (You should not expect the technicalities to be trivial.)

Exercise 8 (*Harder*) Find the implicit solution of the equation

$$zp + q + \lambda z = 0, \qquad (0 < \lambda < 1)$$

of Exercise 2, satisfying $z(x, 0) = f(x)$, for every x. Show that, when $f(x) = \sin x$, the first positive value y_0 of y for which $p = \partial z / \partial x$ is *not* finite is

$$y_0 = -\frac{1}{\lambda} \log(1 - \lambda).$$

[HINT: Use the auxiliary equations technique to derive two families of surfaces, $u = c_1$ and $v = c_2$, which give rise to the characteristics. Insert the boundary values directly and eliminate x to derive a relationship $F(c_1, c_2) = 0$ between the constants c_1 and c_2. Substitute $c_1 = u$, $c_2 = v$ to obtain the general solution $F(u, v) = 0$. (Implicitly) differentiate this equation partially with respect to x without more ado, to give an expression for p.]

Exercise 9 Find the solution of

$$yp - 2xyq = 2xz$$

that satisfies $z = y^3$ when $x = 0$ and $1 \leq y \leq 2$. What is the domain of definition of your solution (in terms of x and y)?

Exercise 10 Find the solutions of the equation

$$xzp - yzq = x^2 - y^2$$

in $xy \geq 0$ which pass through

(a) $x = y = z$;

(b) $x = y, \quad z = 0$.

Describe each solution geometrically.

Exercise 11 (from probability theory) The functions $p_k(t)$ satisfy the simultaneous ordinary differential equations

$$\frac{dp_k}{dt} = kp_{k-1} - (2k + 1)p_k + (k + 1)p_{k+1}, \qquad (k = 0, 1, 2, \ldots).$$

Obtain *formally* a first-order partial differential equation for their generating function

$$P(x, t) = \sum_{k=0}^{\infty} p_k(t)x^k, \qquad (|x| < 1).$$

Solve the partial differential equation. Hence, evaluate $p_k(t)$ for $k \geq 0$ and $t > 0$, when $p_0(0) = 1$ and $p_k(0) = 0$ for $k > 0$.

5.5   Parametric solutions and domains of definition

In this section, we investigate the problem of when a solution surface of (1) can be found which contains a given boundary curve. The discussion gives rise, in our view, to one of the best examples in elementary mathematics where an abstract analytic condition translates to a vivid geometrical one (see section 5.6).

In contrast to the procedure described in section 5.4, where a general solution was found first and the boundary curve fitted after, we shall be concerned now with the 'building' of a solution surface directly on the boundary curve.

We again consider the partial differential equation

$$(1) \qquad\qquad Pp + Qq \;=\; R,$$

where P, Q, R are continuous functions of x, y, z and satisfy the Lipschitz conditions of the form (e) of section 2.2 (see our comments at the end of the introduction to this chapter). The boundary curve Γ will have the notation

$$\mathbf{r} \;=\; \mathbf{r}(t) \;=\; (x(t), y(t), z(t)), \qquad (t \in A)$$

with $\mathbf{r}(t)$ continuously differentiable and $(\dot{x}(t), \dot{y}(t))$ never zero for any value of t in $A \subseteq \mathbb{R}$. (Throughout this chapter the dot will denote differentiation with respect to t. Further, A will normally denote a finite or infinite interval in $\mathbb{R}$.)

For each *fixed* t in A the results of Chapter 2 allow us uniquely to solve the characteristic equations

$$(2) \qquad \frac{dx}{ds} \;=\; P(x, y, z), \qquad \frac{dy}{ds} \;=\; Q(x, y, z), \qquad \frac{dz}{ds} \;=\; R(x, y, z),$$

when subject to the condition of passing through the *point*

$$x \;=\; x(t), \quad y \;=\; y(t), \quad z \;=\; z(t)$$

of the boundary curve. This t-th characteristic may be expressed in the form

$$\mathbf{r} \;=\; \mathbf{r}^t(s) \;=\; (x^t(s), y^t(s), z^t(s))$$

which, if we suppose that s can be chosen to be zero at the point where the characteristic meets the boundary curve, satisfies

$$\mathbf{r}^t(0) \;=\; \mathbf{r}(t) \;=\; (x(t), y(t), z(t)).$$

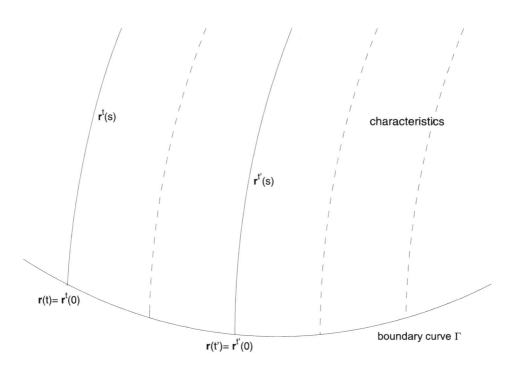

We now allow t to vary in A, so that $\mathbf{r}(t)$ varies along Γ. Writing $\mathbf{r}(s, t)$ for $\mathbf{r}^t(s)$, the equation

$$(8) \qquad\qquad\qquad \mathbf{r} \;=\; \mathbf{r}(s, t)$$

determines a geometrical object in terms of two parameters (which we hope is a solution surface of (1)). By construction, it contains the boundary curve Γ (when $s = 0$) and each characteristic emanating from it. (For given s, t, the point $\mathbf{r}(s, t)$ is found 'by going t along the boundary curve and s up the characteristic that "starts" there'.) By Proposition 1 of section 5.1, if there exists a solution surface containing the boundary curve, then it must contain the object given by (8).

We shall *not* display the existence of a solution surface containing the boundary curve. But we will discuss now when (8) can be written as a continuously differentiable surface in the form $z = z(x, y)$. In this case, when the given functions have power series expansions, a famous result, the Cauchy–Kowalewski Theorem, guarantees a unique solution, at least in a neighbourhood of each point on the boundary curve. The reader will discover that when such a function $z = z(x, y)$ can be constructed, it is often the solution for a wider range of values of its independent variables. This range, known as the *domain of definition* of the (unique) solution, will appear in examples and exercises below. On deriving a solution, one should always try to determine its domain of definition.

Suppose now that we have determined

$$\mathbf{r} = \mathbf{r}(s,t) = (x(s,t), y(s,t), z(s,t))$$

as above for a certain range of s and t with

$$\mathbf{r}(0,t) = \mathbf{r}(t) = (x(t), y(t), z(t)).$$

Working only formally for the present, we solve

(9)
$$x = x(s,t), \quad y = y(s,t)$$

to give

(10)
$$s = s(x,y), \quad t = t(x,y),$$

and then substitute in $z = z(s,t)$, to obtain

(11)
$$z = z(x,y) \equiv z(s(x,y), t(x,y)).$$

For this $z = z(x,y)$ to be continuously differentiable, it is sufficient, by the chain rule, for $s = s(x,y)$ and $t = t(x,y)$ to be continuously differentiable. So, to justify our procedure, we need to justify the 'inversion' from equations (9) to equations (10). Note that, as P and Q are continuous, the partial derivatives x_s, y_s of the solution (9) exist and are continuous. *Assuming the existence and continuity of x_t, y_t, we now invoke the Inverse Function Theorem*[1] of multi-variable calculus which allows inversion in a neighbourhood $N(s_0, t_0)$ of a point (s_0, t_0), to give *continuously differentiable* functions of form (10), *provided* the Jacobian

(12)
$$J \equiv \frac{\partial(x,y)}{\partial(s,t)} \equiv x_s y_t - x_t y_s \neq 0,$$

when evaluated at (s_0, t_0). The resulting function $z = z(x,y) \equiv z(x,y,s_0,t_0)$ of (11) then defines a continuously differentiable surface in a neighbourhood of (x_0, y_0), where

$$x_0 = x(s_0, t_0), \quad y_0 = y(s_0, t_0).$$

Note A corresponding 'one-dimensional' example may clarify the need for the condition (12). Consider the function

$$y = x^3.$$

---

[1]See, for example, T.M. Apostol, *Mathematical Analysis*, 2nd Edition, Addison–Wesley, 1975, Theorem 13.6, page 272.

This function is a (1–1) onto continuously differentiable function on the whole line $\mathbb{R}$, with inverse

$$x = y^{\frac{1}{3}}.$$

However, this latter function fails to be differentiable at $y = 0$ (since $h^{1/3}/h$ has no limit as h tends to 0). This corresponds to

$$\frac{dy}{dx} = 3x^2$$

being zero at $x = 0$. Thus, $y = x^3$ only has a continuously differentiable inverse at points where $\dfrac{dy}{dx}$ is non-zero. In general, one can only achieve an inverse locally, and the 'two-dimensional analogue' of this result is the Inverse Function Theorem used above.

The hope in the above analysis is that one might at least be able to find one function $z = z(x, y)$ which will define a continuously differentiable solution in an (as large as possible) neighbourhood of the boundary curve, and that the surface it defines, containing, as it does, sections of characteristics meeting the boundary curve, is a solution to our partial differential equation (1). So, we should require that the Jacobian $J = J(s, t)$ satisfy (12) along the boundary curve; that is

$$J(0, t) \neq 0,$$

for every $t \in A$. Our hope may then be expressed as a desire that there is a solution surface $z = z(x, y)$ of (1) which agrees with $z = z(x, y, 0, t)$ on $N(0, t)$, for every $t \in A$.

There are two things that can go wrong with the geometrical construction outlined in this section. One is that the boundary curve Γ is tangential to the characteristics. If this occurs at an isolated point, then the point is just a 'singularity' of the surface. However, if it occurs over an interval (a rather special case), then Γ *is* the characteristic and there is no surface at all. The other difficulty is that we may derive a perfectly good surface, but its tangent plane is parallel to the z-axis and so is *not* given locally by an equation of the form $z = z(x, y)$ with continuously differentiable $z(x, y)$. The discussion on the Jacobian above bears on these difficulties and section 5.6 throws further light on them.

The reader may assume in the following examples and exercises that, if $\mathbf{r} = \mathbf{r}(s, t)$ describes, as above, a family of characteristics passing through a boundary curve in a neighbourhood of the boundary curve in which condition (12) holds, then $\mathbf{r} = \mathbf{r}(s, t)$ represents a solution surface of (1) in parametric form in that neighbourhood.

Example 7 Show that there is a solution surface of

$$zp + q + \lambda z = 0, \qquad (0 < \lambda < 1)$$

containing the boundary curve given by

$$y = 0, \quad z = \sin x,$$

provided

$$\log((1+\lambda)^{-\frac{1}{\lambda}}) < y < \log((1-\lambda)^{-\frac{1}{\lambda}}).$$

Choosing the parametrisation $\mathbf{r}(t) = (t, 0, \sin t)$ for the boundary curve, we solve the (corresponding) characteristic equations

$$\frac{dx}{ds} = z, \quad \frac{dy}{ds} = 1, \quad \frac{dz}{ds} = -\lambda z,$$

parametrically. The second and third equations integrate immediately to give

$$y = s + B(t), \quad z = C(t)e^{-\lambda s},$$

where the 'constants of integration' $B(t), C(t)$ vary with the point $\mathbf{r}(t)$ on the boundary curve which the t-th characteristic contains. Substituting in the first equation, we find that

$$\frac{dx}{ds} = C(t)e^{-\lambda s}.$$

On integration (with respect to s with t fixed), we thus have characteristics

$$x = -\frac{1}{\lambda}C(t)e^{-\lambda s} + A(t), \quad y = s + B(t), \quad z = C(t)e^{-\lambda s}.$$

Taking $s = 0$ on the boundary, where $\mathbf{r}(t) = (t, 0, \sin t)$, this reduces to

$$x = t + \frac{1}{\lambda}(1 - e^{-\lambda s})\sin t, \quad y = s, \quad z = e^{-\lambda s}\sin t.$$

The Jacobian

$$J(s,t) \equiv x_s y_t - x_t y_s = -(1 + \frac{1}{\lambda}(1 - e^{-\lambda s})\cos t)$$

is certainly non-zero on the boundary (when $s = 0$). As $|\sec t| \geq 1$ and $\lambda > 0$, it remains non-zero when

$$|1 - e^{-\lambda s}| < \lambda,$$

that is, when

$$1 - \lambda < e^{-\lambda s} < 1 + \lambda.$$

As $y = s$, the result follows. $\qquad\qquad\qquad\qquad\qquad\qquad\qquad\qquad\square$

Example 8 Discuss the existence of a solution surface to the equation

$$p + q = z$$

containing the curve $x = y$, $z = 1$.

Choosing the parametrisation $\mathbf{r}(t) = (t, t, 1)$ for the boundary curve, we solve the characteristic equations

$$\frac{dx}{ds} = 1, \quad \frac{dy}{ds} = 1, \quad \frac{dz}{ds} = z$$

parametrically, to obtain

$$x = s + A(t), \quad y = s + B(t), \quad z = C(t)e^s.$$

Taking $s = 0$ (or any value) on the boundary, where $r(t) = (t, t, 1)$, we see that

$$x = s + t, \quad y = s + t, \quad z = e^s.$$

The variables x, y are clearly *not* independent, it is impossible to solve for s or t in terms of x, y, and $J(s, t)$ is always zero.

The auxiliary equations route is just as fruitless. In this case, those equations are

$$\frac{dx}{1} = \frac{dy}{1} = \frac{dz}{z},$$

leading to characteristics as intersections of the families

$$y - x = c_1,$$

$$ze^{-y} = c_2.$$

A trial solution of the form

$$y - x = f(ze^{-y}),$$

with f continuously differentiable and $x = y$ when $z = 1$, gives $f(u) = 0$ when $u = e^{-y}$; so, $x = y$ when $z > 0$ (all because e^{-y} is always positive).

Note that in the parametric 'solution', $z = e^s$ is always positive. 'Geometrically', the 'solution' is $x = y$ ($z > 0$), but this cannot, of course, be represented in the form $z = z(x, y)$.

Notice also that, in this example, the projection of the boundary curve on to the (x, y)-plane coincides with the characteristic given by $c_1 = c_2 = 0$. In these circumstances, it is unrealistic to expect to be able to 'build a surface on the boundary curve' using the characteristics. This point is discussed further in section 5.6.   □

Example 9 Find a function $z = z(x, y)$ which solves

(13) $$p + zq = 1,$$

where $z = \frac{1}{2}x$ on $x = y$ whenever $0 \leq x \leq 1$. Discuss the domain of definition of your solution.

Choosing the parametrisation $\mathbf{r}(t) = (t, t, \frac{1}{2}t)$, $0 \leq t \leq 1$, for the boundary curve, we solve the characteristic equations

$$\frac{dx}{ds} = 1, \quad \frac{dy}{ds} = z, \quad \frac{dz}{ds} = 1$$

parametrically to give

$$x = s + A(t), \quad y = \frac{s^2}{2} + sC(t) + B(t), \quad z = s + C(t),$$

where A, B, C are arbitrary (suitably differentiable) functions of t. Incorporating the boundary conditions at $s = 0$ yields

$$x = s + t, \quad y = \frac{s^2}{2} + \frac{st}{2} + t, \quad z = s + \frac{t}{2}, \qquad (0 \leq t \leq 1).$$

Solving for s, t in terms of x, y then gives

$$s = \frac{2(y - x)}{x - 2}, \quad t = \frac{x^2 - 2y}{x - 2}, \qquad (x \neq 2).$$

Hence,

(14) $$z = \frac{x^2 - 4x + 2y}{2(x - 2)}, \qquad \left(0 \leq \frac{x^2 - 2y}{x - 2} \leq 1, \ x \neq 2\right).$$

Notice, that the Jacobian

$$J(s, t) \equiv \frac{\partial(x, y)}{\partial(s, t)} \equiv x_s y_t - x_t y_s = 1 - \frac{s + t}{2},$$

is zero only when $x = s + t = 2$.

Second, we find that z given in (14) satisfies the partial differential equation (13) for all values of x, y, save $x = 2$.

Third, the boundary curve lies in the half-space $x < 2$. *Any* solution surface in $x > 2$ (corresponding to arbitrary continuously differentiable $A(t)$, $B(t)$, $C(t)$ such that $J(s,t) \neq 0$) may be matched with z given, for $x < 2$, in (14). The reader is asked to ponder whether any other continuously differentiable solutions exist for $x < 2$. (Certainly, there is the unique solution (14) given in $0 < 2y - x^2 \leq 2 - x$.) ☐

Exercise 12 Fill in the details in Example 9, giving also an alternative solution by means of the general solution/auxiliary equations method.

Exercise 13 Solve parametrically

$$(x - z)p + q + z = 0,$$

where $z = 1$ on $x = y$ for $0 < x < \frac{1}{2}$, showing that your solution must be unique in the domain defined by

$$-\sinh y < x < \frac{e^{\frac{1}{2} - y}}{2}.$$

Exercise 14 Find solutions $z = z(x,y)$ of the following problems, each valid in the positive quadrant $(x > 0, y > 0)$ save on a curve C which should in each case be specified:

(a)　　$p - zq + z = 0,$　　where $z(0,y) = \min(1,y)$,

(b)　　$p + 2zq = 1,$　　　where $z = 0$ on $y = 0$ for $x > 0$, and $z = y$ on $x = 0$ for $y > 0$.

Show further that in each case the partial derivatives p, q are discontinuous across a characteristic C, but that the solution to (b) requires an extra condition (such as continuity of z across C) to ensure that it is unique. (For practice, find solutions by both characteristic and auxiliary equations methods.)

Note Discontinuities of z and its partial derivatives can occur only across characteristic curves, as in Exercise 14. Finite ('jump') discontinuities in z are known as *shocks* because of an application of first-order partial differential equations with such discontinuities to gas dynamics.

Suppose that the partial differential equation is in *conservation form*

(15)
$$\frac{\partial L}{\partial x} + \frac{\partial M}{\partial y} = R,$$

making the equation a *conservation law*, where L, M, N are functions of x, y, z (for example, $2zz_x + z_y \equiv (z^2)_x + z_y = 0$).

One can then define a *weak solution* $z = z(x, y)$ as one which satisfies

$$(16) \qquad \iint_D (Lw_x + Mw_y - Rw)\, dx dy \;=\; \int_\Gamma w(L\, dy - M\, dx)$$

for all appropriate 'test functions' $w = w(x, y)$, where the boundary conditions for solutions z of (1) are given on the curve Γ which forms part of the boundary ∂D of the region D in $\mathbb{R}^2$. The functions w must be continuously differentiable in D and zero on $\partial D \setminus \Gamma$, the part of the boundary of D outside Γ. Using Green's Theorem in the Plane (see page 144), it can easily be shown that this weak solution not only agrees with the usual solution of (1) when z is continuously differentiable, but exists on characteristics in D across which discontinuities produce shocks.

5.6   A geometric interpretation of an analytic condition

The Jacobian condition

$$(12) \qquad\qquad J(s, t) \;\equiv\; x_s y_t - y_s x_t \;\neq\; 0,$$

when applied along the boundary curve, where we have taken $s = 0$, allows an interesting interpretation in geometrical terms. To see this, note that x_s, the partial derivative with respect to s (keeping t constant), must, by definition of $x(s, t)$, be

$$\frac{dx}{ds} \;=\; P$$

of the characteristic equations. Similarly, $y_s = Q$. Further, using the boundary condition $\mathbf{r}(0, t) = \mathbf{r}(t)$, we see that $x_t = \dot{x}(t), y_t = \dot{y}(t)$ when the partial derivatives are evaluated at $s = 0$. So, on $s = 0$, (12) may be re-written as

$$P\dot{y}(t) - Q\dot{x}(t) \;\neq\; 0$$

and is equivalent to saying that the directions determined by the ratios

$$\frac{dx}{ds} : \frac{dy}{ds} \quad \text{and} \quad \dot{x}(t) : \dot{y}(t)$$

are different. We have thus shown that the analytic condition (12) may be expressed by saying that the tangents to the projections on the (x, y)-plane of characteristic and boundary curves must, where they meet, be always in different directions.

We conclude with a set of exercises, which covers a range of the more important ideas and techniques introduced in this chapter.

Exercise 15 Find the solution $z = z(x, y)$ of the equation

$$yzp - xzq \ = \ x - y$$

which satisfies the conditions

$$z \ = \ 1 - x, \quad \text{when} \quad 1 < x < 2 \quad \text{and} \quad y = 1.$$

On what domain is the solution uniquely determined by these conditions?

Exercise 16 Consider the equation

$$ap + zq \ = \ 1,$$

where a is a real number. For $a \neq 1$, find, in explicit form, the unique solution $z = z(x, y)$ of this equation which satisfies $z = -1$ when $x + y = 1$, and determine the region in the (x, y)-plane where the solution is defined. Show that, if $a = 1$, there are two solutions of this equation satisfying $z = -1$ when $x + y = 1$, and explain why uniqueness fails in this case.

Exercise 17 Determine the characteristics of the equation

$$3p + zq \ = \ 2$$

which intersect the curve Γ given by $(x, y, z) = (t, t, 1)$ for $t \leq 0$. Sketch the projection in the (x, y)-plane of the characteristic through $(t, t, 1)$ and show that it touches the line $x = y + 1$. Find the solution surface $z = z(x, y)$ which passes through Γ, determine the region of the (x, y)-plane in which it is defined and indicate this region on your sketch.

Exercise 18 For the equation

$$p + zq \ = \ 0,$$

show that the projections of the characteristics on the (x, y)-plane are straight lines and obtain the solution satisfying $z = f(y)$ when $x = 0$ in the implicit form $F(x, y, z) = 0$. Deduce that $|q|$ becomes infinite when $1 + xf'(y - xz) = 0$.

Solve explicitly for $z = z(x, y)$ when $f(y) = y^2$, showing carefully that the solution $z \to y^2$ as $x \to 0$. Show also that $|q| \to \infty$ as we approach the curve with equation $4xy = -1$ and that each characteristic projection touches this curve once.

Exercise 19 Use the method of characteristics to show that

$$z(x, y) \ = \ \log(y + \exp(\tanh(x - y^2/2)))$$

solves the problem consisting in the equation

$$yp + q \ = \ e^{-z}, \quad (-\infty < x < \infty, \ y \geq 0)$$

together with the boundary condition

$$z(x, 0) \ = \ \tanh x, \qquad (-\infty < x < \infty).$$

Sketch the projections of the characteristics in the (x, y)-plane passing through the points $(1, 0)$, $(0, 0)$ and $(-1, 0)$. For fixed $y > 0$, does the solution have limits as $x \to \pm\infty$? What is the first positive value y_0 of y for which the solution is positive for all $y > y_0$?

6 Second-Order Partial Differential Equations

This chapter discusses the equation

$$(1) \qquad Ar + 2Bs + Ct = \phi(x, y, z, p, q),$$

where p, q, r, s, t denote the partial derivatives

$$p = z_x = \frac{\partial z}{\partial x}, \qquad q = z_y = \frac{\partial z}{\partial y},$$

$$r = z_{xx} = \frac{\partial^2 z}{\partial x^2}, \quad s = z_{xy} = z_{yx} = \frac{\partial^2 z}{\partial x \partial y} = \frac{\partial^2 z}{\partial y \partial x}, \quad t = z_{yy} = \frac{\partial^2 z}{\partial y^2}$$

of the twice continuously differentiable real-valued function $z = z(x, y)$ (which implies $z_{xy} = z_{yx}$) and A, B, C are real-valued functions of (x, y) in an appropriate domain in $\mathbb{R}^2$. The function ϕ is a continuous function of the 'lower order' (and necessarily continuous) functions z, p, q, as well as of the variables x, y. Note immediately that A, B, C are *not* functions of z (unlike the functions P, Q, R of Chapter 5). An equation such as (1), where the terms involving the highest-order derivatives are linear in those derivatives (as here, on the left-hand side of (1)), is said to have *linear principal part*. Equation (1) is *linear* if the function ϕ is given in the form

$$\phi(x, y, z, p, q) = Dp + Eq + Fz + G,$$

where D, E, F, G are functions of x and y.

By an appropriate change of independent variables

$$\xi \;=\; \xi(x,y), \qquad \eta \;=\; \eta(x,y),$$

equation (1) in $z = z(x,y)$ can be transformed to an equation in $Z = Z(\xi, \eta)$ *in canonical form*, where

$$z(x,y) \;=\; Z(\xi(x,y), \eta(x,y)).$$

In fact, there are three posssible canonical forms, according to when the functions A, B, C of (x,y) satisfy $B^2 > AC$, $B^2 < AC$ or $B^2 = AC$. In this way, one *classifies* the partial differential equation (1). Details of this *classification* now follow.

$\boldsymbol{B^2 > AC}$ **at** $\boldsymbol{(x,y)}$: equation (1) is then **hyperbolic at** $\boldsymbol{(x,y)}$ and the canonical form is

$$(2) \qquad\qquad \frac{\partial^2 Z}{\partial \xi \partial \eta} \;=\; \Psi(\xi, \eta, Z, P, Q),$$

where

$$P \;=\; Z_\xi \;=\; \frac{\partial Z}{\partial \xi}, \qquad Q \;=\; Z_\eta \;=\; \frac{\partial Z}{\partial \eta}.$$

A classic example of such an equation is the (one-dimensional) *wave equation*

$$(3) \qquad\qquad \frac{\partial^2 z}{\partial x^2} \;=\; \frac{1}{c^2} \frac{\partial^2 z}{\partial t^2},$$

for a function $z = z(x,y)$ and positive constant c, modelling the transverse motion of a string performing small oscillations.

$\boldsymbol{B^2 < AC}$ **at** $\boldsymbol{(x,y)}$: equation (1) is then **elliptic at** $\boldsymbol{(x,y)}$ and the canonical form is

$$(4) \qquad\qquad \frac{\partial^2 Z}{\partial \xi^2} + \frac{\partial^2 Z}{\partial \eta^2} \;=\; \Psi(\xi, \eta, Z, P, Q).$$

An example of such an equation is (two-dimensional) *Laplace's equation*

$$(5) \qquad\qquad \frac{\partial^2 z}{\partial x^2} + \frac{\partial^2 z}{\partial y^2} \;=\; 0,$$

for a function $z = z(x,y)$, modelling the gravitational, electromagnetic or fluid potential in steady state. Solutions of Laplace's equation are called *harmonic functions*.

$B^2 = AC$ at (x, y): equation (1) is then **parabolic at (x, y)** and the canonical form is

$$(6) \qquad \frac{\partial^2 Z}{\partial \eta^2} = \Psi(\xi, \eta, Z, P, Q).$$

An example of a parabolic equation is the (one-dimensional) *diffusion* (or *heat*) *equation*

$$(7) \qquad k\frac{\partial^2 T}{\partial x^2} = \frac{\partial T}{\partial t}$$

satisfied by the temperature function $T = T(x, t)$, where k represents constant thermal conductivity.

This chapter concerns itself with the reduction to canonical form and with some important techniques for finding solutions. One important topic, as in Chapter 5, will be to discuss which boundary conditions are appropriate and lead to the existence of a unique solution. We leave this discussion to the next chapter.

6.1 Characteristics

As with first-order equations, certain curves, this time in the (x, y)-plane but still carrying the name of *characteristic*, are central to the discussion. We emphasise at once that the characteristic curves involved in the reduction of the second-order equation to canonical form have no relation to those three-dimensional characteristic curves lying on a solution surface of equations of first-order.

Suppose that the curve Γ ($\mathbf{r} = \mathbf{r}(\sigma) = (x(\sigma), y(\sigma))$) is *regular* in the sense that it has a continuous second derivative, non-zero at every point of its domain and hence with well-defined and non-parallel tangent and normal lying in the (x, y)-plane. Suppose we also know the values of a solution $z = z(x, y)$ of (1) and its normal derivative

$$(8) \qquad \frac{\partial z}{\partial n} = \mathbf{n} \cdot \operatorname{grad} z = n_1 p + n_2 q,$$

where $\mathbf{n} = (n_1, n_2)$ is the unit normal to the curve, at each point of Γ (that is, 'we know z and $\partial z/\partial n$ along Γ' as functions of σ). Such boundary conditions are called *Cauchy data* and are often appropriate in examples from physics. By applying the chain rule to $z = z(x(\sigma), y(\sigma))$, we see that

$$(9) \qquad \frac{dz}{d\sigma} = p\frac{dx}{d\sigma} + q\frac{dy}{d\sigma}$$

as well as (1) and (8) along Γ. Since $x = x(\sigma)$ and $y = y(\sigma)$ are given functions and since we have assumed the curve has non-parallel tangents and normals (and hence the determinant of the coefficients of p, q on the right-hand side of (8) and (9) is non-zero), we can solve (8) and (9) for p and q, which are themselves therefore known along Γ and thus as functions $p = p(\sigma)$, $q = q(\sigma)$ of σ. Again, using the chain rule and noting that

$$r = \frac{\partial p}{\partial x}, \qquad s = \frac{\partial p}{\partial y} = \frac{\partial q}{\partial x}, \qquad t = \frac{\partial q}{\partial y},$$

differentiation of p and q gives

(10)
$$\frac{dp}{d\sigma} = r\frac{dx}{d\sigma} + s\frac{dy}{d\sigma}$$

and

(11)
$$\frac{dq}{d\sigma} = s\frac{dx}{d\sigma} + t\frac{dy}{d\sigma}$$

along Γ. We have therefore three equations in (r, s, t)

(1)
$$Ar + 2Bs + Ct = \phi,$$

(10)
$$\frac{dx}{d\sigma}r + \frac{dy}{d\sigma}s = \frac{dp}{d\sigma},$$

(11)
$$\frac{dx}{d\sigma}s + \frac{dy}{d\sigma}t = \frac{dq}{d\sigma},$$

where r, s and t are the only unknowns.

Definition 1 The curve Γ is a *characteristic* if

(*)   we *cannot* determine r, s, t along Γ from (1), (10) and (11).

A necessary and sufficient condition for (*) is that the determinant of the coefficients of r, s, t in (1), (10) and (11) is zero:

$$\begin{vmatrix} A & 2B & C \\ \dfrac{dx}{d\sigma} & \dfrac{dy}{d\sigma} & 0 \\ 0 & \dfrac{dx}{d\sigma} & \dfrac{dy}{d\sigma} \end{vmatrix} = 0,$$

that is,

$$(12) \qquad A\left(\frac{dy}{d\sigma}\right)^2 - 2B\frac{dx}{d\sigma}\frac{dy}{d\sigma} + C\left(\frac{dx}{d\sigma}\right)^2 = 0,$$

which determines the characteristics when they are given in the parametric form $x = x(\sigma)$, $y = y(\sigma)$. Notice that, whilst we retain the coefficients on the left-hand side of (1), the sign of the second term has become reversed.

Remark The reader should notice that if Γ is *not* a characteristic, one can, in view of the fact that all the functions involved are continuous, determine r, s, t even in a neighbourhood of Γ, where the determinant remains non-zero.

We now seek a differential equation for the characteristics when they are expressed in explicit form $y = y(x)$. In that form, dy/dx is given in terms of the parametric equations $x = x(\sigma), y = y(\sigma)$ by

$$\frac{dy}{dx} = \frac{dy}{d\sigma} \bigg/ \frac{dx}{d\sigma}$$

whenever $dx/d\sigma \neq 0$. Equation (12) then reduces to

$$(13) \qquad A\left(\frac{dy}{dx}\right)^2 - 2B\frac{dy}{dx} + C = 0,$$

the equation giving the characteristics in explicit form.

Notice that the equations (12) and (13) are quadratic in $dy/d\sigma : dx/d\sigma$ and dy/dx, respectively; and so, there are two *families* of characteristics, corresponding to varying the constants of integration in the two solutions to the equations.

Suppose these two families, real or possibly co-incident (in the parabolic case) or complex conjugate (when the partial differential equation (1) which has real coefficients is elliptic), are given in the implicit form

$$\xi(x, y) = c_1, \qquad \eta(x, y) = c_2,$$

where c_1, c_2 are constants (which can be varied to give the different curves in each characteristic family). Then, since $x = x(\sigma)$, $y = y(\sigma)$ along a characteristic, we may use the chain rule to show that

$$\xi_x \frac{dx}{d\sigma} + \xi_y \frac{dy}{d\sigma} = 0 = \eta_x \frac{dx}{d\sigma} + \eta_y \frac{dy}{d\sigma}.$$

In this case, (12) gives rise to

$$(14) \qquad\qquad K \equiv A\xi_x^2 + 2B\xi_x\xi_y + C\xi_y^2 = 0,$$

$$(15) \qquad\qquad M \equiv A\eta_x^2 + 2B\eta_x\eta_y + C\eta_y^2 = 0,$$

which are the differential equations for the characteristics when expressed in implicit form and which define the expressions K and M. Note that we have now recovered the positive sign for the second term in both equations.

Notes

(a) Whereas the equations for the characteristics in implicit form will be invaluable to us in deriving canonical forms in the next section, the most convenient equation for calculating the characteristic families is (13): one just solves for dy/dx and integrates each of the two solutions with respect to x.

(b) When $A = 0 \neq B$, one may be led to suppose from (13) that there is just one family of characteristics, found by integrating $dy/dx = C/2B$. However, we have to remember that in deriving (13), we have assumed that $dx/d\sigma \neq 0$. If we go back to (what can be taken to be) the defining equation (12), we see immediately that $A = 0$ implies that $dx/d\sigma$ is a factor of the left-hand side of (12) and $x = c_1$ has to be one family of characteristics. Of course, the lines with these equations (as c_1 varies) cannot be expressed in the form $y = y(x)$.

Example 1 The curve Γ in the (x, y)-plane has equation $y = x^2$. Find the unit tangent and the unit 'upward-drawn' normal at each point of Γ. If the function $z = z(x, y)$ satisfies

$$z = x \quad \text{and} \quad \frac{\partial z}{\partial n} = -\frac{2x}{\sqrt{1 + 4x^2}}$$

along Γ, determine $p = \partial z/\partial x$ and $q = \partial z/\partial y$ along Γ.

The curve Γ can be given the parametrisation $x = \sigma$, $y = \sigma^2$. The tangent $(dx/d\sigma, dy/d\sigma)$ is $(1, 2\sigma)$, so that the unit tangent $\mathbf{t}$ (that is, the tangent of length one) is given by

$$\mathbf{t} = \left(\frac{1}{\sqrt{1 + 4\sigma^2}}, \frac{2\sigma}{\sqrt{1 + 4\sigma^2}} \right).$$

By inspection, the unit upward-drawn normal $\mathbf{n}$ must be

$$\mathbf{n} = \left(-\frac{2\sigma}{\sqrt{1 + 4\sigma^2}}, \frac{1}{\sqrt{1 + 4\sigma^2}} \right)$$

– 'upward' determines where the minus sign must be put (here, in order that the y-component is positive). Equations (8) and (9) above give

$$-\frac{2\sigma}{\sqrt{1 + 4\sigma^2}} p + \frac{1}{\sqrt{1 + 4\sigma^2}} q = -\frac{2\sigma}{\sqrt{1 + 4\sigma^2}},$$

$$p + 2\sigma q = 1,$$

as $z = \sigma$ on Γ. The reader will see that these simultaneous linear equations are solved by $p = 1$, $q = 0$. $\square$

Example 2 Determine where each of the following two equations are hyperbolic and, where this is the case, find the characteristic families.

(i)
$$x^2 r - y^2 t = yq - xp,$$

(ii)
$$y^2 s + yt = q + y^3.$$

In each case, it is only the terms on the left-hand side of the equation, involving as they do the coefficients of r, s, t, that interest us.

In case (i), the condition $B^2 - AC > 0$, necessary for the equation to be hyperbolic, is satisfied at any point not on the co-ordinate axes. Equation (13) for the characteristics is

$$x^2 \left(\frac{dy}{dx}\right)^2 - y^2 = 0$$

which factorises as

$$\left(x\frac{dy}{dx} - y\right)\left(x\frac{dy}{dx} + y\right) = 0.$$

Hence the characteristic families are

$$xy = c_1, \qquad \frac{y}{x} = c_2.$$

For case (ii), $B^2 - AC > 0$ implies that the equation is hyperbolic when $y \neq 0$. This time, equation (13) is

$$-y^2\frac{dy}{dx} + y = 0.$$

Note (b) above reminds us, as $A = 0$, that one family of characteristics is $x = c_1$. When the differential equation is hyperbolic we may cancel through by the non-zero quantity y and solve the resulting ordinary differential equation to give the second characteristic family $y^2 - 2x = c_2$. $\qquad \square$

Exercise 1 Determine whether the equation

$$x^2 r + 2xys + y^2 t - x^2 p - xyq = 0$$

is hyperbolic, elliptic or parabolic (that is, *classify* the equation), and find its characteristics.

Exercise 2 Consider the problem consisting of the equation

$$xr + (x - y)s - yt = \left(\frac{x - y}{x + y}\right)(p + q)$$

together with the boundary conditions $z = 0$, $p = 3y$ along the curve Γ with equation $x = 2y$ when $y > 0$. Determine where the equation is hyperbolic and find the characteristic families. Also, determine q and $\partial z/\partial n$ along Γ as functions of y.

6.2   Reduction to canonical form

The reduction of the second-order equation

(1) $$Ar + 2Bs + Ct \;=\; \phi(x, y, z, p, q)$$

to canonical form is to be achieved by a transformation

$$\xi \;=\; \xi(x, y), \qquad \eta \;=\; \eta(x, y)$$

from independent real variables x, y to independent real variables ξ, η. So, as well as requiring $\xi(x, y)$ and $\eta(x, y)$ to be real-valued functions (and as the reader may be familiar from the theory of change of variable in double integration), it is necessary for the change to be *proper*, that is for the Jacobian J of the transformation to be non-zero; in symbols,

$$J \;\equiv\; \frac{\partial(\xi, \eta)}{\partial(x, y)} \;\equiv\; \frac{\partial \xi}{\partial x}\frac{\partial \eta}{\partial y} - \frac{\partial \xi}{\partial y}\frac{\partial \eta}{\partial x} \;\neq\; 0.$$

This assures us that the new variables ξ, η are independent. We also want to be confident that the second-order partial derivatives exist and are continuous, in particular so that $\xi_{xy} = \xi_{yx}$, $\eta_{xy} = \eta_{yx}$. So, for the rest of this section we require the following.

> The functions $\xi = \xi(x, y)$, $\eta = \eta(x, y)$ are twice continuously differentiable and real-valued in a domain D of $\mathbb{R}^2$ where the Jacobian $\partial(\xi, \eta)/\partial(x, y)$ is never zero.

Of course, the change of variables changes the functional form of z: the point is made immediately by the example $z = x^2 - y^2$ with the change of variable $\xi = x + y$, $\eta = x - y$ (with Jacobian -2). The function z is changed to the function Z given by $Z(\xi, \eta) = \xi\eta$. In general, when $z = z(x, y)$ is a twice continuously differentiable and real-valued function of the independent real variables x, y in a domain D, we define $Z = Z(\xi, \eta)$ by

$$Z(\xi(x, y), \eta(x, y)) \;=\; z(x, y), \qquad (x, y) \in D.$$

In these circumstances, Z is also twice continuously differentiable, and so $Z_{\xi\eta} = Z_{\eta\xi}$.

For convenience, we adopt the notation

$$P = Z_\xi = \frac{\partial Z}{\partial \xi}, \qquad Q = Z_\eta = \frac{\partial Z}{\partial \eta},$$

$$R = Z_{\xi\xi} = \frac{\partial^2 Z}{\partial \xi^2}, \quad S = Z_{\xi\eta} = Z_{\eta\xi} = \frac{\partial^2 Z}{\partial \xi \partial \eta} = \frac{\partial^2 Z}{\partial \eta \partial \xi}, \quad T = Z_{\eta\eta} = \frac{\partial^2 Z}{\partial \eta^2}.$$

We now use the change of variables $\xi = \xi(x,y)$, $\eta = \eta(x,y)$ to convert the partial differential equation (1) in z to an equation in Z. It is important that a reader fresh to the theory perform the details of the following calculations: we give them in an appendix to this chapter. Differentiating $z = z(x,y)$ with respect to x and y, the calculations give

$$p = \xi_x P + \eta_x Q,$$

$$q = \xi_y P + \eta_y Q,$$

$$\text{(16)} \qquad r = (\xi_{xx}P + \eta_{xx}Q) + (\xi_x^2 R + 2\xi_x \eta_x S + \eta_x^2 T),$$

$$s = (\xi_{xy}P + \eta_{xy}Q) + (\xi_x\xi_y R + (\xi_x\eta_y + \xi_y\eta_x)S + \eta_x\eta_y T),$$

$$t = (\xi_{yy}P + \eta_{yy}Q) + (\xi_y^2 R + 2\xi_y\eta_y S + \eta_y^2 T).$$

Hence,

$$\text{(17)} \qquad Ar + 2Bs + Ct = (KR + 2LS + MT)$$
$$+ (A\xi_{xx} + 2B\xi_{xy} + C\xi_{yy})P + (A\eta_{xx} + 2B\eta_{xy} + C\eta_{yy})Q,$$

where

$$\text{(18)} \qquad L = A\xi_x\eta_x + B(\xi_x\eta_y + \xi_y\eta_x) + C\xi_y\eta_y$$

and K and M are the expressions defined in (14), (15) in section 6.1. This suggests that if the functions $\xi(x,y), \eta(x,y)$ giving the characteristics can be taken to define the change of variables (so that $K = M = 0$), and if the expression L is non-zero (which we show below), then (1) is hyperbolic. That $L \neq 0$ cannot surprise, for we should not expect a proper change of variables to leave no second-order partial derivatives in the reduced equation!

First note that the ratios $dy/d\sigma : dx/d\sigma$ for the characteristic families in parametric form $x = x(\sigma)$, $y = y(\sigma)$, as given by (12) in section 6.1, are distinct, if and only if

$B^2 \neq AC$ (a fact that can also be deduced by considering this tangent direction dy/dx from equation (13), in the case $dx/d\sigma \neq 0$). If $\xi(x, y) = c_1$ is one characteristic family, we must have $\xi(x(\sigma), y(\sigma)) = c_1$ and hence, using the chain rule,

$$\xi_x \frac{dx}{d\sigma} + \xi_y \frac{dy}{d\sigma} = 0.$$

So, for this family,

$$\frac{dy}{d\sigma} : \frac{dx}{d\sigma} = -\xi_x : \xi_y.$$

Similarly, for the other family $\eta(x, y) = c_2$,

$$\frac{dy}{d\sigma} : \frac{dx}{d\sigma} = -\eta_x : \eta_y.$$

As the ratios for the two families are distinct, $\xi_x \eta_y \neq \xi_y \eta_x$; that is, $\partial(\xi, \eta)/\partial(x, y) \neq 0$. So, the change of variables $\xi = \xi(x, y), \eta = \eta(x, y)$, when $\xi(x, y) = c_1$ and $\eta(x, y) = c_2$ are the characteristic families and $B^2 \neq AC$, is a proper change which, further, gives $dy/d\sigma : dx/d\sigma$ real and hence real-valued functions $\xi = \xi(x, y)$, $\eta = \eta(x, y)$ in the hyperbolic case $B^2 > AC$.

A further calculation, which the reader should perform and detail of which will be found in the chapter's Appendix, shows that

$$(19) \qquad L^2 - KM = (B^2 - AC) \left(\frac{\partial(\xi, \eta)}{\partial(x, y)} \right)^2.$$

So, in the **hyperbolic case** with ξ, η derived from the equations of characteristic functions, both bracketed terms on the right-hand side are non-zero, whereas $K = M = 0$ from (14), (15). Hence, L is non-zero and, using (17), equation (1) reduces to

$$(2) \qquad S = \Psi(\xi, \eta, Z, P, Q)$$

which is known as *the canonical form for the hyperbolic equation*. The function Ψ is given by

$$(20) \quad 2L\Psi(\xi, \eta, Z, P, Q) = \phi(x(\xi, \eta), y(\xi, \eta), Z, \xi_x P + \eta_x Q, \xi_y P + \eta_y Q)$$
$$- (A\xi_{xx} + 2B\xi_{xy} + C\xi_{yy})P - (A\eta_{xx} + 2B\eta_{xy} + C\eta_{yy})Q,$$

once one uses equations (16) and solves $\xi = \xi(x, y)$, $\eta = \eta(x, y)$, for x, y as continuously differentiable functions of ξ, η. (This is possible by the Inverse Function Theorem of the calculus, as $\partial(\xi, \eta)/\partial(x, y)$ is non-zero.)

Example 3 Reduce each of the equations (i), (ii) of Example 2 of section 6.1 to canonical form wherever they are hyperbolic.

The reduction of

(i) $$x^2 r - y^2 t = yq - xp$$

is to be performed at all points (x, y) not on the co-ordinate axes in $\mathbb{R}^2$. The characteristic families were shown to be $xy = c_1$, $y/x = c_2$; so, we make the change of variables

$$\xi = xy, \qquad \eta = \frac{y}{x}$$

(permissible, as x is never zero). One soon calculates $\phi = 2\eta Q$ and deduces that $\Psi \equiv 0$ by (20). The canonical form for (i) is therefore $S = 0$.

Similarly, the appropriate change of variables for

(ii) $$y^2 s + yt = q + y^3$$

when $y \neq 0$ (and (ii) is hyperbolic) is

$$\xi = x, \qquad \eta = y^2 - 2x.$$

One calculates
$$L = y^3, \qquad \phi = 2yQ + y^3, \qquad \Psi = \frac{1}{2}$$

and hence the canonical form is $S = 1/2$. □

Notes

(a) Faced with a particular partial differential equation to reduce to canonical form, I would suggest attacking the problem *ab initio*, performing the calculations from which (16), (17) above are derived, for the functions given.

(b) Beware that the second-order partial derivatives in the original equation (1) do in general contribute both *first-* and second-order terms to the reduced equation: it is a common error to forget some first-order terms.

(c) It is often convenient, during the working, at first to leave the coefficients of P, Q, R, S, T as functions of x and y (rather than of ξ and η), converting to functions of ξ and η at the last stage, when writing down the reduced equation.

In order to find the (real) canonical form in the **elliptic case** $B^2 < AC$, we note first that, as the functions A, B, C are real-valued, the two solutions for the ratio $dy/d\sigma$: $dx/d\sigma$ in (12) must be complex conjugates. Thus, the functions $\xi = \xi(x, y)$, $\eta = \eta(x, y)$ solving (14), (15) can be taken to be complex conjugates. As $B^2 \neq AC$, they must also be distinct and the change of variables $(x, y) \to (\xi, \eta)$ must be proper.

This suggests making the further change of variables $(\xi, \eta) \to (\xi', \eta')$, where ξ', η' are twice the real and imaginary parts of ξ, η

$$\xi' \;=\; (\xi + \eta), \qquad \eta' \;=\; i(\eta - \xi),$$

in order to find real functions ξ', η' and hence to make the change of variables $(x, y) \to (\xi', \eta')$, defined as the composition of the changes $(x, y) \to (\xi, \eta)$ followed by $(\xi, \eta) \to (\xi', \eta')$, a real change. Defining $Z' = Z'(\xi', \eta')$ by

$$Z(\xi, \eta) \;=\; Z'((\xi + \eta), i(\eta - \xi))$$

and with an obvious extension of our notation, the reader will quickly show that

$$P \;=\; P' - iQ',$$

$$Q \;=\; P' + iQ',$$

$$S \;=\; R' + T'.$$

In this way, we can convert the hyperbolic canonical form (2) to

(4′) $$R' + T' \;=\; \Psi'(\xi', \eta', Z', P', Q')$$

where Ψ' is defined by

$$\Psi((\xi, \eta, Z, P, Q)) \;=\; \Psi'(\xi + \eta, i(\eta - \xi), Z, P' - iQ', P' + iQ').$$

Equation (4′) is known as *the canonical form for the elliptic equation*.

The reader will want to be assured that the composite change of variables $(x, y) \rightarrow (\xi', \eta')$ is itself both real and proper. Justification of the above procedure in going from the hyperbolic canonical form to the elliptic through complex variables, though not given here, is relatively straightforward if there exist analytic functions[1] of two complex independent variables x, y which reduce to $P = P(x, y)$, $Q = Q(x, y)$, $R = R(x, y)$ when x, y are real. When such an extension of P, Q, R to complex variables is not possible, proof is quite hard and subtle.

Example 4 Reduce the equation

(iii) $$y^2 r + x^2 t = 0$$

to canonical form.

Clearly, the equation (iii) is elliptic at all points (x, y) not on the co-ordinate axes. The characteristics can be quickly determined from (13): we write down

$$y^2 \left(\frac{dy}{dx} \right)^2 + x^2 = 0$$

to find the complex conjugate families of solutions

$$\xi(x, y) \equiv y^2 + ix^2 = c_1, \qquad \eta(x, y) \equiv y^2 - ix^2 = c_2.$$

Proceeding as above, with $\xi' = \xi + \eta$, $\eta' = i(\eta - \xi)$, we make the change of variables

$$\xi' = 2y^2, \qquad \eta' = 2x^2.$$

Hence,

$$p = 4xQ',$$

$$q = 4yP',$$

$$r = 4Q' + 16x^2 T',$$

$$t = 4P' + 16y^2 R',$$

and using (iii),

$$16x^2 y^2 (R' + T') = -4x^2 P' - 4y^2 Q'.$$

---

[1]A function of two independent complex variables is *analytic* if it can be expressed as a convergent power series expansion in the pair of variables x, y about every point of its domain.

Thus, the canonical form for (iii) is

$$R' + T' = -\frac{1}{2}\left(\frac{1}{\xi'}P' + \frac{1}{\eta'}Q'\right). \qquad \square$$

Finally, for the **parabolic case** where $B^2 = AC$, when there is just one distinct (real) family $\xi(x, y) = c_1$ of characteristics, we need to find a (twice continuously differentiable) function $\eta = \eta(x, y)$ which, together with the function $\xi = \xi(x, y)$, will give us a proper real change of variables $(x, y) \to (\xi, \eta)$. But this is easy: just pick *any* real-valued twice continuously differentiable $\eta = \eta(x, y)$ such that $\partial(\xi, \eta)/\partial(x, y) \neq 0$. Note that $\eta(x, y) = c_2$ cannot then be a characteristic family, and so the expression M in (15) is non-zero, whereas $\xi(x, y) = c_1$ representing a characteristic family still ensures that K in (14) remains zero. Therefore, from (19) we see that $L = 0$ and, using (17), that (1) reduces to *the canonical form in the parabolic case*

(6) $$T = \Psi''(\xi, \eta, Z, P, Q).$$

In the same way as we defined Ψ in the hyperbolic case, the function Ψ'' is given by letting $M\Psi''(\xi, \eta, Z, P, Q)$ be the expression on the right-hand side of (20) for the functions ξ, η defined for this parabolic case.

Example 5 Reduce the equation

(iv) $$r - 2s + t = 0$$

to canonical form.

Equation (iv) is immediately parabolic everywhere in the plane. Equation (13) gives

$$\left(\frac{dy}{dx}\right)^2 + 2\frac{dy}{dx} + 1 = \left(\frac{dy}{dx} + 1\right)^2 = 0$$

and hence the only characteristic family is $x + y = c_1$. So, we take

$$\xi = x + y, \qquad \eta = x - y,$$

the function η having been chosen so that $\partial(\xi, \eta)/\partial(x, y) \neq 0$. Hence,

$$p \;=\; P + Q,$$

$$q \;=\; P - Q,$$

$$r \;=\; R + 2S + T,$$

$$s \;=\; R - T,$$

$$t \;=\; R - 2S + T,$$

from which equation (iv) reduces to the canonical form $T = 0$ at once.

Alternatively, and even more simply, we could take

$$\xi \;=\; x + y, \qquad \eta \;=\; x$$

giving $p = P + Q$, $q = P$, $r = R + 2S + T$, $s = R + S$, $t = R$ and $T = 0$ very quickly indeed. $\qquad\qquad\qquad\qquad\qquad\qquad\qquad\qquad\qquad\qquad\qquad\qquad\qquad\qquad\qquad\square$

Exercise 3 Reduce the equations in Exercises 1 and 2 to canonical form.

Exercise 4 Reduce the following two equations to canonical form:

(a) $\quad s + t + xp + yq + (1 + xy - x^2)z \;=\; 0,$

(b) $\quad 5r + 8xs + 5x^2 t \;=\; 0.$

6.3   General solutions

According to the definition of section 5.3, a general solution of a second-order partial differential equation is a solution containing two arbitrary (suitably differentiable) functions. We now proceed by example to find general solutions to some widely occurring hyperbolic and parabolic equations, where the method is most useful.

Example 6 For each of the following second-order equations satisfied by the function $Z = Z(\xi, \eta)$, find the general solution:

(a) $\quad S = 0,$

(b) $\quad S = \dfrac{1}{2},$

(c) $\quad S = \dfrac{1}{4} + \dfrac{1}{2\xi} Q, \qquad (\xi > 0)$

(d) $\quad T = 0,$

where $S = \partial^2 Z/\partial\xi\partial\eta$, $T = \partial^2 z/\partial\eta^2$ and $Q = \partial z/\partial\eta$.

Equation (a) has already been discussed in section 5.3. The general solution is

$$Z = f(\xi) + g(\eta),$$

valid for arbitrary (suitably differentiable – a phrase we shall henceforth omit) functions $f = f(\xi)$, $g = g(\eta)$.

We specify two methods for tackling (b). The first uses the complementary function/ particular integral method common to the study of ordinary differential equations: we add to a solution, the complementary function, of the corresponding homogeneous equation $S = 0$ one particular solution, found 'by inspection', of the given non-homogeneous equation (b). Here, by inspection, it is clear that $Z = \xi\eta/2$ is a particular solution; so, using the complementary function determined for (a) above, we can already write down the general solution of (b),

(21) $$Z = f(\xi) + g(\eta) + \frac{1}{2}\xi\eta,$$

valid for arbitrary f, g.

The second method uses what might be called 'partial integration'. It gives us a more general way of tackling not only such differential equations as (b), but also such others as (c) and (d) (see below).

Write $W = \partial Z/\partial\eta$. Then (b) can be written

$$\frac{\partial W}{\partial\xi} = \frac{1}{2}.$$

Integrate with respect to ξ (thinking of it, if you like, as if it were an ordinary differential equation in ξ) to give

$$W \equiv \frac{\partial Z}{\partial \eta} = h(\eta) + \frac{1}{2}\xi,$$

where the arbitrary constant of ordinary differential equation integration is replaced by an arbitrary (integrable) function h of η. Notice that the differential of this function with respect to ξ is zero. Now integrate again, this time with respect to η (thinking of the equation as if it were an ordinary differential equation in η, with ξ constant). This again gives the solution (21) above, where f is an arbitrary function of ξ and $g(\eta)$ is the indefinite integral

$$g(\eta) = \int^\eta h(t)\, dt$$

or equivalently, g is a differentiable function satisfying $g'(\eta) = h(\eta)$. Note that, as (21) is a solution with two arbitrary functions f and g (just differentiate, to check), it is the general solution.

We proceed in the same way, but now more briefly, to solve (c). With $W = \partial Z/\partial \eta$, re-write the equation as

$$\frac{\partial W}{\partial \xi} - \frac{1}{2\xi}W = \frac{1}{4}.$$

Viewed as an equation in ξ, the integrating factor (Appendix (1)) is

$$\exp\left(-\int^\xi \frac{dt}{2t}\right) = \exp\left(-\frac{1}{2}\log \xi\right) = \exp\left(\log \xi^{-1/2}\right) = \frac{1}{\xi^{1/2}}.$$

This is valid, as $\xi > 0$. Hence,

$$\frac{\partial}{\partial \xi}\left(\frac{1}{\xi^{\frac{1}{2}}}W\right) = \frac{1}{4\xi^{\frac{1}{2}}},$$

and so, integrating with respect to ξ,

$$\frac{1}{\xi^{\frac{1}{2}}}W = \frac{1}{\xi^{\frac{1}{2}}}\frac{\partial Z}{\partial \eta} = \frac{1}{2}\xi^{\frac{1}{2}} + h(\eta),$$

where h is an arbitrary integrable function. Integrating now with respect to η (keeping ξ constant)

$$\frac{1}{\xi^{\frac{1}{2}}}Z = \frac{1}{2}\xi^{\frac{1}{2}}\eta + g(\eta) + f(\xi),$$

where f is arbitrary and $g'(\eta) = h(\eta)$. Thus, the general solution is

$$(22) \qquad\qquad Z = \xi^{\frac{1}{2}}(f(\xi) + g(\eta)) + \frac{\xi\eta}{2},$$

for arbitrary f, g and $\xi > 0$.

For (d), we just integrate twice with respect to η to obtain a general solution in terms of two arbitrary functions f, g of ξ:

$$\frac{\partial Z}{\partial \eta} = f(\xi)$$

and hence

$$(23) \qquad\qquad Z = \eta f(\xi) + g(\xi). \qquad\qquad\qquad \square$$

We are now ready to consider how we can take account of given boundary conditions, the subject of the next section.

Exercise 5 Determine general solutions to the equations specified in Exercise 1 and 4(a). Notice that, in the latter exercise, the canonical form can be written, when $\xi = x$ and $\eta = x - y$, as

$$\frac{\partial W}{\partial \xi} - \eta W = 0 \qquad \text{where} \quad W \equiv \frac{\partial Z}{\partial \eta} - \xi Z.$$

Note Using operator notation, the last equation above may be expressed as

$$\left(\frac{\partial}{\partial \xi} - \eta \right) \left(\frac{\partial}{\partial \eta} - \xi \right) Z = 0.$$

The term

$$\left(\frac{\partial}{\partial \xi} - \eta \right) \left(\frac{\partial}{\partial \eta} - \xi \right)$$

is known as a *factorisation* of this canonical form operator.

6.4 Problems involving boundary conditions

We have already commented on the desirability of determining not only where a solution to a problem consisting of a differential equation with appropriate boundary conditions can be found, but also where it can be shown to be unique. Existence and uniqueness will be addressed in the next chapter in the context of some celebrated problems originating from mathematical physics. In this section, we shall concern ourselves solely with the technique of 'fitting' boundary conditions in some cases which 'work'. We again proceed by example.

Example 7 Find the solution $z = z(x, y)$ of the equation

(iv) $$r - 2s + t = 0$$

when subject to the boundary conditions

$$z = p = y \qquad \text{at } x = 0, \quad \text{for } 0 < y < 1,$$

specifying the domain in which the solution is determined.

In Example 5, we showed that the canonical form for this parabolic equation is $T = 0$, where $\xi = x + y$ and $\eta = x$, and in Example 6, we deduced that the general solution of $T = 0$ is

$$Z(\xi, \eta) = \eta f(\xi) + g(\xi).$$

Hence, in (x, y) co-ordinates, the general solution to (iv) can be written

(24) $$z(x, y) = x f(x + y) + g(x + y).$$

Partial differentiation with respect to x gives

$$z_x(x, y) = f(x + y) + x f'(x + y) + g'(x + y).$$

Applying the boundary conditions, we have

$$g(y) = y = f(y) + g'(y), \qquad \text{for } 0 < y < 1,$$

and hence $f(y) = y - 1$ and $g(y) = y$ for $0 < y < 1$. Therefore,

$$f(x + y) = x + y - 1 \quad \text{and} \quad g(x + y) = x + y, \qquad \text{for } 0 < x + y < 1.$$

So, the solution of the problem consisting of equation (iv), together with the given boundary conditions, is, from (24),

$$z(x, y) = x(x + y - 1) + (x + y) = x^2 + xy + y,$$

which is determined in

$$\{(x, y) : 0 < x + y < 1\},$$

the shaded area in the diagram below.   □

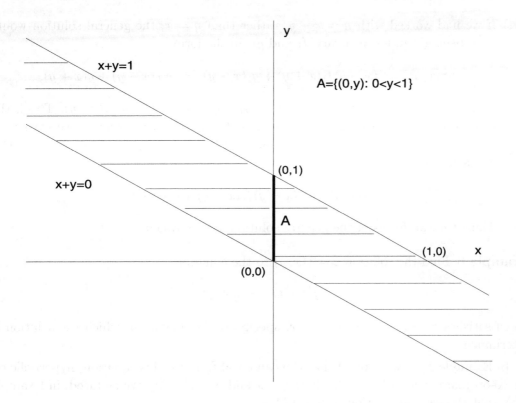

Notes

(a) The reader should note how the boundary condition valid for $0 < y < 1$ translates to a solution 'determined' for $0 < x + y < 1$.

(b) To say that a solution is 'determined' in a domain A means that the solution exists and is unique in A.

(c) Notice that the set of points

$$\{(x, y) : x = 0, \ 0 < y < 1\}$$

in $\mathbb{R}^2$ where the boundary conditions are given is contained in the domain

$$\{(x, y) : x = 0, \ 0 < x + y < 1\}$$

in which the solution is determined (see the diagram).

(d) If we had worked with $\eta = x - y$, rather than $\eta = x$, the general solution would have been given, for arbitrary f_1 and g_1, in the form

$$z(x, y) \;=\; (x - y)f_1(x + y) + g_1(x + y) \;=\; xf(x + y) + g(x + y),$$

if we put $f(x+y) = 2f_1(x+y)$ and $g(x+y) = g_1(x+y) - (x+y)f_1(x+y)$. This is the form of the general solution given in (24). Going 'the other way', if we start with (24) and put $f_1(x+y) = f(x+y)/2$ and $g_1(x+y) = ((x+y)f(x+y) + 2g(x+y))/2$, we easily recover

$$z(x, y) \;=\; (x - y)f_1(x + y) + g_1(x + y).$$

Thus, the two forms of the general solution are equivalent.

Example 8 Find the solution $z = z(x, y)$ to the equation

$$y^2 s + yt \;=\; q + y^3$$

which satisfies $z = q = 0$ when $y^2 = x$, specifying the domain in which the solution is determined.

In Example 3(ii), we showed that the canonical form for this equation, hyperbolic off the co-ordinate axes, is $S = 1/2$, where $\xi = x$ and $\eta = y^2 - 2x$. We deduced, in Example 6(b), that the general solution of $S = 1/2$ is

$$Z(\xi, \eta) \;=\; f(\xi) + g(\eta) + \frac{\xi\eta}{2}$$

which can be expressed in Cartesian co-ordinates as

$$z(x, y) \;=\; f(x) + g(y^2 - 2x) + \frac{x(y^2 - 2x)}{2}.$$

Hence,

$$q \;=\; 2yg'(y^2 - 2x) + xy.$$

By applying the boundary conditions,

$$f(x) + g(-x) - \frac{x^2}{2} \;=\; 0 \;=\; x^{\frac{1}{2}}(2g'(-x) + x).$$

These equations are valid only for $x \geq 0$, as $x = y^2$ on the boundary. When $y^2 = x > 0$, we can cancel $x^{\frac{1}{2}}$ to give $g'(-x) = -x/2$ and hence, with c constant,

$$-g(-x) \;=\; -\frac{x^2}{4} + c, \qquad f(x) \;=\; \frac{x^2}{4} + c,$$

for $x > 0$. So,

$$g(x) \;=\; \frac{x^2}{4} - c,$$

for $x < 0$ and the solution

$$z(x,y) \;=\; \frac{x^2}{4} + \frac{(y^2 - 2x)^2}{4} + \frac{x(y^2 - 2x)}{2} \;=\; \frac{(x - y^2)^2}{4}$$

is therefore determined when $x > 0$ and $y^2 - 2x < 0$, the shaded area in the next diagram ($x > 0$ is clearly redundant). $\qquad\square$

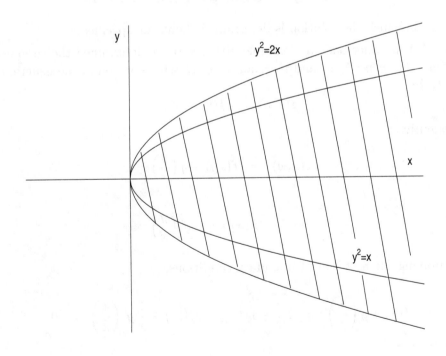

Example 9 Find a continuous solution $z = z(x, y)$ to the equation

$$x^2 r - y^2 t = f(x, y, p, q)$$

in each of the following two cases:

(a) where the solution is required in

$$A = \{(x, y) : x \geq 1, y \geq 1\}$$

and satisfies

$$z(x, 1) = 2(x - 1)^3, \quad q(x, 1) = 0 \text{ for } x \geq 1, \quad \text{and} \quad z(1, y) = 0 \text{ for } y \geq 1,$$

and the function f is given by

$$f(x, y, p, q) = yq - xp,$$

(b) where $f(x, y, p, q) = x^2$ and the solution satisfies

$$z = x^2, \quad q = 0 \quad \text{on } y = 1, \text{ for } 0 < x < 1.$$

The domain in which the solution is determined should be specified.

In case (a), we refer back to Example 3(i), where we determined the canonical form $S = 0$ when $\xi = xy, \eta = y/x$, and to Example 6(a), where we found the general solution of $S = 0$ to be

$$z(\xi, \eta) = f(\xi) + g(\eta)$$

or, equivalently,

(25) $$z(x, y) = f(xy) + g\left(\frac{y}{x}\right).$$

Hence,

$$q = xf'(xy) + \frac{1}{x} g'\left(\frac{y}{x}\right)$$

and so, applying the first pair of boundary conditions,

(26) $$f(x) + g\left(\frac{1}{x}\right) = 2(x - 1)^3, \qquad xf'(x) + \frac{1}{x} g'\left(\frac{1}{x}\right) = 0,$$

for $x \geq 1$. Hence,

$$f'(x) + \frac{1}{x^2} g'\left(\frac{1}{x}\right) = 0,$$

from which it follows that

(27) $$f(x) - g\left(\frac{1}{x}\right) = 2c,$$

where c is a constant, still for $x \geq 1$. From (26) and (27),

$$f(x) = (x-1)^3 + c, \qquad g\left(\frac{1}{x}\right) = (x-1)^3 - c,$$

for $x \geq 1$, and so,

$$f(x) = (x-1)^3 + c, \qquad \text{for } x \geq 1,$$

and

$$g(x) = \left(\frac{1}{x} - 1\right)^3 - c, \qquad \text{for } 0 < x \leq 1.$$

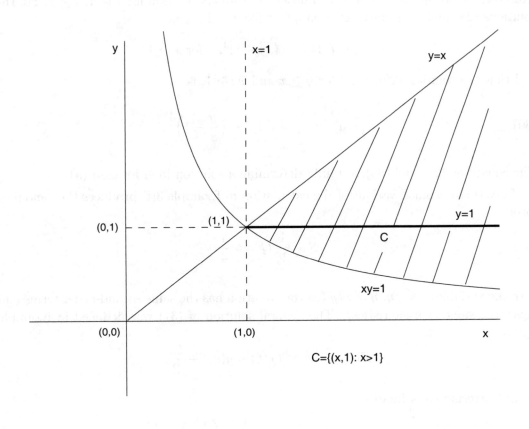

$C = \{(x,1): x > 1\}$

The function

(28) $$z(x, y) = (xy - 1)^3 + \left(\frac{x}{y} - 1\right)^3$$

provides a solution to the problem when both $xy \geq 1$ and $0 < y/x \leq 1$, shown as the shaded region in the diagram above.

To determine the solution 'above' the line $y = x$, we apply the last boundary condition to the general solution (25) to give

$$f(y) + g(y) = 0, \quad \text{for } y \geq 1.$$

Hence, on substituting for g,

(29) $$z(x, y) = f(xy) - f\left(\frac{y}{x}\right), \quad \text{for } \frac{y}{x} \geq 1.$$

However, we are specifically asked to find a *continuous* solution for $x \geq 1, y \geq 1$. So, the solutions (28) and (29) must agree on $y = x$ for $x \geq 1$. Hence,

$$f(x^2) - f(1) = (x^2 - 1)^3, \quad \text{for } x \geq 1$$

and thus the solution (29), valid for $y \geq x$ and $xy \geq 1$, is

(30) $$z(x, y) = (xy - 1)^3 - \left(\frac{y}{x} - 1\right)^3.$$

The equations (28) and (30) certainly determine a solution in A for case (a).

In case (b), a small variant of the calculation in Example 3(i) produces the canonical form

(31) $$S - \frac{1}{2\xi} Q = \frac{1}{4}$$

with the variables $\xi = xy$, $\eta = x/y$ (as the equation has the same second-order terms and hence the same characteristics). The general solution of (31) was deduced in Example 6(c) as

$$Z(\xi, \eta) = \xi^{\frac{1}{2}}(f(\xi) + g(\eta)) + \frac{\xi\eta}{2}$$

or, in Cartesian co-ordinates,

$$z(x, y) = (xy)^{\frac{1}{2}}\left(f(xy) + g\left(\frac{x}{y}\right)\right) + \frac{x^2}{2}.$$

Hence,

$$q = \frac{1}{2}\left(\frac{x}{y}\right)^{\frac{1}{2}}\left(f(xy) + g\left(\frac{x}{y}\right)\right) + (xy)^{\frac{1}{2}}\left(xf'(xy) - \frac{x}{y^2}g'\left(\frac{x}{y}\right)\right)$$

and, applying the boundary conditions, both

$$x^{\frac{1}{2}}(f(x) + g(x)) + \frac{x^2}{2} = x^2$$

and

$$\frac{x^{\frac{1}{2}}}{2}(f(x) + g(x)) + x^{\frac{1}{2}}(xf'(x) - xg'(x)) = 0,$$

for $0 < x < 1$. These simultaneous equations are easily solved:

$$f(x) = \frac{x^{\frac{3}{2}}}{6} + c, \qquad g(x) = \frac{x^{\frac{3}{2}}}{3} - c, \qquad \text{for } 0 < x < 1.$$

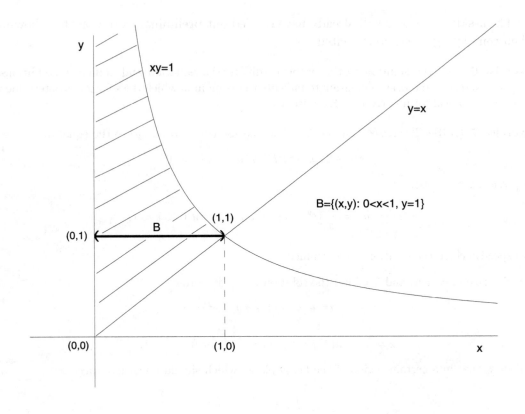

Hence, it would, at a first glance, seem that the solution

$$z(x,y) \; = \; \frac{x^2}{2} + (xy)^{\frac{1}{2}} \left(\frac{1}{6} (xy)^{\frac{3}{2}} + \frac{1}{3} \left(\frac{x}{y} \right)^{\frac{3}{2}} \right) \; = \; \frac{x^2}{6y} (y^3 + 3y + 2)$$

would be determined when both $0 < xy < 1$ and $0 < x/y < 1$. However, the solution is not valid on $y = 0$ and so, there is no way in which the boundary values given on $y = 1$ can determine the values of the functions f and g for $y < 0$ and hence give a unique solution there. Thus, the solution is determined only for $y > 0$, $0 < xy < 1$, $0 < x/y < 1$ which give the shaded region in the diagram above. $\qquad\qquad\qquad\qquad\qquad\square$

Notes

(a) The reader should note the argument in the last paragraph, restricting the solution to $y > 0$.

(b) As in the examples considered, a diagram can often help to clarify one's thinking.

The assiduous reader will already have carried out preliminary work on the following when completing the exercises cited.

Exercise 6 Find the solution to the problem (differential equation and boundary conditions) given in Exercise 2 and draw a diagram to indicate the domain in which the solution is determined. (Use the canonical form derived in Exercise 3.)

Exercise 7 (i) (See Exercises 1, 3 and 5.) Find the solution $z = z(x,y)$ to the equation

$$x^2 r + 2xys + y^2 t - x^2 p - xyq \; = \; 0$$

in $\{(x,y) : y > x > 0\}$,

$$z \; = \; q \; = \; \frac{1}{x} (e^x - 1) \quad \text{on } y = 1, \text{ for } 0 < x < 1,$$

and specify where the solution is determined.

(ii) (See Exercises 4(a) and 5.) Find the solution $z = z(x,y)$ to

$$s + t + xp + yq + (1 + xy - x^2)z \; = \; 0,$$

$$z = 1 \quad \text{and} \quad q = 0 \quad \text{on } y = 0, \text{ for } 0 < x < 1,$$

and show that in a certain region of the (x,y)-plane, which should be clearly specified,

$$z \; = \; \frac{1}{2} (e^{-xy} + e^{xy - y^2}).$$

6.5   Appendix: technique in the use of the chain rule

In this Appendix, we fulfil promises made in section 6.2 to go through the calculations which establish equations (17) and (19) in detail. The reader is strongly encouraged to try this now before reading on. What follows is essential technique and involves repeated use of the chain rule from multivariable calculus. The operator idea is central to our presentation.

To obtain (17), repeated differentiation of

$$z(x,y) \ = \ Z(\xi(x,y), \eta(x,y))$$

is required. Partial differentiation, first with respect to x, then y, gives (once one uses the chain rule)

(32)
$$\frac{\partial z}{\partial x} \ = \ p \ = \ \xi_x P + \eta_x Q = \left(\xi_x \frac{\partial}{\partial \xi} + \eta_x \frac{\partial}{\partial \eta} \right) Z,$$

(33)
$$\frac{\partial z}{\partial y} \ = \ q \ = \ \xi_y P + \eta_y Q = \left(\xi_y \frac{\partial}{\partial \xi} + \eta_y \frac{\partial}{\partial \eta} \right) Z.$$

Notice that the operator $\partial/\partial x$ acting on the function z, given as a function of x and y, is equivalent to the operator $\xi_x \frac{\partial}{\partial \xi} + \eta_x \frac{\partial}{\partial \eta}$ acting on Z, the expression of z in (ξ, η)-coordinates. We use this idea to obtain r, now differentiating (32) with respect to x.

$$r \ = \ \frac{\partial \xi_x}{\partial x} P + \frac{\partial \eta_x}{\partial x} Q + \xi_x \frac{\partial P}{\partial x} + \eta_x \frac{\partial Q}{\partial x}$$

$$= \ \xi_{xx} P + \eta_{xx} Q + \xi_x \left(\xi_x \frac{\partial P}{\partial \xi} + \eta_x \frac{\partial P}{\partial \eta} \right) + \eta_x \left(\xi_x \frac{\partial Q}{\partial \xi} + \eta_x \frac{\partial Q}{\partial \eta} \right)$$

$$= \ \xi_{xx} P + \eta_{xx} Q + \xi_x^2 R + 2\xi_x \eta_x S + \eta_x^2 T.$$

Similarly, to find an expression for s, we differentiate (32) with respect to y (or, equivalently – and the reader should check – (33) with respect to x).

$$s = \xi_{xy}P + \eta_{xy}Q + \xi_x\left(\xi_y\frac{\partial P}{\partial\xi} + \eta_y\frac{\partial P}{\partial\eta}\right) + \eta_x\left(\xi_y\frac{\partial Q}{\partial\xi} + \eta_y\frac{\partial Q}{\partial\eta}\right)$$

$$= \xi_{xy}P + \eta_{xy}Q + \xi_x\xi_y R + (\xi_x\eta_y + \xi_y\eta_x)S + \eta_x\eta_y T.$$

And differentiating q with respect to y to obtain t:

$$t = \xi_{yy}P + \eta_{yy}Q + \xi_y\left(\xi_y\frac{\partial P}{\partial\xi} + \eta_y\frac{\partial P}{\partial\eta}\right) + \eta_y\left(\xi_y\frac{\partial Q}{\partial\xi} + \eta_y\frac{\partial Q}{\partial\eta}\right)$$

$$= \xi_{yy}P + \eta_{yy}Q + \xi_y^2 R + 2\xi_y\eta_y S + \eta_y^2 T.$$

Equation (17) now follows quickly.

To establish (19) is straightforward algebraic manipulation.

$$L^2 - KM = (A\xi_x\eta_x + B(\xi_x\eta_y + \xi_y\eta_x) + C\xi_y\eta_y)^2$$

$$- (A\xi_x^2 + 2B\xi_x\xi_y + C\xi_y^2)(A\eta_x^2 + 2B\eta_x\eta_y + C\eta_y^2).$$

The coefficients of A^2, C^2, AB and BC are easily seen to be zero, leaving

$$L^2 - KM = B^2((\xi_x\eta_y + \xi_y\eta_x)^2 - 4\xi_x\xi_y\eta_x\eta_y) + AC(2\xi_x\xi_y\eta_x\eta_y - \xi_x^2\eta_y^2 - \xi_y^2\eta_x^2)$$

$$= (B^2 - AC)(\xi_x\eta_y - \xi_y\eta_x)^2$$

which is (19).

7 The Diffusion and Wave Equations and the Equation of Laplace

The material in this chapter is in many ways of considerable significance. The equations under consideration arise from classical, and classic, problems in mathematical physics. They have provoked some of the finest work in classical analysis. The range of theory and techniques that have been invented to tackle the various instances that arise is extraordinary and has occupied some of the most celebrated figures in the history of mathematics.

We have necessarily had to have rather limited objectives here. The aim is to find some important, if elementary, solutions to the equations which in turn will expose some of their main properties. Our main tool is indeed an elementary one which may already be familiar to the reader: the method of separation of variables. However, even with this tool, we find ourselves working with such well-known analytical objects as Bessel functions and Legendre polynomials. Later chapters discuss other aspects of the subject in the context of the calculus of variations, of the Sturm–Liouville equation, of complex analysis and of transform theory.

The final section of this chapter fulfils a promise made in the last: we shall discuss existence and uniqueness of solutions and, what is so important these days when numerical analysis and the computer provide approximate solutions to otherwise intractable equations, whether a problem is 'well-posed', that is, whether its possible solutions are 'continuously dependent on the boundary conditions', and so, approximations are likely to be accurate.

7.1   The equations to be considered

This section specifies the notation and co-ordinate systems we shall use in the main text. Some of the exercises will lead the reader further afield.

We make two purely notational points at the outset.

(a) We shall consistently use u (or U) as the dependent variable (rather than z (or Z) as in the last two chapters) because of the extension beyond the context of Chapter 6 to the three-dimensional Laplace equation, where we use z as an independent variable.

(b) For the diffusion and wave equations, we shall use t, rather than x, y or z, as the second independent variable, to stress the fact that our interest is restricted to $t \geq 0$. Of course, the reason is that many classical applications traditionally employ t as the time variable and attempt to discover subsequent motion in problems containing initial conditions at $t = 0$. The range of values taken by the space co-ordinate, or co-ordinates, will vary.

Taking the above comments into account, two of the equations for functions $u = u(x, t)$ may be written down at once; namely, the parabolic equation

(1)
$$k \frac{\partial^2 u}{\partial x^2} = \frac{\partial u}{\partial t}$$

the one-dimensional diffusion, or heat, equation

and hyperbolic equation

(2)
$$\frac{\partial^2 u}{\partial x^2} = \frac{1}{c^2} \frac{\partial^2 u}{\partial t^2}$$

the one-dimensional wave equation

where k and c are positive constants.

Laplace's equation, an elliptic equation, will be considered in polar co-ordinates. In two dimensions, the equation for $u = u(x, y)$,

(3)
$$\nabla^2 u \equiv \frac{\partial^2 u}{\partial x^2} + \frac{\partial^2 u}{\partial y^2} = 0$$

the two-dimensional Laplace's equation in Cartesian co-ordinates

translates, once one changes variables, $(x, y) \to (r, \theta)$ from Cartesian to circular polar co-ordinates defined by

$$x = r \cos \theta,$$
$$y = r \sin \theta$$

for $r \geq 0$, $0 \leq \theta < 2\pi$ and

$$U(r, \theta) = u(r \cos \theta, r \sin \theta)$$

to

(4)
$$\frac{\partial^2 U}{\partial r^2} + \frac{1}{r} \frac{\partial U}{\partial r} + \frac{1}{r^2} \frac{\partial^2 U}{\partial \theta^2} = 0$$

the two-dimensional Laplace's equation in circular polar co-ordinates

for $r \neq 0$. Similarly, in three dimensions, moving from

(5)
$$\nabla^2 u \equiv \frac{\partial^2 u}{\partial x^2} + \frac{\partial^2 u}{\partial y^2} + \frac{\partial^2 u}{\partial z^2} = 0$$

the three-dimensional Laplace's equation in Cartesian co-ordinates

for $u = u(x, y, z)$, by the change $(x, y, z) \to (r, \theta, \phi)$ of Cartesian to spherical polar co-ordinates defined by

$$x = r \sin \theta \cos \phi,$$
$$y = r \sin \theta \sin \phi, \qquad (r \geq 0,\ 0 \leq \theta \leq \pi,\ 0 \leq \phi < 2\pi)$$
$$z = r \cos \theta$$

and, when

$$U(r, \theta, \phi) = u(r \sin\theta \cos\phi, r \sin\theta \sin\phi, r \cos\theta),$$

gives, for $r \neq 0$ and $\theta \neq 0, \pi$,

---

(6)
$$\frac{1}{r^2} \frac{\partial}{\partial r}\left(r^2 \frac{\partial U}{\partial r}\right) + \frac{1}{r^2 \sin\theta} \frac{\partial}{\partial \theta}\left(\sin\theta \frac{\partial U}{\partial \theta}\right) + \frac{1}{r^2 \sin^2\theta} \frac{\partial^2 U}{\partial \phi^2} = 0$$

the three-dimensional Laplace's equation in spherical polar co-ordinates

---

In the same way, using circular polar co-ordinates to replace the space variables in the two-dimensional wave equation for $u = u(x, y, t)$,

---

(7)
$$\nabla^2 u \equiv \frac{\partial^2 u}{\partial x^2} + \frac{\partial^2 u}{\partial y^2} = \frac{1}{c^2} \frac{\partial^2 u}{\partial t^2}$$

the two-dimensional wave equation in Cartesian co-ordinates

---

one obtains, for $r \neq 0$,

---

(8)
$$\frac{\partial^2 U}{\partial r^2} + \frac{1}{r} \frac{\partial U}{\partial r} + \frac{1}{r^2} \frac{\partial^2 U}{\partial \theta^2} = \frac{1}{c^2} \frac{\partial^2 U}{\partial t^2}$$

the two-dimensional wave equation in circular polar co-ordinates

---

Notes

(a) We leave the derivation of (4) from (3) (and thus (8) from (7)) and of (6) from (5) as exercises for the reader. The first follows quickly from the initial result proved in the Appendix to Chapter 6. The others are deduced similarly.

(b) The reader should remember that the symbols r and θ, used for both circular and spherical co-ordinates here (as is commonplace), represent quite different quantities in the two cases. Some authors will make this point by using ρ and ψ, say, for circular polars.

(c) Bessel's, respectively Legendre's, equation follows from using the separation of variables technique on equation (8), respectively (6), as we shall show in later sections of this chapter.

We shall be assuming some elementary working knowledge of the use of Fourier series below. The Bibliography provides references for revision of this material.

7.2   One-dimensional heat conduction

The temperature $u = u(x,t)$ of a thin rod, or bar, of constant cross-section and homogeneous material, lying along the x-axis and perfectly insulated laterally, may be modelled by the one-dimensional heat equation

$$(1) \qquad k\frac{\partial^2 u}{\partial x^2} = \frac{\partial u}{\partial t}.$$

Rather than describe the separation of variables technique in any abstract sense, we will now apply it to find a solution to (1). We seek such a solution in the (separated) form

$$(9) \qquad u(x,t) = X(x)T(t),$$

where X, T are suitably differentiable functions – we will denote the derivatives with dashes: $X' = dX/dx$, $X'' = d^2X/dx^2$, etc. Then, for u to satisfy (1), we need $kX''T = XT'$ or

$$(10) \qquad \frac{X''}{X} = \frac{T'}{kT}.$$

The left-hand side of (10) is a function only of x, where the right-hand side is only of t. Hence, both must be equal to a (real) *separation constant* λ, say. Thus, we have separate differential equations satisfied by $X = X(x)$ and $T = T(t)$; namely,

$$(11) \qquad X'' - \lambda X = 0,$$

$$(12) \qquad T' - \lambda kT = 0.$$

Finite rod   Suppose that the rod lies along the part of the x-axis between $x = 0$ and $x = L > 0$ is kept at zero temperature at its ends and that, at each of its points x, starts off with temperature $f(x)$ arising from a twice continuously differentiable function $f = f(x)$. Thus, we have boundary conditions

$$(13) \qquad u(0,t) = 0, \qquad u(L,t) = 0 \qquad \text{for } t \geq 0$$

and an initial condition

(14) $u(x,0) = f(x),$ for $0 \leq x \leq L.$

From (9), the boundary conditions (13) imply $X(0)T(t) = X(L)T(t) = 0$ for all $t \geq 0$. To exclude the possibility that T is identically zero, we must have

(15) $X(0) = X(L) = 0.$

The reader will easily show that the general solutions $X = ax + b$ (for $\lambda = 0$) and $X = c \exp(\sqrt{\lambda}x) + d \exp(-\sqrt{\lambda}x)$ (for $\lambda > 0$) of (11), where a, b, c, d are constants, lead to $X \equiv 0$ (and hence $u \equiv 0$) once the boundary conditions (15) are applied. So, to find an interesting solution, we must insist that λ is negative, say $\lambda = -\mu^2$, where $\mu > 0$. Then

$$X = A \sin \mu x + B \cos \mu x,$$

where A and B are constants. Conditions (15) then imply $B = 0$ and (again excluding $A = 0$ giving $X \equiv 0$) $\sin \mu L = 0$ and hence $\mu L = n\pi$, where $n \in \mathbb{Z}$, the set of all integers. Thus, for each integer n, we have shown that

(16) $X(x) = X_n(x) \equiv A_n \sin \dfrac{n\pi x}{L},$

with A_n constant, is a solution of (11) subject to (15), where

$$\lambda = \lambda_n \equiv -\frac{n^2 \pi^2}{L^2},$$

derived from $\lambda = -\mu^2$ and $\mu = n\pi/L$.

    Solving (12) with $\lambda = \lambda_n$ and hence integrating factor $\exp(\lambda_n kt)$, we obtain, for each integer n,

(17) $T(t) = T_n(t) \equiv C_n e^{-\frac{n^2 \pi^2 kt}{L^2}}.$

Combining (16) and (17), we have found, for each integer n, a separated solution

(18n) $u(x,t) = D_n \sin \dfrac{n\pi x}{L} e^{-\frac{n^2 \pi^2 kt}{L^2}}$

with D_n constant.

    To incorporate the initial condition (14), we will try to fit a certain combination of solutions (14) to it. This will be justified by its success. The combination we choose is that infinite sum corresponding to positive n; namely,

$$(19) \qquad u(x,t) = \sum_{n=1}^{\infty} D_n \sin \frac{n\pi x}{L} e^{-\frac{n^2\pi^2 kt}{L^2}}.$$

Condition (14) then implies

$$\sum_{n=1}^{\infty} D_n \sin \frac{n\pi x}{L} = f(x).$$

For a fixed positive integer m, we multiply through by $\sin(m\pi x/L)$ and integrate along the rod, that is, from 0 to L, so that

$$\int_0^L \sum_{n=1}^{\infty} D_n \sin \frac{m\pi x}{L} \sin \frac{n\pi x}{L} \, dx = \int_0^L f(x) \sin \frac{m\pi x}{L} \, dx.$$

We now interchange the sum and integral (which may be justified by uniform convergence – but see our Note at the very end of Chapter 0) and, using the trigonometric identity

$$2 \sin P \sin Q = \cos(P - Q) - \cos(P + Q),$$

the reader will quickly deduce that

$$\int_0^L \sin \frac{m\pi x}{L} \sin \frac{n\pi x}{L} \, dx = \begin{cases} 0, & (m \neq n), \\ \dfrac{L}{2}, & (m = n). \end{cases}$$

Hence,

$$(20\text{m}) \qquad D_m = \frac{2}{L} \int_0^L f(x) \sin \frac{m\pi x}{L} \, dx$$

for every positive integer m. We have thus found a solution to the problem consisting of the expression (19) with constants D_m given by (20m).

Notes

(a) Those readers familiar with Fourier series will see that we have sufficient continuity and differentiability assumptions to justify convergence and the interchangeability of sum and integral.

(b) A natural question is: are there solutions which are not in this separated form? The answer is in the negative, as there is a uniqueness theorem for this problem. Indeed, such a theorem pertains when $u(0, t)$ and $u(L, t)$ are any continuous functions of t.

(c) Another question is: why should we choose an infinite number of solutions of the form (18n), but only those corresponding to positive n? A full answer to this question would lead us naturally into a study of vector spaces of functions and which collections of vectors form spanning sets, that is, allow any function in the space (here, any solution of the finite bar problem) to be expressed as a linear combination of basis vectors (here, in (18n)).

(d) It is worth noting that separated solutions again allow us to tackle the problem of the infinite bar, though boundary conditions need especial attention. An appropriate place for discussion of the existence of solutions is in the application of transform methods and complex analysis (see Chapter 14).

Exercise 1 Suppose that the boundary conditions (13) are replaced by

$$u_x(0, t) \; = \; 0, \qquad u_x(L, t) \; = \; 0$$

(corresponding to the ends being insulated), but that in all other respects the problem remains the same. Obtain an expression for $u(x, t)$.

Exercise 2 Solve the heat equation (1) for the finite bar when subject to

$$u(0, t) \; = \; u_x(L, t) \; = \; 0 \qquad \text{for } t \geq 0$$

and

$$u(x, 0) \; = \; T_0 x \qquad \text{for } 0 \leq x \leq L,$$

where T_0 is a non-zero constant.

Exercise 3 Solve the heat equation (1) for the finite bar when subject to

$$u(0, t) \; = \; 0, \qquad u(L, t) \; = \; 1 \qquad \text{for } t \geq 0$$

and

$$u(x, 0) \; = \; h(x) \qquad \text{for } 0 \leq x \leq L,$$

where $h(0) = 0$ and $h(L) = 1$.

[HINT: First find a substitution of the form $v(x, t) = u(x, t) + f(x)$, where $v(x, t)$ satisfies (1) and $v(0, t) = v(L, t) = 0$.]

7.3 Transverse waves in a finite string

A string of constant density, performing small transverse vibrations and with constant tension along its length may be modelled by a function $u = u(x,t)$ satisfying the one-dimensional wave equation

$$\text{(2)} \qquad \frac{\partial^2 u}{\partial x^2} = \frac{1}{c^2} \frac{\partial^2 u}{\partial t^2}, \qquad (c > 0).$$

For the case of a finite string, lying when in equilibrium along the x-axis for $0 \le x \le L$, fixed at its ends and with given initial position and transverse velocity, appropriate conditions on the boundary are

$$\text{(21)} \qquad u(0,t) = 0 \quad \text{and} \quad u(L,t) = 0, \qquad \text{for } t \ge 0$$

and initially both

$$\text{(22)} \qquad u(x,0) = \phi(x) \quad \text{and} \quad \frac{\partial u}{\partial t}(x,0) = \psi(x), \qquad \text{for } 0 \le x \le L,$$

where ϕ and ψ are twice continuously differentiable functions.

Separation of variables is again useful and, as for the heat equation in section 7.2, putting

$$u(x,t) = X(x)T(t)$$

leads to

$$\frac{X''}{X} = \frac{T''}{c^2 T}$$

and, with separation constant λ, to the equations

$$\text{(23)} \qquad X'' - \lambda X = 0,$$

$$\text{(24)} \qquad T'' - \lambda c^2 T = 0.$$

Again, as with the heat equation, exclusion of the case $u \equiv 0$ implies first that

$$\text{(25)} \qquad X(0) = X(L) = 0$$

from the boundary conditions (21), and then that the only solutions of (23) we can consider are of the form

$$X(x) \;=\; A \sin \mu x + B \cos \mu x,$$

where $\lambda = -\mu^2$ is negative and A and B are constants. The conditions (25) then give solutions to (23) in the form

$$X(x) \;=\; X_n(x) \;\equiv\; A_n \sin \frac{n\pi x}{L}$$

for $0 \le x \le L$ and each integer n, where each A_n is constant. Correspondingly, the solution of (24) is

$$T \;=\; T_n(t) \;\equiv\; C_n \sin \frac{n\pi ct}{L} + D_n \cos \frac{n\pi ct}{L},$$

for $t \ge 0$ and constants C_n, D_n. We seek a solution $u = u(x,t)$ as a sum,

$$(26) \qquad\qquad u(x,t) \;=\; \sum_{n=1}^{\infty} u_n(x,t),$$

of the *normal modes*

$$(27) \qquad u_n(x,t) \;=\; \left(E_n \sin \frac{n\pi ct}{L} + F_n \cos \frac{n\pi ct}{L} \right) \sin \frac{n\pi x}{L}$$

of frequency $(n\pi/L)(1/2\pi) = n/2L$, where E_n, F_n are constants. Hence,

$$\frac{\partial u}{\partial t}(x,t) \;=\; \sum_{n=1}^{\infty} \frac{n\pi c}{L} \left(E_n \cos \frac{n\pi ct}{L} - F_n \sin \frac{n\pi ct}{L} \right) \sin \frac{n\pi x}{L}$$

and the initial conditions (22) provide

$$\sum_{n=1}^{\infty} F_n \sin \frac{n\pi x}{L} \;=\; \phi(x) \quad \text{and} \quad \sum_{n=1}^{\infty} \frac{n\pi c}{L} E_n \sin \frac{n\pi x}{L} \;=\; \psi(x)$$

for $0 \le x \le L$. Referring to our work in the last section, we can immediately write down values for E_n and F_n:

$$(28) \qquad E_n \;=\; \frac{L}{n\pi c} \cdot \frac{2}{L} \int_0^L \psi(x) \sin \frac{n\pi x}{L} \, dx, \qquad F_n \;=\; \frac{2}{L} \int_0^L \phi(x) \sin \frac{n\pi x}{L} \, dx$$

valid for every positive integer n. The assumption we have made can easily be shown (from the theory of Fourier series) to establish rigorously that $u(x,t)$, as given by (26), (27) and (28), is a solution of the problem consisting of equation (2) with boundary conditions (21) and initial conditions (22).

Exercise 4 The function $u = u(x,t)$ solves equation (2) when subject to the boundary condition (21). Find u in each of the following cases:

(a)
$$u(x,0) = 0 \qquad \text{for } 0 \leq x \leq L$$

and

$$\frac{\partial u}{\partial t}(x,0) = \begin{cases} 0 & \text{for } 0 \leq x \leq \dfrac{L}{4}, \\[2mm] V_0 & \text{for } \dfrac{L}{4} < x < \dfrac{3L}{4}, \\[2mm] 0 & \text{for } \dfrac{3L}{4} \leq x \leq L, \end{cases}$$

where V_0 is a small positive constant. (Assume the above analysis remains valid, despite the discontinuities of $\partial u/\partial t$.)

(b)
$$u(x,0) = \varepsilon x(L-x) \quad \text{and} \quad \frac{\partial u}{\partial t}(x,0) = 0$$

for $0 \leq x \leq L$, where ε is a small positive constant.

Exercise 5 By first transforming the equation to canonical form (as in Chapter 6), show that the solution to (2) when subject only to the initial conditions (22) may be written in d'Alembert's form

$$(29) \qquad u(x,t) = \frac{1}{2}\left(\phi(x+ct) + \phi(x-ct)\right) + \frac{1}{2c} \int_{x-ct}^{x+ct} \psi(u)\, du$$

and specify the domain in the (x,t)-plane in which this solution is determined.

Now consider the special case $\psi \equiv 0$. Show that, if in addition we apply the boundary conditions (21), then ϕ must be an odd function ($\phi(x) = -\phi(-x)$, all x) of period $2L$. Also, by first re-writing the solution (26) in this special case, using the identity

$$\cos P \sin Q = \frac{1}{2}\left(\sin(P+Q) - \sin(P-Q)\right),$$

compare (26) with (29).

Note Further discussion of examples, such as in Exercise 4, and the application of d'Alembert's formula (29) in Exercise 5 (including to the propagation of waves) may be found in any book on wave motion, examples of which may be found in the Bibliography.

7.4 Separated solutions of Laplace's equation in polar co-ordinates and Legendre's equation

Our real interest in this section is in applying the method of separation of variables to three-dimensional Laplace's equation in spherical polar co-ordinates and in the derivation of Legendre's equation. As a curtain-raiser, which in a simpler way shows some of the techniques involved, we shall look at the two-dimensional Laplace equation in circular polar co-ordinates (r, θ)

$$
(4) \qquad \frac{\partial^2 U}{\partial r^2} + \frac{1}{r}\frac{\partial U}{\partial t} + \frac{1}{r^2}\frac{\partial^2 U}{\partial \theta^2} \;=\; 0 \qquad\qquad (r \neq 0)
$$

and seek separated solutions $U = U(r, \theta)$ in the form

$$
U(r, \theta) \;=\; R(r)\Theta(\theta),
$$

where $R = R(r)$ and $\Theta = \Theta(\theta)$ have continuous second derivatives. This leads quickly to

$$
\frac{r^2}{R}\left(R'' + \frac{1}{r}R'\right) \;=\; -\frac{\Theta''}{\Theta} \;=\; \lambda
$$

for a separation constant λ, and hence

$$
(30) \qquad\qquad r^2 R'' + rR - \lambda R \;=\; 0
$$

and

$$
(31) \qquad\qquad \Theta'' + \lambda\Theta \;=\; 0.
$$

Requirement For the purposes of many applications, especially in physics, it is natural to insist, as (r, θ) and $(r, \theta + 2\pi)$ represent the same points in the plane, that the function Θ be periodic of period divisible by 2π. This requires $\lambda = n^2$ where n is an integer – a requirement we now adopt.

The solutions of (31) are then

$$\Theta(\theta) \;=\; \begin{cases} C_0\theta + D_0 & (n = 0), \\[2mm] C_n \sin n\theta + D_n \cos n\theta & (n \neq 0), \end{cases}$$

for integers n. The constant C_0 will be taken to be zero for the same reason as we put $\lambda = n^2$. Equation (30) becomes

$$(32) \qquad\qquad r^2 R'' + rR' - n^2 R \;=\; 0 \qquad (r > 0)$$

which is of Euler type (see Appendix (7)). When $n = 0$, $R = a + b\log n$ (a, b constants) is a solution. For $n \neq 0$, the trial solution $R = r^\mu$ leads to the auxiliary equation $\mu^2 = n^2$ and hence the general solution of (32) is given by

$$R(r) \;=\; \begin{cases} A_0 + B_0 \log r & (n = 0), \\[3mm] A_n r^n + \dfrac{B_n}{r^n} & (n \neq 0). \end{cases} \qquad (r > 0)$$

As solutions for $n < 0$ are already present in the list for $n > 0$, we ignore them and, to match boundary conditions, seek solutions to (4) in the form

$$(33) \qquad U(r,\theta) \;=\; A_0 + B_0 \log r + \sum_{n=1}^{\infty} \left(A_n r^n + \frac{B_n}{r^n} \right) (C_n \sin n\theta + D_n \cos n\theta).$$

In practice, with the simpler boundary conditions, one may try to fit the conditions by choosing only the (seemingly) appropriate terms in (33). The following example gives a case in point.

Example 1 Find a continuous function $u = u(x, y)$ which satisfies Laplace's equation inside the circle $\Gamma = \{(x, y) : x^2 + y^2 = 1\}$ but equals $1 + 2y^2$ on Γ.

Noting that on Γ, as $y = \sin\theta$ and hence

$$(34) \qquad\qquad 1 + 2y^2 \;=\; 1 + 2\sin^2\theta = 2 - \cos 2\theta,$$

we seek solutions in the form

$$U(r,\theta) \;=\; (A_0 + B_0 \log r) + \left(A_2 r^2 + \frac{B_2}{r^2} \right) (C_2 \sin 2\theta + D_2 \cos 2\theta).$$

As we require a continuous solution, valid at the origin, we must have $B_0 = B_2 = 0$. ($C_2 = D_2 = 0$ cannot give a solution which satisfies the boundary condition on Γ.) Hence, having regard to (34), a natural 'trial solution' is

$$(35) \qquad\qquad U(r, \theta) \;=\; a + br^2 \cos 2\theta,$$

where a and b are constants. To equal $1 + 2y$ on Γ, we must by (34) have

$$a + b \cos 2\theta \;=\; 2 - \cos 2\theta \qquad (0 \le \theta < 2\pi).$$

For this to hold when $\theta = \pi/4$, a has to be 2, and hence $b = -1$. With these values for a and b, the trial solution (35) becomes

$$U(r, \theta) \;=\; 2 - r^2 \cos 2\theta = 2 - r^2(1 - 2\sin^2\theta) = 2 - r^2 + 2(r \sin\theta)^2$$

and thus

$$u(x, y) \;=\; 2 - (x^2 + y^2) + 2y^2 = 2 - x^2 + y^2. \qquad \square$$

Exercise 6 The change of variables $(x, y, z) \to (r, \theta, z)$ into cylindrical polar co-ordinates given by

$$x \;=\; r \cos\theta,$$
$$y \;=\; r \sin\theta,$$
$$z \;=\; z$$

for $r \ge 0$, $0 \le \theta < 2\pi$, and $U(r, \theta, z) = u(r \cos\theta, r \sin\theta, z)$ translates the three-dimensional Laplace equation (5) for $u = u(x, y, z)$ into

$$(36) \qquad\qquad \frac{1}{r}\frac{\partial}{\partial r}\left(r\frac{\partial U}{\partial r}\right) + \frac{1}{r^2}\frac{\partial^2 U}{\partial \theta^2} + \frac{\partial^2 U}{\partial z^2} \;=\; 0 \qquad (r \ne 0).$$

Show that separated solutions of the form $U(r, \theta) = R(r)\Theta(\theta)$ which are independent of z again give rise to equations (30) and (31), and hence to the solution (33) if we require $\lambda = n^2$.

Find a solution of (36) inside the cylinder $r = a$ (a is a positive constant) which satisfies

$$\lim_{r \to 0} U(r, \theta) \text{ finite}, \qquad \lim_{r \to a} \frac{\partial U}{\partial r}(r, \theta) \;=\; \cos^3\theta$$

for $0 \le \theta < 2\pi$. Is the solution unique?

For three-dimensional Laplace's equation in spherical polar co-ordinates and $r \ne 0$, $0 \ne \theta \ne \pi$,

$$(6) \qquad \frac{1}{r^2}\frac{\partial}{\partial r}\left(r^2\frac{\partial U}{\partial r}\right) + \frac{1}{r^2 \sin\theta}\frac{\partial}{\partial \theta}\left(\sin\theta\frac{\partial U}{\partial \theta}\right) + \frac{1}{r^2 \sin^2\theta}\frac{\partial^2 U}{\partial \phi^2} \;=\; 0,$$

we shall only seek separated solutions $U = U(r, \theta, \phi)$ independent of the co-ordinate ϕ and therefore in the form

$$U(r, \theta) = R(r)\Theta(\theta).$$

Substitution leads directly to

$$\frac{1}{R}(r^2 R'' + 2rR') = -\frac{1}{\sin\theta \cdot \Theta}(\sin\theta \cdot \Theta'' + \cos\theta \cdot \Theta') = \lambda,$$

for a separation constant λ, and hence to

(37) $$r^2 R'' + 2rR' - \lambda R = 0$$

and

(38) $$\Theta'' + \cot\theta \cdot \Theta' + \lambda\Theta = 0.$$

Requirement In order that a solution $\Theta = \Theta(\theta)$ of (38) is continuously differentiable, it is necessary (but we shall not show here) that $\lambda = n(n+1)$ where n is a positive integer or zero. We make this a requirement.

With $\lambda = n(n+1)$, equation (37) becomes

(39) $$r^2 R'' + 2rR' - n(n+1)R = 0,$$

an equation again of Euler type: the trial solution $R = r^\mu$ leading to the general solution

$$R = R_n(r) \equiv A_n r^n + \frac{B_n}{r^{n+1}}$$

valid for n a positive integer or zero.

The corresponding equation in Θ is

(40) $$\Theta'' + \cot\Theta \cdot \Theta' + n(n+1)\Theta = 0.$$

The substitution $x = \cos\theta$ transforms (40) into an equation in $L = L(x)$ by means of

$$\Theta(\theta) = L(\cos\theta), \qquad \Theta'(\theta) = -\sin\theta \cdot L'(\cos\theta)$$

$$\Theta''(\theta) = \sin^2\theta \cdot L''(\cos\theta) - \cos\theta \cdot L'(\cos\theta)$$

and resulting, after minor re-arrangement, in *Legendre's equation*

$$(41) \qquad\qquad (1 - x^2)L'' - 2xL' + n(n+1)L \;=\; 0.$$

As we shall see in Chapter 13, solutions to this equation, the *Legendre functions*, come in two kinds. Corresponding to each n, there is a polynomial solution $L = P_n(x)$, called a *Legendre polynomial* and, linearly independent of $P_n = P_n(x)$, a solution which is a *Legendre function of the second kind* $Q_n = Q_n(x)$. The Legendre functions of the second kind are often not useful in analysing practical problems, as they are undefined at $x = 1$ (corresponding to $\theta = 0$). For example,

$$Q_0(x) \;=\; \frac{1}{2}\log\left(\frac{1+x}{1-x}\right), \qquad Q_1(x) \;=\; \frac{1}{2}x\log\left(\frac{1+x}{1-x}\right) - 1.$$

Before taking up a more general discussion later on, and in order to tackle some realistic problems, we specify here the first four Legendre polynomials, which the reader may easily check solve equation (41):

$$(42) \quad P_0(x) \;=\; 1, \quad P_1(x) \;=\; x, \quad P_2(x) \;=\; \frac{1}{2}(3x^2 - 1), \quad P_3(x) \;=\; \frac{1}{2}(5x^3 - 3x).$$

Of course, as (41) is a homogeneous equation, constant multiples of these functions also solve it. It is, however, usual to choose the multiples, as we have here, so that $P_n(1) = 1$ (corresponding to $\Theta(0) = 1$).

Exercise 7 For integers m, n ranging over $0, 1, 2, 3$, verify the following identities:

(a)
$$P_n(x) \;=\; \frac{1}{2^n n!}\frac{d^n}{dx^n}(x^2 - 1)^n,$$

(b)
$$\int_{-1}^{1}(P_n(x))^2\, dx \;=\; \frac{2}{2n+1},$$

(c)
$$\int_{-1}^{1} P_m(x)P_n(x)\, dx \;=\; 0 \qquad (m \neq n).$$

Note Equation (a) is known as Rodrigues's Formula and (c) is referred to as the orthogonality property of Legendre polynomials (corresponding to orthogonality of vectors in the appropriate inner product space of functions, as studied in linear algebra).

As with circular polars in the plane discussed above, we might in the general case seek solutions of three-dimensional Laplace's equation, which are independent of the angle ϕ, in the form

$$U(r,\theta) = \sum_{n=0}^{\infty} \left(A_n r^n + \frac{B_n}{r^{n+1}} \right) (C_n P_n(\cos\theta) + D_n Q_n(\cos\theta)).$$

For simple problems, such as in the example now following, it may be possible to limit ourselves to sums of the following four functions:

(43)

$$\left(A_0 + \frac{B_0}{r} \right) P_0(\cos\theta) = A_0 + \frac{B_0}{r},$$

$$\left(A_1 r + \frac{B_1}{r^2} \right) P_1(\cos\theta) = \left(A_1 r + \frac{B_1}{r^2} \right) \cos\theta,$$

$$\left(A_2 r^2 + \frac{B_2}{r^3} \right) P_2(\cos\theta) = \left(A_2 r^2 + \frac{B_2}{r^3} \right) \frac{1}{2}(3\cos^2\theta - 1),$$

$$\left(A_3 r^3 + \frac{B_3}{r^4} \right) P_3(\cos\theta) = \left(A_3 r^3 + \frac{B_3}{r^4} \right) \frac{1}{2}(5\cos^3\theta - 3\cos\theta).$$

Example 2 In three-dimensional Euclidean space, $u = u(x,y,z)$ satisfies Laplace's equation outside the sphere $S = \{(x,y,z) : x^2 + y^2 + z^2 = 1\}$ and is equal to $3 + 4z + 5(x^2 + y^2)$ on S. Find a solution to this problem, given that $u \to 0$ as $x^2 + y^2 + z^2 \to \infty$.

In spherical polars (r, θ, ϕ), the problem may be expressed as seeking to find a solution U of Laplace's equation for $r > 1$, which is independent of ϕ, equals $3 + 4\cos\theta + 5\sin^2\theta$ when $r = 1$ and tends to zero as $r \to \infty$. Now, the reader will quickly show that

(44) $$3 + 4\cos\theta + 5\sin^2\theta = \frac{19}{3} P_0(\cos\theta) + 4P_1(\cos\theta) - \frac{10}{3} P_2(\cos\theta).$$

It is natural to seek a solution involving just the first three functions in (43). Further, we shall discard terms involving r^n ($n = 0, 1, 2$) which do not tend to zero as $r \to \infty$.

So, our trial solution is

$$U(r, \theta) \;=\; \frac{a}{r} P_0(\cos \theta) + \frac{b}{r^2} P_1(\cos \theta) + \frac{c}{r^3} P_2(\cos \theta)$$

for constants a, b, c. Comparing this with (44) when $r = 1$ gives $a = 19/3$, $b = 4$, $c = -10/3$ (justified, for example, by using (b) and (c) of Exercise 7). A solution to the problem is hence found to be

$$u(x, y, z) \;=\; \frac{19}{3(x^2 + y^2 + z^2)^{\frac{1}{2}}} + \frac{4z}{(x^2 + y^2 + z^2)^{\frac{3}{2}}} + \frac{5(x^2 + y^2 - 2z^2)}{3(x^2 + y^2 + z^2)^{\frac{5}{2}}},$$

once one uses $z = r \cos \theta$ and $r^2 = x^2 + y^2 + z^2$.   □

Exercise 8 Find a continuous function $u = u(x, y, z)$ which satisfies Laplace's equation within the sphere $S = \{(x, y, z) : x^2 + y^2 + z^2 = 1\}$ and equals z^3 on S.

Exercise 9 Investigate separated solutions *not* independent of ϕ,

$$U(r, \theta, \phi) \;=\; R(r)\Theta(\theta)\Phi(\phi),$$

of three-dimensional Laplace's equation in spherical polar co-ordinates (r, θ, ϕ) as follows. First write

$$U(r, \theta, \phi) \;=\; R(r)K(\theta, \phi)$$

and choose your separation constant so that the equation in R is (39). Then write

$$K(\theta, \phi) \;=\; \Theta(\theta)\Phi(\phi)$$

choosing a second separation constant so that the equation in Φ is

$$\Phi'' + m^2 \Phi \;=\; 0.$$

Show that if we write $x = \cos \theta$ and $\Theta(\theta) = M(\cos \theta)$, then the equation satisfied by $M = M(x)$ is

(45) $$(1 - x^2)M'' - 2xM' + \left\{ n(n+1) - \frac{m^2}{1 - x^2} \right\} M \;=\; 0.$$

Note Solutions of (45) are linear combinations of the so-called *associated Legendre functions* $P_n^m = P_n^m(x)$ and $Q_n^m = Q_n^m(x)$ not discussed further here.

7.5 The Dirichlet problem and its solution for the disc

In a number of examples and exercises in the last section, solutions of Laplace's equation, *harmonic functions*, were sought which satisfied specified conditions on the boundary ∂D of a region D in which the equation was valid. For instance, in Example 1, a solution of two-dimensional Laplace's equation in the interior of the unit disc, which equalled a given function on the unit circle, was required, whereas in Example 2, the solution of three-dimensional Laplace's equation outside the unit sphere had to have specific values on the sphere.

Definition 1 The problem of finding a harmonic function in a region D of Euclidean space, which equals a given function on the boundary ∂D of D, is called the *Dirichlet problem* for D.

Because of its great importance in applications, the Dirichlet problem has received considerable attention, especially in a number of areas of classical mathematical physics, such as gravitation, electromagnetism and hydrodynamics. In this section, we show that the Dirichlet problem for the disc has a solution in integral form. We work in circular polar co-ordinates, so that the problem can be expressed as follows: to find $U = U(r, \theta)$ for $0 \leq r \leq R$, $0 \leq \theta < 2\pi$, such that

$$(46) \qquad \frac{\partial^2 U}{\partial r^2} + \frac{1}{r} \frac{\partial U}{\partial r} + \frac{1}{r^2} \frac{\partial^2 U}{\partial \theta^2} = 0, \qquad (r \neq 0)$$

and

$$\lim_{r \to R} U(r, \theta) = f(\theta), \qquad (0 \leq \theta < 2\pi)$$

where we assume that f is a continuous function of θ. As we require our solution to exist in the whole disc, and in particular at the origin, we reject solutions containing $\log r$ and r^{-n} for all positive integers n, and use the trial solution (33) in the restricted form

$$U(r, \theta) = A_0 + \sum_{n=1}^{\infty} r^n (C_n \sin n\theta + D_n \cos n\theta).$$

Hence,

$$f(\theta) = A_0 + \sum_{n=1}^{\infty} R^n (C_n \sin n\theta + D_n \cos n\theta),$$

for $0 \leq \theta < 2\pi$,

where

$$A_0 = \frac{1}{2\pi} \int_0^{2\pi} f(\phi)\, d\phi$$

and

$$C_n = \frac{1}{\pi R^n} \int_0^{2\pi} f(\phi) \sin n\phi\, d\phi, \qquad D_n = \frac{1}{\pi R^n} \int_0^{2\pi} f(\phi) \cos n\phi\, d\phi$$

for every positive integer n. Therefore, using the uniform convergence of the sum for $f(\theta)$ to interchange sum and integral, we have

$$U(r,\theta) = \frac{1}{2\pi} \int_0^{2\pi} f(\phi)\, d\phi + \frac{1}{\pi} \int_0^{2\pi} \left(\sum_{n=1}^{\infty} \left(\frac{r}{R}\right)^n \cos n(\theta - \phi) \right) f(\phi)\, d\phi$$

and hence

$$(47) \qquad U(r,\theta) = \frac{1}{2\pi} \int_0^{2\pi} \left(1 + 2\sum_{n=1}^{\infty} \left(\frac{r}{R}\right)^n \cos n(\theta - \phi) \right) f(\phi)\, d\phi,$$

all for $0 < r < R$, $0 \le \theta < 2\pi$. Two elementary lemmas now aid our discussion.

Lemma 1 (Mean Value Theorem) Under the conditions of the given Dirichlet problem, the value of U at the origin is

$$\frac{1}{2\pi} \int_0^{2\pi} f(\phi)\, d\phi,$$

that is, the average value of f over the circle on which it is defined.

The proof of Lemma 1, which relies on Green's Theorem in the Plane (the two-dimensional Divergence Theorem) and techniques not central to this chapter, is given in an appendix.

Lemma 2 For $0 \le \rho < 1$

$$(48) \qquad \sum_{n=1}^{\infty} \rho^n \cos nt = \frac{\rho \cos t - \rho^2}{1 - 2\rho \cos t + \rho^2}$$

for all real t.

Proof Noting that, as a convergent geometric series,

$$(49) \qquad \sum_{n=1}^{\infty} (\rho e^{it})^n = \frac{\rho e^{it}}{1 - \rho e^{it}} = \frac{\rho e^{it}(1 - \rho e^{-it})}{(1 - \rho e^{it})(1 - \rho e^{-it})}$$

$$= \frac{\rho e^{it} - \rho^2}{1 - \rho(e^{it} + e^{-it}) + \rho^2},$$

all we need to do is recall that

$$\cos t = \mathrm{Re}\,(e^{it}) = \frac{1}{2}(e^{it} + e^{-it})$$

and take real parts.   □

Corollary 1 With the same conditions,

$$1 + 2 \sum_{n=1}^{\infty} \rho^n \cos nt = \frac{1 - \rho^2}{1 - 2\rho \cos t + \rho^2}.$$

Lemma 1 tells us that equation (47) represents $U = U(r, \theta)$ even when $r = 0$, and hence Lemma 2 allows us to deduce, when writing $\rho = r/R$ and $t = \theta - \phi$, that

$$(50) \qquad U(r, \theta) = \frac{1}{2\pi} \int_0^{2\pi} \frac{R^2 - r^2}{R^2 - 2rR\cos(\theta - \phi) + r^2} f(\phi)\, d\phi$$

for $0 \le r < R$, $0 \le \theta < 2\pi$, once one multiplies both numerator and denominator of the integrand by R^2.

Thus, equation (50) solves the Dirichlet problem for the disc, presenting the harmonic function $U = U(r, \theta)$ as *Poisson's integral*. The function

$$\frac{R^2 - r^2}{2\pi(R^2 - 2rR\cos(\theta - \phi) + r^2)}$$

multiplying $f(\phi)$ in the integrand is known as *Poisson's kernel*.

Note The reader will have noticed that we have re-formulated the usual form of the boundary condition, using $\lim_{r \to R} u(r, \theta) = f(\theta)$ rather than $u(R, \theta) = f(\theta)$, to ensure that Poisson's integral will continue to give us a solution as we approach the boundary.

7.6   Radially symmetric solutions of the two-dimensional wave equation and Bessel's equation

We have noted in the introduction to this chapter that when we change the Cartesian space variables (x, y) to circular polar co-ordinates (r, θ), the two-dimensional wave equation becomes

$$(51) \qquad \frac{\partial^2 U}{\partial r^2} + \frac{1}{r} \frac{\partial U}{\partial r} + \frac{1}{r^2} \frac{\partial^2 U}{\partial \theta^2} = \frac{1}{c^2} \frac{\partial^2 U}{\partial t^2},$$

where c is a positive constant. Here we look for radially symmetric separated solutions

$$U(r, t) = R(r) T(t),$$

that is, separated solutions which are independent of θ. Quickly we find

$$(52) \qquad \frac{1}{R} \left(R'' + \frac{1}{r} R' \right) = \frac{T''}{c^2 T} = \lambda,$$

for a separation constant λ.

   We now impose conditions which correspond to those useful for the analysis of small transverse vibrations of a circular membrane, fixed along the whole of its rim; namely, the boundary condition

$$(53) \qquad\qquad U(R, t) = 0, \qquad \text{for all } t \geq 0$$

and, where $\phi = \phi(r)$ and $\psi = \psi(r)$ are continuous, initial conditions

$$(54) \qquad U(r, 0) = \phi(r), \qquad \frac{\partial U}{\partial t}(r, 0) = \psi(r), \qquad \text{for } 0 \leq r \leq R.$$

Requirement   In order to ensure oscillatory motion, we insist that $\lambda = -\mu^2$ (μ is a positive constant), so that the differential equation in $T = T(t)$ arising from (52)

$$T'' - \lambda c^2 T \equiv T'' + \mu^2 c^2 T = 0,$$

has solutions

$$T(t) = A \sin \mu c t + B \cos \mu c t, \qquad (A, \ B \ \text{constants}).$$

The corresponding equation in $R = R(r)$, again from (52), is

$$r R'' + R' + \mu^2 r R = 0,$$

which, after making the 'normalising' substitution $x = \mu r$, gives an equation in $y = y(x)$, through

$$R(r) = y(\mu r), \qquad R'(r) = \mu y'(\mu r), \qquad R'' = \mu^2 y''(\mu r).$$

This equation is *Bessel's equation of order zero*:

(55)
$$xy'' + y' + xy = 0.$$

Solutions of this equation (discussed in Chapters 12, 13 and 14) are linear combinations of the *zero order Bessel functions* $J_0 = J_0(x)$, respectively $Y_0 = Y_0(x)$, of the *first*, respectively, *second kind*. As $Y_0 \to \infty$ as $x \to 0$, for the purposes of our problem Y_0 must be excluded. Thus, we want constant multiples of

$$J_0 = J_0(x) = J_0(\mu r)$$

and hence have found a separated solution of the two-dimensional wave equation in the form

$$U(r,t) = J_0(\mu r)(A\sin\mu ct + B\cos\mu ct).$$

In order for this to satisfy the boundary condition (53), we must have

$$J_0(\mu R) = 0,$$

as otherwise $T \equiv 0$ implying $U \equiv 0$, which is only possible for the uninteresting initial conditions given by $\phi \equiv \psi \equiv 0$. It is known (but not established here) that $J_0 = J_0(x)$ has a countably infinite number of positive zeros which we arrange as an increasing sequence (α_n), where

$$\alpha_1 < \alpha_2 < \ldots < \alpha_n < \ldots .$$

Correspondingly, we may write $\mu = \mu_n = \alpha_n/R$, for each n. So, in line with earlier work in this chapter and in order to fit the initial conditions, we choose a trial solution in the form

$$U(r,\theta) = \sum_{n=1}^{\infty} J_0\left(\frac{\alpha_n r}{R}\right)\left(A_n \sin\frac{\alpha_n ct}{R} + B_n \cos\frac{\alpha_n ct}{R}\right).$$

The Fourier analysis (valid for differentiable ϕ, ψ) which we used in sections 7.2 and 7.3 may then be employed to determine the constants A_n, B_n as integrals involving $J_0(\alpha_n r/R)$.

Exercise 10 Investigate the same problem (including the same boundary and initial conditions) when the (non-zero) solution $U = U(r, \theta, t)$ sought varies with θ, as well as with r and t.

[*Plan of action:* first write $U(r, \theta, t) = f(r, \theta)T(t)$ and, as above, write down equations for $f = f(r, \theta)$ and $T = T(t)$ involving $f(r, \theta) = R(r)\Theta(\theta)$ and, for consistency, choose a separation constant so that the resulting equation in Θ is $\Theta'' + n^2\Theta = 0$, for positive integers n. Supposing that $R(r) = J_n(\mu r)$, where $J_n = J_n(x)$ is the n-th order Bessel function, 'well-behaved' at $x = 0$, and that $\psi(r) \equiv 0$, write down a natural trial solution.]

Exercise 11 Investigate the two-dimensional wave equation in Cartesian co-ordinates

$$\frac{\partial^2 u}{\partial x^2} + \frac{\partial^2 u}{\partial y^2} = \frac{1}{c^2} \frac{\partial^2 u}{\partial t^2},$$

seeking solutions $u = u(x, y, t)$ satisfying the boundary conditions

$$u(x, 0, t) = u(x, b, t) = 0 \qquad \text{for } 0 \leq x < a,$$

$$u(0, y, t) = u(a, y, t) = 0 \qquad \text{for } 0 \leq y < b,$$

both for all $t \geq 0$, and the initial conditions

$$u(x, y, 0) = \phi(x, y),$$

$$\frac{\partial u}{\partial t}(x, y, 0) = \psi(x, y),$$

both for $0 \leq x \leq a$, $0 \leq y \leq b$. This corresponds to small transverse vibrations of a rectangular membrane fixed at every point of its rim.

[You should expect your solution to be expressed in terms of a 'double' Fourier series.]

7.7   Existence and uniqueness of solutions, well-posed problems

The topics heading this section represent an area where pure mathematics has the most profound and useful application, to assure applied mathematicians that their models of the real world produce problems where solutions may be found (existence), or where a solution, once found, covers all possibilities (uniqueness), or where a computer approximation of a problem produces solutions near the true solution of the real (well-posed) problem.

Our remarks split naturally along the lines of the classification of second-order partial differential equations with linear principal part given in Chapter 6. However, we make one general comment concerning well-posed problems first.

If a problem is well-posed, then the continuity of a solution dependent on continuously varying the boundary conditions necessitates that r, s, t can be defined as continuous functions along the boundary curve. Hence, the boundary cannot, by definition, be a characteristic.

It is only for **hyperbolic equations** that one can say something general and positive. The *Cauchy problem*, consisting of the hyperbolic equation in canonical form

$$(56) \qquad \frac{\partial^2 u}{\partial x \partial y} = \phi \left(x, y, u, \frac{\partial u}{\partial x}, \frac{\partial u}{\partial y} \right)$$

together with *Cauchy boundary data*

$$(57) \quad u(x(\sigma), y(\sigma)) = f_1(\sigma), \quad \frac{\partial u}{\partial x}(x(\sigma), y(\sigma)) = f_2(\sigma), \quad \frac{\partial u}{\partial y}(x(\sigma), y(\sigma)) = f_3(\sigma)$$

given along a regular curve C with equation $\mathbf{r} = \mathbf{r}(\sigma) = \mathbf{r}(x(\sigma), y(\sigma))$, where f_1, f_2, f_3 are continuous functions of σ, permits an elegant solution *locally*, in the following manner.

Theorem 1 Suppose that $\phi = \phi(x, y, u, p, q)$ is continuous in a neighbourhood N of

$$(x_0, y_0, f_1(\sigma_0), f_2(\sigma_0), f_3(\sigma_0))$$

in $\mathbb{R}^5$, where $\mathbf{r}(\sigma_0) = (x(\sigma_0), y(\sigma_0)) = (x_0, y_0)$, a point on C. Further, suppose that no tangent of C is parallel to either the x- or y-axis and that ϕ satisfies Lipschitz conditions in each of the variables u, p, q (see Chapter 2). Then, in a sufficiently small neighbourhood of (x_0, y_0), there exists a unique solution of (56) satisfying the Cauchy data (57).

Notes

(a) This Cauchy problem is 'stable', in the sense that slight changes in the Cauchy data (57) produce only slight changes in the solution. Thus, the problem is well-posed.

(b) The proof[1] of Theorem 1 parallels, with its use of Lipschitz conditions and the Weierstrass M-test, the proof of the Cauchy–Picard Theorem given in Chapter 2.

(c) When the equation is linear and $\phi(x, y, u, p, q)$ may be written in the form

$$\phi(x, y, u, p, q) = a(x, y)p + b(x, y)q + c(x, y)u + d(x, y),$$

one can use the particularly elegant method of Riemann to write down a global solution of the Cauchy problem in terms of the values along the boundary curve and the integral of $d = d(x, y)$ over a region bounded by characteristic and boundary curves. The exercises at the end of this section consider special cases.

---

[1] See, for example, Epstein, *Partial Differential Equations*, Chapter 3.

We can often find solutions of the **parabolic equation** and, when the solution is unique, this can be expected to arise alongside a well-posed problem in a bounded region.

Example 3 The one-dimensional heat equation

$$(1) \qquad\qquad k\frac{\partial^2 u}{\partial x^2} = \frac{\partial u}{\partial t},$$

for $u = u(x,t)$, taken together with the appropriate Cauchy data

$$u(0,t) = f_1(t), \qquad u(L,t) = f_2(t), \qquad u(x,0) = f_3(x)$$

(compare section 7.2) is a well-posed problem producing a unique solution.

However, in infinite domains, uniqueness can fail. To furnish an example, we allow x to range over the whole real line and specify an initial condition in the limit form

$$\lim_{t\to 0+} u(x,t) = f(x), \qquad \text{for all } x.$$

If $u_i = u_i(x,t)$, for $i = 1,2$, are two solutions of the problem, then $u = u_1 - u_2$ clearly also solves (1) and $u(x,t) \to 0$ as $t \to 0+$, for all x. If the problem had a unique solution, u would have to be the identically zero function. However, the reader may easily show that if we take $u = u(x,t)$ to be the function defined by

$$u(x,t) = \frac{x}{t^{3/2}} \exp\left(-\frac{x^2}{4kt}\right), \qquad (t > 0, \text{ all } x)$$

then

$$\lim_{t\to 0+} u(x,t) = 0, \qquad (x \neq 0)$$

whereas $u(0,t) = 0$, for all t. Note that $u(x,t)$ goes to infinity as (x,t) approaches $(0,0)$ along the curve $x^2 = 4t$. $\qquad\qquad\qquad \square$

Note Uniqueness is retrieved if we know that u must remain bounded; but the proof is technically demanding (and not given here).

The **elliptic equation** is far less tractable; hence the importance of solutions to the Dirichlet problem. We now give some classical examples to show that Cauchy data may be prescribed for which Laplace's equation has no solution, and also for which, though a solution can be found, it is extremely unstable.

Example 4 We consider two-dimensional Laplace's equation

$$\frac{\partial^2 u}{\partial x^2} + \frac{\partial^2 u}{\partial y^2} = 0$$

with Cauchy data

$$u(x,0) = \frac{\partial u}{\partial x}(x,0) = 0, \qquad \frac{\partial u}{\partial y}(x,0) = |x|^3.$$

It can be shown (but not here) that any solution must have derivatives of every order on $y = 0$. However, the third derivative of $|x|^3$ does not exist at $x = 0$. □

Example 5 The same as Example 4, save this time

$$\frac{\partial u}{\partial y}(x,0) = \frac{\sin nx}{n},$$

for positive integers n. The right-hand side of this equation tends uniformly to zero as $n \to \infty$, but the solution to the problem, given by

$$u(x,y) = \frac{\sinh ny \sin nx}{n^2},$$

does *not* tend to zero as $n \to \infty$. However, if

$$\frac{\partial u}{\partial y}(x,0) = 0,$$

then the solution is $u \equiv 0$. The problem is not well-posed. □

It is conjectured that *all* problems involving elliptic equations are ill-posed.

Exercise 12 (The Dirichlet problem is not soluble for the hyperbolic equation.) Show that if $u = u(x,y)$ is a solution of the hyperbolic equation $u_{xy} = 0$, valid in the closed rectangle, the sides of which are given by the equations

$$x = 0, \qquad x = a, \qquad y = 0, \qquad y = b,$$

where a and b are positive constants, then

$$u(0,0) + u(a,b) = u(a,0) + u(0,b).$$

Hence, show that it is not in general possible to solve a hyperbolic equation with prescribed values on the boundary of a given region.

Exercise 13 Show that $v(x,y) = J_0((xy)^{\frac{1}{2}})$, with $J_0(0) = 1$, is a *Riemann function* for the *telegraphy equation*

$$Lu \equiv \frac{\partial^2 u}{\partial x \partial y} + \frac{1}{4} u = f(x,y),$$

that is, that it satisfies

$$Lv = 0, \qquad \frac{\partial v}{\partial x}(x,0) = \frac{\partial v}{\partial y}(0,y) = 0,$$

for $x \neq 0$, $y \neq 0$ and $v(0,0) = 1$.

[HINT: Recall that the Bessel function $J_0 = J_0(t)$ of zero order satisfies Bessel's equation

$$tJ_0'' + J_0' + tJ_0 = 0.]$$

Note By translation, the origin $(0,0)$ can be replaced by any point (X,Y) in the domain of the problem. The Riemann function would then be $v(x,y) = J_0((x-X)^{\frac{1}{2}}(y-Y)^{\frac{1}{2}})$ and would correspond to boundary conditions

$$Lv = 0, \qquad \frac{\partial v}{\partial x}(x,Y) = \frac{\partial v}{\partial y}(X,y) = 0,$$

for $x \neq X, y \neq Y$ and $v(X,Y) = 1$.

Exercise 14 Consider the problem consisting of the hyperbolic equation

$$\frac{\partial^2 u}{\partial x^2} - \frac{\partial^2 u}{\partial t^2} = f(x,t),$$

with Cauchy data

$$u(x,0) = \phi(x), \qquad \frac{\partial u}{\partial t}(x,0) = \psi(x), \qquad (a \leq x \leq b)$$

to be satisfied by a twice continuously differentiable function $u = u(x,t)$ and continuous functions $\phi = \phi(x)$, $\psi = \psi(x)$. By using Green's Theorem in the Plane (see the Appendix to this chapter), as applied to the triangular region A, bounded by segments of the characteristics L_1, respectively L_2, given by $x - t = X - T$, respectively $x + t = X + T$, passing through the point (X,T), and the boundary curve L_3, given by $t = 0$, where $T \neq 0$, $a \leq X - T$ and $X + T \leq b$, show that

$$\iint_A f(x,t)\,dx dt = \int_{L_1+L_2+L_3} \left(\frac{\partial u}{\partial t}\frac{dx}{ds} + \frac{\partial u}{\partial x}\frac{dt}{ds} \right) ds.$$

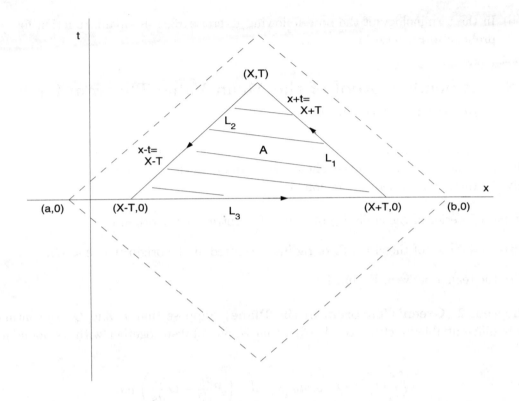

Deduce that

$$2u(X,T) \;=\; \phi(X-T) + \phi(X+T) + \int_{X-T}^{X+T} \psi(x)\,dx - \int_{0}^{T}\int_{X-T+t}^{X+T-t} f(x,t)\,dx\,dt.$$

Hence, show that a solution to the problem both exists and is unique. Show further that if ϕ and ψ are also continuous functions of a parameter λ, then so is u.

[HINT: Notice that $dx/ds = -dt/ds$ on L_1, $dx/ds = dt/ds$ on L_2 and $dt/ds = 0$ on L_3, and hence that the integrand for L_1 is $(-du/ds)$ and for L_2 is du/ds.]

Notes

(a) The reader will have noticed that $u(X,T)$ can be determined by the above method whenever (X,T) lies in the square bounded by the dotted lines in the diagram; that is, by the lines with equations

$$X - T \;=\; a, \quad X - T \;=\; b, \quad X + T \;=\; a, \quad X + T \;=\; b.$$

We only know ϕ and ψ on the interval $[a,b]$. The square is called the *domain of dependence* on the boundary/initial conditions.

(b) In the terminology of the note following Exercise 13, a Riemann function for the problem in Exercise 14 is $v \equiv 1$.

7.8 Appendix: proof of the Mean Value Theorem for harmonic functions

We here deduce Lemma 1 of section 7.5 as a corollary of Green's Theorem in the Plane (the two-dimensional Divergence Theorem), a result which we shall assume. We consider only the three cases where the region A is

(i) the interior of the circle Γ_R of radius R, centred at the origin 0,

(ii) the interior of the circle Γ_ε of radius ε, centred at the origin $(0 < \varepsilon < R)$,

(iii) the region between Γ_ε and Γ_R.

Theorem 2 (Green's Theorem in the Plane) Suppose that P and Q are continuously differentiable functions on $A \cup \partial A$; that is, on A taken together with its boundary ∂A. Then

(58)
$$\iint_A (P_x + Q_y)\, dx dy \;=\; \int_{\partial A} \left(P\frac{dy}{ds} - Q\frac{dx}{ds} \right) ds,$$

where s is arc-length measured along ∂A and the integral on the right-hand side is the corresponding line-integral.

Notice that a curve $\mathbf{r} = \mathbf{r}(s) = (x(s), y(s))$ with unit tangent vector $\mathbf{t} = (dx/ds, dy/ds)$ has (outward-drawn) normal $\mathbf{n} = (dy/ds, -dx/ds)$, as this latter vector is clearly also of unit length and such that $\mathbf{t}.\mathbf{n} = 0$. So, using vector notation, Green's identity (58) may be re-written

(59)
$$\iint_A (P_x + Q_y)\, dx dy \;=\; \int_{\partial A} (P, Q).\mathbf{n}\, ds$$

in 'Divergence Theorem form'.

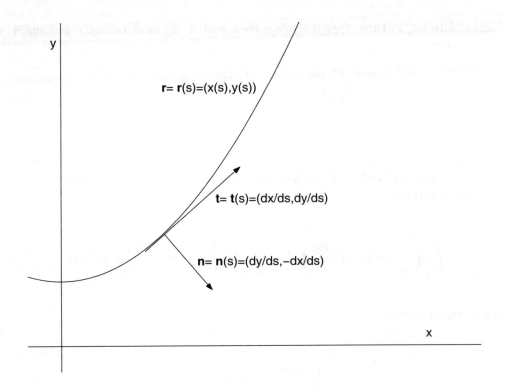

Suppose now that $u = u(x,y)$ and $v = v(x,y)$ are twice continuously differentiable *harmonic* functions in $A \cup \partial A$. We may apply Green's Theorem to the continuously differentiable functions

$$P \;=\; uv_x - vu_x, \qquad Q \;=\; uv_y - vu_y.$$

The reader will quickly calculate that $P_x + Q_y$ is zero and hence, as

$$\frac{\partial u}{\partial n} \;=\; \mathbf{n} \cdot (u_x, u_y), \qquad \frac{\partial v}{\partial n} \;=\; \mathbf{n} \cdot (v_x, v_y),$$

that (59) reduces in this case to

$$(60) \qquad\qquad \int_{\partial A} \left(u \frac{\partial v}{\partial n} - v \frac{\partial u}{\partial n} \right) ds \;=\; 0.$$

For the rest of this section, suppose u is harmonic in all of region (i), and hence in regions (ii) and (iii) as well.

Taking A to be, in turn, region (i) and then region (ii), and taking $v \equiv 1$ on A, we see at once that

$$(61) \qquad \int_{\Gamma_R} \frac{\partial u}{\partial n} \, ds \;=\; \int_{\Gamma_\varepsilon} \frac{\partial u}{\partial n} \, ds = 0.$$

Now take A to be region (iii), so that $\partial A = \Gamma_R \cup \Gamma_\varepsilon$, and let $v = \log r$ which (as we saw in section 7.4) is harmonic in all of A, a region not containing the origin. As $\partial/\partial n$ is $\partial/\partial r$ on Γ_R and $(-\partial/\partial r)$ on Γ_ε (where the outward-drawn normal points *towards* the origin), equation (60) gives

$$(62) \qquad \int_{\Gamma_R} \left(\frac{1}{R} u - \log R \cdot \frac{\partial u}{\partial r} \right) ds + \int_{\Gamma_\varepsilon} \left(-\frac{1}{\varepsilon} u + \log \varepsilon \cdot \frac{\partial u}{\partial r} \right) ds = 0$$

which reduces at once to

$$(63) \qquad \frac{1}{\varepsilon} \int_{\Gamma_\varepsilon} u \, ds \;=\; \frac{1}{R} \int_{\Gamma_R} u \, ds$$

once one applies (61) to the second term in each integral in (62). The length of the circumference of Γ_ε is

$$\int_{\Gamma_\varepsilon} ds \;=\; 2\pi\varepsilon.$$

So, if u_0 is the value of u at the origin,

$$\left| \int_{\Gamma_\varepsilon} u \, ds - 2\pi\varepsilon u_0 \right| \;=\; \left| \int_{\Gamma_\varepsilon} (u - u_0) \, ds \right| \;\le\; 2\pi\varepsilon \sup |u - u_0|,$$

where $\sup |u - u_0|$ is the maximum value of $|u - u_0|$ on Γ_ε. Therefore, the left-hand side of (63) tends to $2\pi u_0$ as $\varepsilon \to 0$ and, if $u = f(\theta)$ on Γ_R, where θ is the circular polar angle,

$$u_0 \;=\; \frac{1}{2\pi R} \int_{\Gamma_R} u \, ds \;=\; \frac{1}{2\pi R} \int_0^{2\pi} f(\theta) R \, d\theta \;=\; \frac{1}{2\pi} \int_0^{2\pi} f(\theta) \, d\theta.$$

Exercise 15 Using Green's Theorem, with $P = uu_x$, $Q = uu_y$, derive another of Green's identities:

$$\iint_A (u\nabla^2 u + (\operatorname{grad} u)^2)\, dx\, dy \;=\; \int_{\partial A} u\frac{\partial u}{\partial n}\, ds.$$

Hence, show that, if u is a twice continuously differentiable harmonic function on $A \cup \partial A$, which is zero on ∂A, then $u \equiv 0$ on $A \cup \partial A$. Hence, show that, if there is a solution to the Dirichlet problem in A, with values given on ∂A, then it is unique.

Exercise 16 Use Exercise 15 to show that the integral with respect to θ, from 0 to 2π, of the Poisson kernel is one (see section 7.5).

[HINT: Note that $U \equiv 1$ solves the Dirichlet problem for the disc if U is given as 1 on its boundary.]

8 The Fredholm Alternative

The Fredholm Alternative analyses the solutions of the non-homogeneous Fredholm equation

$$y(x) \;=\; f(x) + \lambda \int_a^b K(x,t)y(t)\,dt, \qquad (x \in [a,b])$$

where $f : [a,b] \to \mathbb{R}$ and $K : [a,b]^2 \to \mathbb{R}$ are continuous and λ is a constant. We shall only establish it in the special and useful case when $K = K(x,t)$ is *degenerate* (some authors say, *of finite rank*), that is, when

$$K(x,t) \;=\; \sum_{j=1}^n g_j(x)h_j(t), \qquad (x,t \in [a,b])$$

and $g_j : [a,b] \to \mathbb{R}$, $h_j : [a,b] \to \mathbb{R}$ are continuous ($j = 1,\ldots,n$). Note that the summation is a finite one. As with Green's functions in Chapter 4, the method of proof we shall employ also gives a practical method for solving problems.

8.1 A simple case

In order to motivate both the statement and proof of the Fredholm Alternative Theorem, we now discuss an equation with perhaps the simplest degenerate kernel $K = K(x,t)$ of any interest; namely,

$$K(x,t) \;=\; g(x)h(t), \qquad (x,t \in [a,b])$$

where g and h are continuous on $[a,b]$ and neither of which is identically zero on $[a,b]$.

The equation is

(N) $$y(x) \;=\; f(x) + \lambda \int_a^b g(x)h(t)y(t)\,dt$$

$$=\; f(x) + \lambda X g(x), \qquad \text{where } X \;=\; \int_a^b h(t)y(t)\,dt.$$

We note immediately that the constant X depends on y. We discuss the equation in terms of its (non-homogeneous) *transpose*

(N^{T}) $$y(x) \;=\; f(x) + \lambda \int_a^b K(t,x)y(t)\,dt$$

$$=\; f(x) + \lambda \int_a^b g(t)h(x)y(t)\,dt$$

$$=\; f(x) + \lambda Y h(x), \qquad \text{where } Y \;=\; \int_a^b g(t)y(t)\,dt,$$

the corresponding homogeneous equation

(H) $$y(x) \;=\; \lambda \int_a^b g(x)h(t)y(t)\,dt \;=\; \lambda X g(x),$$

and that equation's (homogeneous) transpose

(H^{T}) $$y(x) \;=\; \lambda \int_a^b g(t)h(x)y(t)\,dt \;=\; \lambda Y h(x).$$

Multiplying (N) by $h(x)$ and integrating with respect to x gives

(1) $$\left(1 - \lambda \int_a^b gh\right) X \;=\; \int_a^b fh.$$

Similar consideration of (N^{T}) results in

(2) $$\left(1 - \lambda \int_a^b gh\right) Y \;=\; \int_a^b fg.$$

Provided $\lambda \int_a^b gh \neq 1$, these equations allow the determination of (unique) X and Y and hence unique continuous solutions

$$y(x) = f(x) + \frac{\lambda \int_a^b fh}{1 - \lambda \int_a^b gh} g(x), \qquad (x \in [a, b])$$

of (N) and

$$y(x) = f(x) + \frac{\lambda \int_a^b fg}{1 - \lambda \int_a^b gh} h(x), \qquad (x \in [a, b])$$

of (N^T). However, if $\lambda \int_a^b gh = 1$, then neither (N) nor (N^T) permits a unique solution. Indeed, from (1), (N) cannot then have a solution unless

(3) $$\int_a^b fh = 0,$$

and, from (2), (N^T) cannot have a solution unless

$$\int_a^b fg = 0.$$

Now, every solution of (H), from its very statement, must be of the form

(4) $$y(x) = cg(x), \qquad (x \in [a, b])$$

where c is a constant. Further, if $y = cg$ on $[a, b]$, for some constant c, and $\lambda \int_a^b gh = 1$, then

$$\lambda X g(x) = c\lambda g(x) \int_a^b h(t)g(t)\, dt = cg(x) = y(x)$$

for each x in $[a, b]$; and so, (4) is the (only) solution of (H). Similarly, if $\lambda \int_a^b gh = 1$,

$$y(x) = dh(x), \qquad (x \in [a, b])$$

where d is a constant, is the (only) solution of (H^T). Thus, if there does *not* exist a unique solution to (N) (which occurs here if $\lambda \int_a^b gh = 1$) then there are non-zero solutions of (H) and (H^T); and there can only be solutions of (N) provided 'the integral of f times the solution of (H^T)' is zero (here $\int_a^b fh = 0$).

In the case of this particularly simple kernel and when the 'consistency condition' $\int_a^b fh = 0$ is met, it is clear (by simple substitution) that $y = f$ is a particular solution of (N). Further, if $y = Y$ is any solution of (N), then $y = Y - f$ is a solution of (H). So, provided the consistency condition is met and there is no unique solution of (N), the complete solution of (N) is given, with arbitrary constant c, by

$$y(x) \;=\; f(x) + cg(x), \qquad (x \in [a, b])$$

(a particular solution plus the complete solution of (H)).

Example 1 Find the complete solution to each of the following Fredholm equations:

(a) $y(x) \;=\; x + \lambda \int_0^1 e^{x+t} y(t)\, dt,$ $\hspace{4cm}$ $(x \in [0, 1]);$

(b) $y(x) \;=\; x - \dfrac{2e^x}{e^2 - 1} + \lambda \int_0^1 e^{x+t} y(t)\, dt,$ $\hspace{2.5cm}$ $(x \in [0, 1]).$

Consider first the equation (a): the kernel $e^{x+t} = e^x e^t$ is symmetric, and so

(N) = (N^{T}) $\hspace{2cm}$ $y(x) \;=\; x + \lambda X e^x, \qquad$ where $X \;=\; \int_0^1 e^t y(t)\, dt.$

Since

$$\int_0^1 x e^x\, dx \;=\; 1, \qquad \int_0^1 e^{2x}\, dx \;=\; \tfrac{1}{2}(e^2 - 1)$$

(we always find it useful to carry out such manipulations first), multiplying (N) by e^x and integrating gives

(5) $\hspace{4cm}$ $(1 - \lambda(e^2 - 1)/2)X \;=\; 1.$

So, provided $\lambda \neq 2/(e^2 - 1)$ and substituting for X, (N) has the unique continuous solution

$$y(x) \;=\; x + \frac{2\lambda}{2 - (e^2 - 1)\lambda}\, e^x, \qquad (x \in [0, 1]).$$

When $\lambda = 2/(e^2 - 1)$, (5) shows that there can be no solution to (N). That is the end of the story for (a)!

Now turning to (b), we have

(N) = (N^{T}) $\hspace{2cm}$ $y(x) \;=\; x - \dfrac{2e^x}{e^2 - 1} + \lambda X e^x, \qquad$ where $X \;=\; \int_0^1 e^t y(t)\, dt.$

Corresponding to (5) there is

(6) $$(1 - \lambda(e^2 - 1)/2)X = 0,$$

and hence, provided $\lambda \neq 2/(e^2 - 1)$, (N) has the unique continuous solution

$$y(x) = x - \frac{2e^x}{e^2 - 1}, \qquad (x \in [0, 1]).$$

When $\lambda = 2/(e^2 - 1)$, (6) provides no restriction on X and the complete solution to (N) is given as

$$y(x) = x - \frac{2e^x}{e^2 - 1} + ce^x = x + c'e^x, \qquad (x \in [0, 1])$$

where c, c' are arbitrary (though, of course, related) constants.

For both (a) and (b) the homogeneous/transposed-homogeneous equation is

$$(\mathrm{H}) = (\mathrm{H}^T) \qquad\qquad y(x) = \lambda X e^x,$$

having the unique solution $y \equiv 0$ when $\lambda \neq 2/(e^2 - 1)$, and when $\lambda = 2/(e^2 - 1)$ the complete solution

$$y(x) = ce^x$$

where c is an arbitrary constant. So, the consistency condition $\int_a^b fh = 0$ in the above discussion is not met in case (a), as

$$\int_0^1 x e^x \, dx \neq 0,$$

whereas it is met in case (b), where

$$\int_0^1 \left(x - \frac{2e^x}{e^2 - 1} \right) e^x \, dx = 0.$$

These correspond to the 'right-hand sides' of (5) and (6) being 1 and 0 respectively. $\square$

Exercise 1 Solve the equation

$$y(x) = (e - 1)x^n - x + \lambda \int_0^1 x t e^{x^3 + t^3} y(t) \, dt, \qquad (x \in [0, 1])$$

for the cases $n = 1$ and $n = 4$.

8.2　Some algebraic preliminaries

In our proof of the Fredholm Alternative Theorem, we shall need to rely on the reader's acquaintance with some further elementary results from linear algebra.

Suppose A is the $n \times n$ matrix $A = (a_{ij})$ and that $A^T = (a'_{ij})$, where

$$a'_{ij} = a_{ji}, \qquad \text{for } i = 1, \ldots, n \text{ and } j = 1, \ldots, n,$$

is the transpose of A. The system of linear equations

$$
\begin{array}{ccccc}
a_{11}x_1 & + & \ldots & + & a_{1n}x_n & = & b_1 \\
\vdots & & & & \vdots & & \vdots \\
a_{n1}x_1 & + & \ldots & + & a_{nn}x_n & = & b_n
\end{array}
$$

may be conveniently expressed as

$$\sum_{j=1}^{n} a_{ij}x_j = b_i, \qquad (i = 1, \ldots, n)$$

or more compactly as

$$Ax = b,$$

where $x = (x_i)$ and $b = (b_i)$ are column vectors.

Proposition 1

(a) $\text{Rank}(A) = \text{Rank}(A^T)$.

(b) If $\text{Rank}(A) = n$, then $\det A \neq 0$ and the non-homogeneous systems

$$Ax = b, \qquad A^T y = b$$

have unique solutions $x = (x_i)$, $y = (y_i)$.

(c) If $1 \leq \text{Rank}(A) = n - r \leq n - 1$, then $\det A = 0$ and the systems of homogeneous equations

$$Ax = 0, \qquad A^T y = 0$$

both have maximal linearly independent sets of solutions consisting of r elements. If $y^k = (y_i^k)$, $k = 1, \ldots, r$, is such a set for $A^T y = 0$, then $Ax = b$ has solutions if and only if

$$b^T y^k = b_1 y_1^k + \ldots + b_n y_n^k = 0, \qquad (k = 1, \ldots, r).$$

Proposition 2 The number of distinct solutions μ of the characteristic equation of A,

$$\det(\mu I - A) = 0,$$

is at most n.

8.3   The Fredholm Alternative Theorem

The theorem analyses the solutions of the equation

$$(\text{N}) \qquad y(x) = f(x) + \lambda \int_a^b K(x,t)y(t)\,dt, \qquad\qquad (x \in [a,b])$$

in terms of the solutions of the corresponding equations

$$(\text{N}^{\text{T}}) \qquad y(x) = f(x) + \lambda \int_a^b K(t,x)y(t)\,dt, \qquad\qquad (x \in [a,b])$$

$$(\text{H}) \qquad y(x) = \lambda \int_a^b K(x,t)y(t)\,dt, \qquad\qquad (x \in [a,b])$$

$$(\text{H}^{\text{T}}) \qquad y(x) = \lambda \int_a^b K(t,x)y(t)\,dt, \qquad\qquad (x \in [a,b])$$

where $f : [a,b] \to \mathbb{R}$ and the kernel $K : [a,b]^2 \to \mathbb{R}$ are continuous and λ is a constant.

Theorem 1 (The Fredholm Alternative) For each (fixed) λ exactly one of the following statements is true:

(F1) The equation (N) possesses a unique continuous solution and, in particular, $f \equiv 0$ on $[a,b]$ implies that $y \equiv 0$ on $[a,b]$. In this case, (N^{T}) also possesses a unique continuous solution.

(F2) The equation (H) possesses a finite maximal linearly independent set of, say, r continuous solutions $y_1, \ldots, y_r$ $(r > 0)$. In this case, (H^{T}) also possesses a maximal linearly independent set of r continuous solutions $z_1, \ldots, z_r$ and (N) has solutions if and only if the 'consistency conditions'

$$\int_a^b f(x)z_k(x)\,dx = 0, \qquad (k = 1, \ldots, r)$$

are all met. When they are, the complete solution of (N) is given by

$$y(x) \;=\; g(x) + \sum_{i=1}^{r} c_i y_i(x), \qquad (x \in [a,b])$$

where $c_1, \ldots, c_r$ are arbitrary constants and $g : [a,b] \to \mathbb{R}$ is any continuous solution of (N).

The above is true for general kernels, but when $K = K(x,t)$ is the degenerate kernel given by

(D) $$\qquad\qquad K(x,t) \;=\; \sum_{j=1}^{n} g_j(x) h_j(t), \qquad (x,t \in [a,b])$$

there are at most n values of λ at which (F2) occurs.

The reader is at once encouraged to commit the statement of Theorem 1 to memory! We prove the result only in an important special case.

Proof of Theorem 1 when K is given by (D)   We may assume that each of the sets of functions $(g_1, \ldots, g_n)$, $(h_1, \ldots, h_n)$ is linearly independent on $[a,b]$; otherwise, we may express each of their elements in terms of linearly independent subsets to reach the required form. (For instance, if $(g_1, \ldots, g_n)$ is linearly independent but h_n is dependent on the linearly independent subset $(h_1, \ldots, h_{n-1})$ and

$$h_n(t) \;=\; \sum_{j=1}^{n-1} d_j h_j(t), \qquad (t \in [a,b])$$

where d_j is a constant $(j = 1, \ldots, n-1)$, then

$$K(x,t) \;=\; \sum_{j=1}^{n-1} g_j(x) h_j(t) + g_n(x) \sum_{j=1}^{n-1} d_j h_j(t)$$

$$=\; \sum_{j=1}^{n-1} (g_j(x) + d_j g_n(x)) h_j(t),$$

for each x and each t in $[a,b]$. The reader may easily verify that each of the sets

(7) $$\qquad\qquad (g_1 + d_1 g_n, \ldots, g_{n-1} + d_{n-1} g_n), \qquad (h_1, \ldots, h_{n-1})$$

is linearly independent on $[a,b]$.)

For K given by (D), the equations (N), (N^T), (H), (H^T) may be written

$$(N_1) \quad y(x) = f(x) + \lambda \sum_{j=1}^{n} X_j g_j(x), \qquad \text{where } X_j = \int_a^b h_j(t) y(t) \, dt, \quad j = 1, \ldots, n,$$

$$(N_1^T) \quad y(x) = f(x) + \lambda \sum_{j=1}^{n} Y_j h_j(x), \qquad \text{where } Y_j = \int_a^b g_j(t) y(t) \, dt, \quad j = 1, \ldots, n,$$

$$(H_1) \quad y(x) = \lambda \sum_{j=1}^{n} X_j g_j(x),$$

$$(H_1^T) \quad y(x) = \lambda \sum_{j=1}^{n} Y_j h_j(x),$$

for x in $[a, b]$. So, a solution of (N) is determined once we know the values of the X_j's (which are, of course, defined in terms of the unknown quantity y).

First, let us note that $\lambda = 0$ corresponds to the unique continuous solution $y = f$ of both (N) and (N^T): (F1) occurs. From this point in the proof, we assume that $\lambda \neq 0$.

We now convert the problem to one of algebra, involving a system of simultaneous linear equations. Multiplying (N_1) through by $h_i(x)$ and integrating with respect to x gives, on slight re-arrangement,

$$\mu X_i - \sum_{j=1}^{n} a_{ij} X_j = b_i, \qquad (i = 1, \ldots, n)$$

once we define

$$\mu = 1/\lambda, \qquad a_{ij} = \int_a^b g_j(x) h_i(x) \, dx, \qquad b_i = \mu \int_a^b f(x) h_i(x) \, dx$$

for $i = 1, \ldots, n$ and $j = 1, \ldots, n$. These n equations in the X_i's may be rewritten in the compact form

$$(N_2) \qquad\qquad\qquad (\mu I - A) X = b,$$

where I is the identity matrix, A is the matrix $A = (a_{ij})$, and $X = (X_i)$, $b = (b_i)$ are column vectors. Similarly, working in turn with (N_1^T), (H_1) and (H_1^T), we have

(N_2^T) $$(\mu I - A)^T Y \;=\; (\mu I - A^T) Y \;=\; c,$$

where $A^T = (a_{ji})$ is the transpose of A, $Y = (Y_i)$, $c = (c_i)$ are column vectors, and $c_i = \mu \int_a^b f(x) g_i(x)\, dx \;\; (i = 1, \dots, n)$,

(H_2) $$(\mu I - A) X \;=\; 0,$$

and

(H_2^T) $$(\mu I - A)^T Y \;=\; 0.$$

By Proposition 1 of section 8.2, if $\mathrm{Rank}(\mu I - A) = n$, (N_2) and (N_2^T) have unique solutions $X = (X_i)$ and $Y = (Y_i)$. Substitution of these values in (N_1) and (N_1^T) gives unique solutions to (N) and (N^T) in this case when $(F1)$ occurs.

We are left with establishing that, when $1 \leq \mathrm{Rank}(\mu I - A) = n - r \leq n - 1$, $(F2)$ must occur. In this case, both (H_2) and (H_2^T) have maximal linearly independent sets $X^k = (X_i^k)$ and $Y^k = (Y_i^k)$, $k = 1, \dots, r$, consisting of r elements. Put

$$y_k(x) \;=\; \lambda \sum_{j=1}^{n} X_j^k g_j(x),$$

and

$$z_k(x) \;=\; \lambda \sum_{j=1}^{n} Y_j^k h_j(x),$$

for each x in $[a, b]$ and $k = 1, \dots, r$. Each y_k is clearly a solution of (H_1) and each z_k a solution of (H_1^T).

We claim now that the set $(y_k : k = 1, \dots, r)$ is linearly independent on $[a, b]$. Suppose that

$$\sum_{k=1}^{r} c_k y_k \;=\; 0$$

for some constants $c_1, \dots, c_r$. Then

$$\lambda \sum_{j=1}^{n} \left(\sum_{k=1}^{r} c_k X_j^k \right) g_j \;=\; 0.$$

But, the g_j's are linearly independent; so

$$\sum_{k=1}^{r} c_k X_j^k = 0 \qquad \text{for } j = 1, \ldots, n;$$

that is,

$$\sum_{k=1}^{r} c_k X^k = 0.$$

Since the X^k's are also linearly independent, $c_k = 0$ for each $k = 1, \ldots, r$ and the y_k's are thus independent also. Similarly, $(z_k : k = 1, \ldots, r)$ is linearly independent. These linearly independent sets (y_k), (z_k) cannot be enlarged whilst remaining independent, for the argument above could then be reversed for the augmented set to contradict the maximality of the independent sets (X^k) and (Y^k). We have thus shown that maximal linearly independent sets of solutions of (H) and (H^T) have the same finite non-zero number of elements when (F1) does not occur.

It remains to consider (N) in this (F2) case. Proposition 1 of section 8.2 asserts in this context that (N_2) has solutions if and only if

$$b^T Y^k = 0, \qquad \text{for } k = 1, \ldots, r,$$

that is, employing the definitions of b and $Y^k = (Y_j^k)$ and noting that Y^k corresponds to the solution z_k of (H_1^T), if and only if

$$\sum_{j=1}^{n} \left(\int_a^b f(x) h_j(x)\, dx \right) \left(\int_a^b g_j(t) z_k(t)\, dt \right) = 0,$$

for $k = 1, \ldots, r$. This is equivalent to

$$\int_a^b \left(\int_a^b \left(\sum_{j=1}^{n} h_j(x) g_j(t) \right) z_k(t)\, dt \right) f(x)\, dx = 0,$$

for $k = 1, \ldots, r$. Using again the fact that z_k is a solution of (H^T), the necessary and sufficient condition for (N) to have solutions reduces to

$$\int_a^b z_k(x) f(x)\, dx = 0, \qquad \text{for } k = 1, \ldots, r.$$

Suppose finally that g is a particular continuous solution of (N) (existing because the consistency conditions are met). Then, if y is any other solution of (N), $y - g$ is a solution of (H) and hence expressible as a linear combination of the elements of the maximal independent set of solutions (y_k):

$$y - g = \sum_{i=1}^{r} c_i y_i,$$

for some constants $c_1, \ldots, c_r$. The Fredholm Alternative Theorem is thus established for degenerate kernels. □

Note The reader will want to remember to distinguish between the 'eigenvalues' λ corresponding to non-zero solutions of (H) in the case (F2) and the eigenvalues $\mu = 1/\lambda$ which satisfy the matrix equation (H$_2$).

8.4   A worked example

We now show how the method used in the proof of the Fredholm Theorem can be used in a practical example.

Example 2 Solve the integral equation

$$y(x) = f(x) + \lambda \int_0^{2\pi} \sin(x + t) y(t)\, dt, \qquad (x \in [0, 2\pi])$$

in the two cases:

(a) $f(x) = 1, \quad (x \in [0, 2\pi])$,   (b) $f(x) = x, \quad (x \in [0, 2\pi])$.

The equation may be written

(N) = (N^T) $y(x) = f(x) + \lambda X_1 \sin x + \lambda X_2 \cos x$

where

$$X_1 = \int_0^{2\pi} y(t) \cos t\, dt, \qquad X_2 = \int_0^{2\pi} y(t) \sin t\, dt.$$

Note that

$$\int_0^{2\pi} \cos^2 x\, dx = \int_0^{2\pi} \sin^2 x\, dx = \pi, \qquad \int_0^{2\pi} x \sin x\, dx = -2\pi,$$

$$\int_0^{2\pi} x \cos x\, dx = \int_0^{2\pi} \cos x\, dx = \int_0^{2\pi} \sin x\, dx = \int_0^{2\pi} \sin x \cos x\, dx = 0.$$

Multiplying (N) through by $\cos x$ and integrating with respect to x and carrying out the same operation with $\sin x$ give

$$X_1 - \lambda\pi X_2 = \int_0^{2\pi} f(x)\cos x\,dx,$$

(8)

$$-\lambda\pi X_1 + X_2 = \int_0^{2\pi} f(x)\sin x\,dx.$$

When the determinant of the coefficients of the X_i's is non-zero,

$$\begin{vmatrix} 1 & -\lambda\pi \\ -\lambda\pi & 1 \end{vmatrix} = 1 - \lambda^2\pi^2 \neq 0,$$

these equations have the unique solutions

$$X_1 = \frac{1}{1 - \lambda^2\pi^2} \int_0^{2\pi} f(x)(\cos x + \lambda\pi \sin x)\,dx,$$

$$X_2 = \frac{1}{1 - \lambda^2\pi^2} \int_0^{2\pi} f(x)(\sin x + \lambda\pi \cos x)\,dx.$$

In case (a), we thus have the solution

$$y(x) = 1, \qquad (x \in [0, 2\pi]),$$

and in case (b), the solution

$$y(x) = x - \frac{2\pi\lambda}{1 - \lambda^2\pi^2}(\lambda\pi \sin x + \cos x),$$

provided $1 - \lambda^2\pi^2 \neq 0$. We have dealt with (F1).

The case (F2) can only occur when $\lambda = \pm\pi^{-1}$. It is easy to check that the corresponding homogeneous equation

(H) = (H^{T}) $y(x) = \lambda X_1 \sin x + \lambda X_2 \cos x$

only has the solutions

$$y(x) = c(\sin x + \cos x) \quad \text{when } \lambda = \pi^{-1}, \qquad\qquad (x \in [0, 2\pi])$$

$$y(x) = d(\sin x - \cos x) \quad \text{when } \lambda = -\pi^{-1}, \qquad\qquad (x \in [0, 2\pi])$$

where c, d are constants. (To check this, consider the equations (8) with the 'right-hand sides' zero, as is the case with the homogeneous equation.) The consistency conditions which need to be met to allow solutions of (N) to exist are therefore

$$\int_0^{2\pi} (\sin x + \cos x) f(x) \, dx \;=\; 0, \qquad \text{when } \lambda = \pi^{-1},$$

and

$$\int_0^{2\pi} (\sin x - \cos x) f(x) \, dx \;=\; 0, \qquad \text{when } \lambda = -\pi^{-1}.$$

In case (a), both conditions are clearly fulfilled. As $y(x) = f(x) = 1$ $(x \in [0, 2\pi])$ is a particular solution when either $\lambda = \pi^{-1}$ or $\lambda = -\pi^{-1}$, the complete solution of (N) is, when $\lambda = \pi^{-1}$,

$$y(x) \;=\; 1 + c(\sin x + \cos x), \qquad (x \in [0, 2\pi])$$

and, when $\lambda = -\pi^{-1}$,

$$y(x) \;=\; 1 + d(\sin x - \cos x), \qquad (x \in [0, 2\pi])$$

where c, d are arbitrary constants.

In case (b), neither condition is met and (N) has, therefore, no solution when either $\lambda = \pi^{-1}$ or $\lambda = -\pi^{-1}$.   □

Exercise 2 Find the eigenvalues and eigenfunctions of the integral equation

$$y(x) \;=\; \lambda \int_0^1 (g(x)h(t) + g(t)h(x)) y(t) \, dt, \qquad (x \in [0, 1])$$

where g and h are continuous functions satisfying

$$\int_0^1 (g(x))^2 \, dx \;=\; \int_0^1 (h(x))^2 \, dx \;=\; 1, \qquad \int_0^1 g(x)h(x) \, dx \;=\; 0.$$

Exercise 3 Let φ and ψ be continuous real-valued functions defined on the interval $[a, b]$ which satisfy

$$\int_a^b \varphi(x) \, dx \;=\; \int_a^b \psi(x) \, dx \;=\; 1 \quad \text{and} \quad \int_a^b \varphi(x)\psi(x) \, dx \;=\; 0.$$

Show that the integral equation

$$y(x) = \lambda \int_a^b K(x,t)y(t)\, dt, \qquad (x \in [a,b])$$

with kernel

$$K(x,t) = \varphi(x) + \psi(t),$$

has a unique eigenvalue, λ_0 say. Find λ_0 and the associated eigenfunctions.

If f is a continuous function, determine the solution of the equation

$$y(x) = f(x) + \lambda \int_a^b K(x,t)y(t)\, dt, \qquad (x \in [a,b])$$

when $\lambda \neq \lambda_0$. If $\lambda = \lambda_0$, show that this equation has no solution unless

$$\int_a^b \psi(x)f(x)\, dx = 0$$

and find all the solutions when f satisfies this condition.

Exercise 4 Show that, for every continuous real-valued function $f = f(x)$ on $[0,1]$, the integral equation

$$y(x) = f(x) + \lambda \int_0^1 (1 + xe^t)y(t)\, dt \qquad (x \in [0,1])$$

has a unique solution provided that

$$\lambda \neq \left(1 \pm \left(\frac{e-1}{2}\right)^{\frac{1}{2}}\right)^{-1}$$

Does there exist a solution of the integral equation

$$y(x) = x + \frac{1}{1 - \left(\frac{e-1}{2}\right)^{\frac{1}{2}}} \int_0^1 (1 + xe^t)y(t)\, dt \ ?$$

Exercise 5 Solve the equation

$$y(x) = 1 - x^2 + \lambda \int_0^1 (1 - 5x^2 t^2)y(t)\, dt, \qquad (x \in [0,1]).$$

Exercise 6 Determine values of the real numbers K, L for which the integral equation

$$y(x) \;=\; 1 + Kx + Lx^2 + \frac{1}{2}\int_{-1}^{1}(1+3xt)y(t)\,dt \qquad (x \in [-1,1])$$

has a solution, and find the solutions of this equation.

Exercise 7 Show that the equation

$$y(x) \;=\; f(x) + \frac{\sqrt{3}}{2a^2}\int_{-a}^{a}(x+t)y(t)\,dt, \qquad (x \in [-a,a])$$

where f is continuous and a is a positive constant, has a solution if and only if

$$\int_{-a}^{a}(x + a/\sqrt{3})f(x)\,dx \;=\; 0,$$

and find all solutions in this case.

Exercise 8   (i) Obtain a non-trivial restriction on $|\lambda|$ which ensures that the sequence (y_n) defined for $x \in [0,1]$ by

$$y_0(x) \;=\; f(x), \qquad y_{n+1}(x) \;=\; f(x) + \lambda\int_{0}^{1}K(x,t)y_n(t)\,dt \qquad (n \geq 1)$$

will converge as $n \to \infty$ to a continuous solution of the integral equation

$$y(x) \;=\; f(x) + \lambda\int_{0}^{1}K(x,t)y(t)\,dt, \qquad (x \in [0,1])$$

provided f and K are continuous functions.
[HINT: Use Picard's Method – see Chapter 1.]

  (ii) Solve

$$y(x) \;=\; f(x) + \lambda\int_{0}^{1}(1-3xt)y(t)\,dt, \qquad (x \in [0,1])$$

without using the above iterative method.

Is the restriction on $|\lambda|$ that you have derived in part (i) the weakest possible restriction that ensures convergence of the iterative method for this particular equation? Give reasons for your answer.

Note We return to the subject of the last paragraph of Exercise 8 in Chapter 10.

9 Hilbert–Schmidt Theory

In this chapter, we shall recount some elements of the Hilbert–Schmidt theory of the homogeneous Fredholm equation

(H) $$y(x) \; = \; \lambda \int_a^b K(x,t) y(t) \, dt, \qquad (x \in [a,b])$$

where λ is a constant and the kernel $K = K(x,t)$ is real-valued, continuous, and symmetric:

$$K(x,t) \; = \; K(t,x), \qquad (x,t \in [a,b]).$$

We recall that an *eigenvalue* of (H) is a value of λ for which there is a continuous solution $y(x)$ of (H), which is not identically zero on $[a,b]$. Such a $y(x)$ is an *eigenfunction* corresponding to the eigenvalue λ. Notice that all eigenvalues must necessarily be non-zero.

Although the material in this chapter fits nicely into our discussion of integral equations in Chapters 8–10, the pay-off comes in Chapter 12, as it is the widely applicable Sturm–Liouville equation that provides an important context where symmetric kernels naturally arise.

For the next section only, we shall consider the case where eigenfunctions can be complex-valued. The reader will want to note how the theory for real-valued functions fits into this general case.

9.1   Eigenvalues are real and eigenfunctions corresponding to distinct eigenvalues are orthogonal

Suppose that y and z are eigenfunctions of (H) corresponding, respectively, to eigenvalues λ and μ. Taking the complex conjugate of

$$z(x) \;=\; \mu \int_a^b K(x,t) z(t)\, dt,$$

we have, as $K(x,t)$ is real,

$$\overline{z(x)} \;=\; \bar{\mu} \int_a^b K(x,t)\overline{z(t)}\, dt;$$

so that $\bar{z}$ is an eigenfunction of (H) corresponding to the eigenvalue $\bar{\mu}$. Hence, as K is symmetric and using Fubini's Theorem ([F] of Chapter 0),

$$\bar{\mu} \int_a^b y(x)\overline{z(x)}\, dx \;=\; \bar{\mu} \int_a^b \left(\lambda \int_a^b K(x,t) y(t)\, dt \right) \overline{z(x)}\, dx$$

$$= \; \lambda \int_a^b \left(\bar{\mu} \int_a^b K(t,x)\overline{z(x)}\, dx \right) y(t)\, dt$$

$$= \; \lambda \int_a^b \overline{z(t)} y(t)\, dt.$$

Therefore,

(1) $$(\lambda - \bar{\mu}) \int_a^b y(x)\overline{z(x)}\, dx \;=\; 0.$$

When $z = y$, (1) reduces to

$$(\lambda - \bar{\lambda}) \int_a^b |y(x)|^2\, dx \;=\; 0.$$

As y is continuous and, as an eigenfunction, not identically zero on $[a,b]$, this implies that $\lambda = \bar{\lambda}$. Thus, each eigenvalue λ is real.

With this result behind us, we may re-write (1) as

$$(\lambda - \mu) \int_a^b y(x)\overline{z(x)} \, dx \;=\; 0.$$

When λ, μ are distinct, this implies that y, z are *complex orthogonal* over $[a, b]$:

(2)
$$\int_a^b y(x)\overline{z(x)} \, dx \;=\; 0.$$

If we had wished, we could have chosen to work only with real-valued eigenfunctions. For, if y is any eigenfunction corresponding to an eigenvalue λ,

$$\mathrm{Re}(y(x)) + i\mathrm{Im}(y(x)) \;=\; \lambda \int_a^b K(x,t)\{\mathrm{Re}(y(t)) + i\mathrm{Im}(y(t))\} \, dt.$$

Hence, as λ and K are real,

$$\mathrm{Re}(y(x)) \;=\; \lambda \int_a^b K(x,t)\mathrm{Re}(y(t)) \, dt$$

and

$$\mathrm{Im}(y(x)) \;=\; \lambda \int_a^b K(x,t)\mathrm{Im}(y(t)) \, dt.$$

One of the two functions $\mathrm{Re}(y(x))$, $\mathrm{Im}(y(x))$ must be a real-valued continuous function satisfying (H) and not identically zero on $[a, b]$.

Choosing y and z both to be real-valued, they are *real orthogonal* over $[a, b]$, as (2) would then reduce to

$$\int_a^b y(x)z(x) \, dx \;=\; 0.$$

Note The reader familiar with the theory of inner product spaces will already have noticed that not only the language of that theory pertains here. We continue in this vein in the next section.

9.2   Orthonormal families of functions and Bessel's inequality

A finite or infinite sequence (y_n) of continuous functions $y_n : [a, b] \to \mathbb{R}$ will be called *orthonormal* over $[a, b]$ if and only if

$$\int_a^b y_m(x) y_n(x) \, dx \;=\; \delta_{mn},$$

where δ_{mn} is the *Kronecker delta* defined, for integers m, n, by

$$\delta_{mn} \;=\; \begin{cases} 1, & \text{if } m = n, \\[2mm] 0, & \text{if } m \neq n. \end{cases}$$

Suppose that $f : [a, b] \to \mathbb{R}$ is continuous and that $(y_1, \ldots, y_N)$ is orthonormal over $[a, b]$. For $n = 1, \ldots, N$, define

$$c_n \;=\; \int_a^b f(t) y_n(t) \, dt.$$

Then,

$$0 \;\leq\; \int_a^b \left(f(t) - \sum_{n=1}^N c_n y_n(t) \right)^2 dt$$

$$=\; \int_a^b (f(t))^2 \, dt - 2 \sum_{n=1}^N c_n \int_a^b f(t) y_n(t) \, dt + \sum_{m=1}^N \sum_{n=1}^N c_m c_n \int_a^b y_m(t) y_n(t) \, dt$$

$$=\; \int_a^b (f(t))^2 \, dt - 2 \sum_{n=1}^N c_n^2 + \sum_{n=1}^N c_n^2$$

$$=\; \int_a^b (f(t))^2 \, dt - \sum_{n=1}^N c_n^2.$$

We have established *Bessel's inequality*

$$(3) \qquad \int_a^b (f(t))^2 \, dt \;\geq\; \sum_{n=1}^{N} c_n^2.$$

9.3 Some results about eigenvalues deducible from Bessel's inequality

We first note that if (λ_n) is a sequence of *distinct* eigenvalues with corresponding eigenfunctions (y_n), then we may assume that the latter sequence is orthonormal. For, we know from section 9.1 that each pair of eigenfunctions is orthogonal; so we can be sure they are orthonormal by replacing each y_n with

$$\frac{y_n}{\left(\int_a^b |y_n|^2 \right)^{\frac{1}{2}}}.$$

In line with our comment in section 9.1 and for the rest of this chapter, *all eigenfunctions will be real-valued.*

Suppose that $(\lambda_1, \ldots, \lambda_N)$ is a finite sequence of eigenvalues with corresponding eigenfunctions $(y_1, \ldots, y_N)$ such that the latter sequence is orthonormal. Fix x in $[a, b]$ and let $f(t) = K(x, t)$ in Bessel's inequality (3); so that, as $\lambda_n \neq 0$ and

$$c_n \;=\; \int_a^b K(x,t) y_n(t) \, dt \;=\; \frac{1}{\lambda_n} y_n(x), \qquad (n = 1, \ldots, N)$$

we have

$$\int_a^b (K(x,t))^2 \, dt \;\geq\; \sum_{n=1}^{N} \frac{(y_n(x))^2}{\lambda_n^2}.$$

Integrating with respect to x,

$$(4) \qquad M \;\equiv\; \int_a^b \int_a^b (K(x,t))^2 \, dt dx \;\geq\; \sum_{n=1}^{N} \frac{1}{\lambda_n^2},$$

using the fact that (y_n) is orthonormal.

It is the repeated use of formula (4) that will establish all the remaining results of the section.

Proposition 1 Each eigenvalue λ of (H) has finite multiplicity (that is, there does not exist an infinite set of eigenfunctions, each corresponding to λ, for which each finite subset is linearly independent).

Proof If there were such a set of eigenfunctions, then there would exist an infinite orthonormal sequence (y_n) of eigenfunctions, each corresponding to λ (exercise: use the Gram–Schmidt process from the theory of inner product spaces). We can then use (4) to deduce, for each positive integer N, that

$$M \geq \frac{N}{\lambda^2}.$$

But M is fixed and this is therefore impossible. So, λ must have finite multiplicity. $\quad\square$

Proposition 2 If a is a limit point of the set

$$\left\{ \frac{1}{\lambda} : \lambda \text{ is an eigenvalue of (H)} \right\},$$

then $a = 0$.

Proof By our hypothesis, there exists an infinite sequence (λ_n) of distinct eigenvalues for which $1/\lambda_n \to a$ as $n \to \infty$. Corresponding to this sequence is an orthonormal sequence of eigenfunctions (y_n). We can therefore use (4) to deduce that

$$\sum_{n=1}^{\infty} \frac{1}{\lambda_n^2}$$

is convergent and hence that $1/\lambda_n^2 \to 0$ as $n \to \infty$. But then $1/\lambda_n \to 0$ as $n \to \infty$ and a must be zero. $\quad\square$

Proposition 3 If p is a positive constant, there are only a finite number of distinct eigenvalues of (H) in $[-p, p]$.

Proof Suppose there were an infinite sequence (λ_n) of distinct eigenvalues in $[-p, p]$. A corresponding orthonormal sequence of eigenfunctions (y_n) exists and we may again apply (4) to give

$$M \geq \sum_{n=1}^{N} \frac{1}{\lambda_n^2} \geq \frac{N}{p^2},$$

for each N. But this is impossible. $\quad\square$

Corollary There can be at most a countable number of distinct eigenvalues.

Proof For each integer n, there can be at most a finite number of eigenvalues in $[-n, n]$. From section 9.1, we know that each eigenvalue is real, and so the totality of eigenvalues is contained in

$$\mathbb{R} \;=\; \bigcup_{n=1}^{\infty} [-n, n].$$

But a countable union of finite sets is countable. $\square$

Note We have *not* established the central result that (H) has at least one eigenvalue. The proof is beyond the scope of this book.

Exercise 1 Find the eigenvalues λ_j and corresponding eigenfunctions of the integral equation

$$y(x) \;=\; \lambda \int_0^\pi \sin(x + t) y(t)\, dt, \qquad (x \in [0, \pi]).$$

Show that these eigenvalues satisfy

$$\sum \frac{1}{\lambda_j^2} = \int_0^\pi \int_0^\pi \sin^2(x + t)\, dx dt.$$

Exercise 2 Show that $\sin^2 \theta < \theta$ for $0 < \theta < \pi$. If $K(x, t) = \sin(xt)$, $a = 0$ and $b = \pi$, prove that (H) has no eigenvalues in the interval

$$\left[-\frac{2}{\pi^2}, \frac{2}{\pi^2} \right].$$

Exercise 3 The integral equation

$$y(x) \;=\; \lambda \int_0^1 K(x, t) y(t)\, dt$$

has kernel $K = K(x, t)$ defined by

$$K(x, t) \;=\; \begin{cases} x(1 - t), & 0 \le x \le t \le 1, \\ t(1 - x), & 0 \le t \le x \le 1. \end{cases}$$

Verify that the equation has eigenfunctions $\sin m\pi x$, where m is a positive integer, and find the corresponding eigenvalues. Hence, show that

$$\sum_{m=1}^{\infty} \frac{1}{m^4} \leq \frac{\pi^4}{90}.$$

Exercise 4 Suppose that the integral equation

$$y(x) = \lambda \int_a^b K(x,t)y(t)\,dt \qquad (x \in [a,b])$$

has only the N distinct eigenvalues $\lambda_1, \ldots, \lambda_N$, each λ_n having multiplicity 1 and corresponding real eigenfunction y_n with $\int_a^b y_n^2 = 1$. Show that any eigenfunction corresponding to the (symmetric) kernel $L = L(x,t)$ defined on $[a,b]^2$ by

$$L(x,t) = K(x,t) - \sum_{n=1}^{N} \frac{y_n(x)y_n(t)}{\lambda_n}$$

is orthogonal to all the y_n and hence show that, on $[a,b]^2$,

$$K(x,t) = \sum_{n=1}^{N} \frac{y_n(x)y_n(t)}{\lambda_n}, \qquad (x,\,t \in [a,b]).$$

[An eigenvalue λ has multiplicity 1 if and only if any two eigenfunctions corresponding to λ are linearly dependent. You may assume, in accordance with the note above, that $L = L(x,t)$ has at least one eigenvalue.]

With $K = K(x,t)$ as above and $f = f(x)$ continuous and real-valued on $[a,b]$, prove that the non-homogeneous equation

$$y(x) = f(x) + \lambda \int_a^b K(x,t)y(t)\,dt \qquad (x \in [a,b])$$

has a unique solution for $a \leq x \leq b$, if $\lambda \notin \{\lambda_1, \ldots, \lambda_N\}$. What conditions must f satisfy for a solution to exist when $\lambda = \lambda_m$ $(1 \leq m \leq N)$? What is the general solution in this case?

Note The expansion of $K(x,t)$ in Exercise 4 may be extended to the case of an infinite number of eigenvalues for appropriate kernels. In this regard, see our Note following Exercise 7 of Chapter 10, where the celebrated theorem of Mercer is stated.

9.4 Description of the sets of all eigenvalues and all eigenfunctions

The important result of this section is best achieved by an induction in which the first step includes a demonstration that (H) actually possesses an eigenvalue. We do not include a complete proof but take the view that a proper understanding of the structure of the sets of eigenvalues and eigenfunctions requires an appreciation of an outline of the proof.

Our discussion involves the integral

$$I(y, L) \;=\; \int_a^b \!\! \int_a^b L(x, t) y(x) y(t) \, dx dt,$$

where L is a continuous, real-valued, symmetric kernel and $y : [a, b] \to \mathbb{R}$ is continuous. The inductive process to be described involves finding maxima and minima of $I(y, L)$ under more and more restrictive hypotheses on the continuous real-valued function y.

(A) Suppose that there is a continuous $y : [a, b] \to \mathbb{R}$ for which $I(y, K) > 0$, where K denotes the kernel of (H).

Step 0 One shows that $I(y, K)$, subject to $\int_a^b y^2 = 1$, achieves a maximum $M_1 > 0$ when $y = y_1$, say. Then y_1 is an eigenfunction of (H) corresponding to the eigenvalue $\lambda_1 = M_1^{-1}$.

Step 1 Putting

$$K_1(x, t) \;=\; K(x, t) - \frac{y_1(x) y_1(t)}{\lambda_1},$$

which is still symmetric, continuous and real-valued, one now shows that if $I(y, K_1) > 0$ for some continuous $y : [a, b] \to \mathbb{R}$, then, subject to $\int_a^b y^2 = 1$, $I(y, K_1)$ achieves a maximum $M_2 > 0$ at the eigenfunction $y = y_2$ of (H) corresponding to the eigenvalue $\lambda_2 = M_2^{-1}$ of (H). M_2 is in fact the maximum value of $I(y, K)$ when subjected to both

$$\int_a^b y^2 \;=\; 1, \qquad \int_a^b y y_1 \;=\; 0.$$

Therefore, $M_1 \geq M_2$ and $\lambda_1 \leq \lambda_2$. (The reader may now care to attempt Exercise 5 to fix ideas.)

Step n Put

$$K_n(x, t) \;=\; K(x, t) - \sum_{i=1}^{n} \frac{y_i(x) y_i(t)}{\lambda_i}$$

and suppose that $I(y, K_n) > 0$ for some continuous $y : [a, b] \to \mathbb{R}$. Proceeding as in Step 2, one finds a maximum value $M_{n+1} \leq M_n$ of $I(y, K)$ when subject to

$$\int_a^b y^2 = 1, \qquad \int_a^b yy_i = 0, \qquad (i = 1, \dots, n).$$

The maximum is achieved at the eigenfunction $y = y_{n+1}$ of (H) corresponding to the eigenvalue $\lambda_{n+1} = M_{n+1}^{-1} \geq \lambda_n$.

In this way, supposing that each $I(y, K_n)$ can take on positive values, one finds a sequence (λ_n) of positive eigenvalues with

$$\lambda_1 \leq \lambda_2 \leq \dots \leq \lambda_n \leq \dots$$

and a corresponding orthonormal sequence (y_n) of eigenfunctions.

(B) If at the p-th stage, say, $I(y, K_p)$ takes on negative, but no positive, values, then one finds a minimum M_{-1} of $I(y, K_p)$ subject to $\int_a^b y^2 = 1$. If M_{-1} is achieved at $y = y_{-1}$, then y_{-1} is an eigenfunction corresponding to the eigenvalue $\lambda_{-1} = M_{-1}^{-1}$. Proceeding by induction for as long as the integrals $I(y, K_m)$ take on negative values, one finds a sequence (λ_{-m}) of negative eigenvalues satisfying

$$\lambda_{-1} \geq \lambda_{-2} \geq \dots \geq \lambda_{-m} \geq \dots$$

together with a corresponding orthonormal sequence (y_{-m}) of eigenfunctions.

(C) If at the q-th stage, say, $I(y, K_q) = 0$ for all continuous $y : [a, b] \to \mathbb{R}$. Then it can be shown that K_q is identically zero on $[a, b]$; so, K is the degenerate kernel

$$K(x, t) = \sum_{i=1}^q \frac{y_i(x)y_i(t)}{\lambda_i}.$$

The process thus produces a sequence of eigenvalues in groups alternately positive and negative, and termination of this sequence is governed by the following proposition.

Proposition 4 The equation (H) possesses only a finite number of eigenvalues if and only if its kernel is degenerate.

By Proposition 1 of section 9.3, the degenerate case allows only a finite orthonormal sequence of distinct eigenfunctions. (The converse is established as in (C) above.)

The process gives a countable number of eigenvalues and we know, by the Corollary to Proposition 3 of section 9.3, that there can be at most a countable number. The picture is completed by the next result. *From now on*, we label the totality of eigenvalues, obtained as in (A) and (B) above, in order of increasing modulus, so that

$$|\lambda_1| \leq |\lambda_2| \leq \cdots \leq |\lambda_n| \leq \cdots,$$

ensuring that the eigenfunctions are correspondingly labelled, y_n corresponding to λ_n.

Proposition 5 If the continuous function $y : [a, b] \to \mathbb{R}$ satisfies both (H) and $\int_a^b y y_i = 0$ for every i, then y is identically zero on $[a, b]$, and so, cannot be an eigenfunction.

Thus, one cannot enlarge the orthonormal set (y_i) of eigenfunctions without losing orthonormality.

Arising from the proof, of which an outline is given above, are the following two interesting results. The second will be used at a vital point of the proof of the Expansion Theorem of the next section.

Proposition 6 $I(y, K)$ takes only positive values if and only if all the eigenvalues of (H) are positive.

Proposition 7 If $y : [a, b] \to \mathbb{R}$ and $z : [a, b] \to \mathbb{R}$ are continuous, then

$$\int_a^b \int_a^b K_n(x, t) y(x) z(t) \, dx dt \to 0$$

as $n \to \infty$, where K_n is defined in Step n of (A).

Exercise 5 With the notation of (A) above and assuming that λ_2 is an eigenvalue of

$$y(x) = \lambda \int_a^b K_1(x, t) y(t) \, dt$$

with corresponding eigenfunction y_2, prove that

(a) $\int_a^b y_1 y_2 = 0$,

(b) λ_2 is an eigenvalue of (H) with corresponding eigenfunction y_2,

(c) if $\int_a^b y y_1 = 0$, then $I(y, K_1) = I(y, K)$, and so, if $I(y, K_1)$ achieves a maximum at y_2, then $I(y, K)$ does also.

9.5   The Expansion Theorem

We now establish a major result of the Hilbert–Schmidt theory, giving a sufficient condition for a function to be expandable in terms of eigenfunctions of (H). The kernel $K = K(x,t)$ remains continuous, real-valued and symmetric, and the notation for the sequence (λ_n) of eigenvalues and corresponding orthonormal sequence (y_n) of eigenfunctions is as established in the section 9.4.

Theorem 1 (Expansion Theorem) If $f : [a,b] \to \mathbb{R}$ is the continuous function defined by

$$(5) \qquad\qquad f(x) \;=\; \int_a^b K(x,t)g(t)\,dt, \qquad (x \in [a,b])$$

for some continuous $g : [a,b] \to \mathbb{R}$, then f may be expanded in the uniformly convergent series

$$f \;=\; \sum_{i=1}^{\infty} c_i y_i, \quad \text{where } c_i \;=\; \int_a^b f y_i, \qquad (i \geq 1)$$

on $[a,b]$.

Note The major part of the proof of this result concerns establishing *uniform* convergence of the expansion.

Proof   An inequality of section 9.3 implies that, for $x \in [a,b]$ and any integer N,

$$\sum_{i=1}^{N} \frac{(y_i(x))^2}{\lambda_i^2} \;\leq\; \sup_{x \in [a,b]} \left(\int_a^b (K(x,t))^2 dt \right) \;\equiv\; P.$$

So, for any positive integers m, n with $m \leq n$,

$$\sum_{i=m}^{n} \frac{(y_i(x))^2}{\lambda_i^2} \;\leq\; P.$$

Defining

$$d_i \;=\; \int_a^b g y_i, \qquad (i \geq 1)$$

Bessel's inequality (section 9.2) shows that

$$\sum_{i=1}^{N} d_i^2 \leq \int_a^b g^2,$$

for every N. So,

$$\sum_{i=1}^{\infty} d_i^2$$

is convergent and hence Cauchy. Given $\epsilon > 0$, we may therefore find N such that

$$\sum_{i=m}^{n} d_i^2 < \frac{\epsilon}{P} \quad \text{whenever } n \geq m \geq N.$$

Now, by the Cauchy–Schwarz inequality,

$$\left(\sum_{i=m}^{n} \frac{d_i y_i(x)}{\lambda_i} \right)^2 \leq \left(\sum_{i=m}^{n} d_i^2 \right) \left(\sum_{i=m}^{n} \frac{(y_i(x))^2}{\lambda_i^2} \right) < \frac{\epsilon}{P} \cdot P = \epsilon,$$

for each $x \in [a, b]$ and whenever $n \geq m \geq N$. Since a uniformly Cauchy series is uniformly convergent, the series

$$\sum_{i=1}^{\infty} \frac{d_i y_i}{\lambda_i}$$

is uniformly convergent on $[a, b]$. But notice that, using the symmetry of K and Fubini's Theorem ([F] of Chapter 0),

$$c_i = \int_a^b f(x) y_i(x) \, dx$$

$$= \int_a^b \left(\int_a^b K(x, t) g(t) \, dt \right) y_i(x) \, dx \qquad \text{(by (5))}$$

$$= \int_a^b g(t) \left(\int_a^b K(t, x) y_i(x) \, dx \right) dt$$

$$= \int_a^b g(t) \frac{y_i(t)}{\lambda_i} \, dt.$$

Hence, $c_i = d_i/\lambda_i$, for every i. So, $s : [a, b] \to \mathbb{R}$ defined by

$$s \;=\; \sum_{i=1}^{\infty} c_i y_i$$

is uniformly convergent on $[a, b]$ and is therefore, by [I] of Chapter 0, continuous.

To complete the proof, we need to show that s is identically equal to f. As in section 9.4, put

$$K_n(x, t) \;=\; K(x, t) - \sum_{i=1}^{n} \frac{y_i(x) y_i(t)}{\lambda_i}, \qquad (x, t \in [a, b])$$

for every positive integer n. Then, by (5),

$$\int_a^b K_n(x, t) g(t)\, dt \;=\; f(x) - \sum_{i=1}^{n} \frac{y_i(x)}{\lambda_i} \int_a^b y_i(t) g(t)\, dt$$

$$=\; f(x) - \sum_{i=1}^{n} \frac{d_i y_i(x)}{\lambda_i}$$

$$=\; f(x) - s_n(x),$$

for each x in $[a, b]$, where s_n is the partial sum

$$s_n \;=\; \sum_{i=1}^{n} c_i y_i, \qquad (n \geq 1).$$

As $f - s$ is continuous, by Proposition 7 of section 9.4,

$$J_n \;\equiv\; \int_a^b (f(x) - s_n(x))(f(x) - s(x))\, dx$$

$$=\; \int_a^b \int_a^b K_n(x, t) g(t)(f(x) - s(x))\, dt dx$$

tends to zero as $n \to \infty$.

But, using the fact that s_n tends uniformly to s and [I] of Chapter 0,

$$J_n \rightarrow \int_a^b (f - s)^2 \quad \text{as } n \rightarrow \infty.$$

The continuity of $f - s$ now ensures that $f = s$, and the proof is complete.   □

Exercise 6 Suppose that $y = Y$ is a solution of

(N) $$y(x) \;=\; f(x) + \lambda \int_a^b K(x,t)y(t)\,dt, \qquad (x \in [a,b])$$

where K is real-valued, continuous and symmetric, and f is continuous. Considering $Y - f$, use the Expansion Theorem to show that Y can be written in the form

$$Y \;=\; f + \lambda \sum_{i=1}^{\infty} \frac{c_i}{\lambda_i - \lambda} y_i, \qquad (x \in [a,b])$$

where $c_i = \int_a^b f y_i$ and whenever $\lambda \neq \lambda_i$ for every i.

10 Iterative Methods and Neumann Series

In this chapter we take a new look at the iterative method introduced in Chapter 1. Our goal will be to introduce the Neumann series, the terms of which involve iterations of the kernel of an integral equation, and use this series to derive the solution of the equation. Throughout the chapter we shall be concerned solely with the non-homogeneous Fredholm equation

$$(N) \qquad\qquad y(x) \;=\; f(x) + \lambda \int_a^b K(x,t) y(t)\, dt, \qquad (x \in [a,b])$$

where K and f are real-valued continuous functions and λ is a constant. The kernel $K = K(x,t)$ need not be symmetric.

10.1 An example of Picard's method

We first suggest that the reader review Picard's method as used in section 1.2 to solve the non-homogeneous Fredholm equation.

We now introduce a simple example to which we shall return at various points in this chapter. It is used here to continue the review of Picard's method and will also be used to exemplify the new procedures which will appear later.

Example Use Picard's method to find a solution of the equation

$$y(x) \;=\; 1 + \lambda \int_0^1 xt y(t)\, dt, \qquad (x \in [0,1]).$$

The method provides the successive iterations ((9) of Chapter 1):

$$y_0(x) \;=\; 1,$$

$$y_1(x) \;=\; 1 + \frac{\lambda x}{2},$$

$$\vdots \quad \vdots \quad \vdots$$

$$y_n(x) \;=\; 1 + \frac{\lambda x}{2}\left(1 + \frac{\lambda}{3} + \ldots + \left(\frac{\lambda}{3}\right)^{n-1}\right),$$

as the reader will readily verify. The kernel $K(x,t) = xt$ is bounded by 1 on $[0,1]^2$. So, using (10) of Chapter 1, y_n tends to the solution of the equation if $|\lambda| < 1$. $\square$

Notes In the above example:

(a) The sequence (y_n) actually converges when the geometric series

$$\sum_{n=0}^{\infty} \left(\frac{\lambda}{3}\right)^n$$

converges; that is, when $|\lambda| < 3$. The solution is then given by

(1) $$y(x) \;=\; 1 + \frac{\lambda x}{2}\left(1 - \frac{\lambda}{3}\right)^{-1} \;=\; 1 + \frac{3\lambda x}{2(3 - \lambda)},$$

for $x \in [0,1]$.

(b) The solution (1), and for the widest possible range of λ, can be found most easily by using the Fredholm Alternative method. For simple degenerate kernels, that method is the one advised. The methods of this chapter come into their own in more complicated situations.

Exercise 1 Provide the details which establish the form of the iterated sequence (y_n) in the above example.

Exercise 2 Use the Fredholm Alternative method to show that the Example has the solution (1) for every $\lambda \neq 3$.

Exercise 3 By using (4) of Chapter 9, show that if λ_1 is the eigenvalue of

$$y(x) = \lambda \int_a^b K(x,t)y(t)\,dt$$

of least modulus, K is symmetric and $b > a$, then

$$\lambda_1 \geq \left(\int_a^b\!\!\int_a^b K^2 \right)^{-\frac{1}{2}} \geq [L(b-a)]^{-1},$$

where L is the least upper bound of $|K|$ on $[a,b]^2$. Show that these inequalities hold when $a = 0$, $b = 1$ and either $K(x,t) = xt$, or $K(x,t) = x^2t^2$, or $K(x,t) = x+t$. Note when the inequalities are strict.

10.2 Powers of an integral operator

The material of this section only provides an alternative way of looking at Picard's method.

The *integral operator* $\mathcal{K} = \mathcal{K}[a,b]$ associates, with every continuous $F : [a,b] \to \mathbb{R}$, the function $\mathcal{K}F$ defined by

$$\mathcal{K}F(x) = \int_a^b K(x,t)F(t)\,dt, \qquad (x \in [a,b]).$$

Powers of $\mathcal{K}$ are defined in the obvious way, using

$$\mathcal{K}^1 = \mathcal{K},$$

$$\mathcal{K}^n F = \mathcal{K}(\mathcal{K}^{n-1}F), \qquad (n \geq 2).$$

(Of course, every $\mathcal{K}^n F$ must be continuous.) In particular,

$$\mathcal{K}^2 F(x) = \int_a^b K(x,u) \int_a^b K(u,t)F(t)\,dt\,du,$$

$$\mathcal{K}^3 F(x) = \int_a^b K(x,v) \int_a^b K(v,u) \int_a^b K(u,t)F(t)\,dt\,du\,dv.$$

The Picard iteration (9) of Chapter 1 may then be written

$$y_0(x) = f(x),$$

$$y_n(x) = f(x) + \lambda \mathcal{K} y_{n-1}(x), \qquad (n \geq 1).$$

By an induction the reader is asked to provide,

(2)
$$y_n(x) = f(x) + \sum_{i=1}^{n} \lambda^i \mathcal{K}^i f(x);$$

so that,

$$y_n(x) - y_{n-1}(x) = \lambda^n \mathcal{K}^n f(x), \qquad (n \geq 1).$$

It is straightforward to show directly that

(3)
$$|\mathcal{K}^n f(x)| \leq L^n M (b-a)^n, \qquad (n \geq 1)$$

where L and M are bounds on $|K|$ and $|f|$, respectively, and the proof that (y_n) converges uniformly to the solution of (N) may then be completed as in Chapter 1. The solution may thus be expressed

(4)
$$y(x) = f(x) + \sum_{n=1}^{\infty} \lambda^n \mathcal{K}^n f(x), \qquad (x \in [a,b])$$

which is certainly valid for $|\lambda| < [L(b-a)]^{-1}$, as it was in section 1.2.

Exercise 4 By a careful induction, establish the equation (2) and the inequality (3).

Exercise 5 If $K(x,t) = xt$ and $f(x) = 1$, for each x and t in $[0,1]$, as in the Example in section 10.1, calculate $\mathcal{K}^n f(x)$ directly when $\mathcal{K} = \mathcal{K}[0,1]$.

10.3 Iterated kernels

Instead of iterating an operator, one can 'iterate under the integral sign' by iterating the kernel as follows. Define, for x, t in $[a,b]$,

$$K_1(x,t) = K(x,t)$$

$$K_n(x,t) = \int_a^b K(x,u) K_{n-1}(u,t) \, du, \qquad (n \geq 2).$$

We show by induction that

$$(5) \qquad\qquad \mathcal{K}^n F(x) \;=\; \int_a^b K_n(x,t) F(t)\, dt, \qquad\qquad (n \geq 1)$$

for any continuous $F : [a, b] \to \mathbb{R}$. The definition of $\mathcal{K}$ gives (5) for $n = 1$. So, suppose that equality holds for $n - 1$. Then, using Fubini's Theorem ([F] of Chapter 0) to interchange the order of integration,

$$\mathcal{K}^n F(x) \;=\; \mathcal{K}(\mathcal{K}^{n-1} F(x))$$

$$= \; \mathcal{K}\left(\int_a^b K_{n-1}(x,t) F(t)\, dt \right)$$

$$= \; \int_a^b K(x,u) \left(\int_a^b K_{n-1}(u,t) F(t)\, dt \right) du$$

$$= \; \int_a^b \left(\int_a^b K(x,u) K_{n-1}(u,t)\, du \right) F(t)\, dt$$

and the formula (5) is thus generally established.

Exercise 6 For the Example in section 10.1, where $K(x,t) = xt$ and integration is over $[0, 1]$, show that

$$K_n(x,t) \;=\; \frac{xt}{3^{n-1}},$$

for each x and t in $[0, 1]$.

Exercise 7 In the case when K is symmetric, and using the notation and Expansion Theorem of section 9.5, show that

$$K_2(x,t) \;=\; \sum_{i=1}^{\infty} \frac{y_i(x) y_i(t)}{\lambda_i^2},$$

for each x, t in $[a, b]$. Hence, prove that

$$(6) \qquad\qquad \sum_{i=1}^{\infty} \frac{y_i(x) y_i(t)}{\lambda_i}$$

converges in the mean to $K(x,t)$; that is,

$$
\text{(7)} \qquad \lim_{n \to \infty} \int_a^b \left[K(x,t) - \sum_{i=1}^n \frac{y_i(x)y_i(t)}{\lambda_i} \right]^2 dt = 0.
$$

Note *If the series* (6) *converged uniformly in t to $J = J(x,t)$, say, then proposition* [I] of Chapter 0 would entail that J is continuous in t, and then it would follow from (7) that $K = J$ by virtue of [I](b) of Chapter 0. Thus, in this case, we would have

$$
\text{(8)} \qquad K(x,t) = \sum_{i=1}^{\infty} \frac{y_i(x)y_i(t)}{\lambda_i}, \qquad (x,t \in [a,b]).
$$

More generally, for a real, continuous, symmetric kernel $K = K(x,t)$, if all but a finite number of eigenvalues are of the same sign, or if $I(y,K)$ of section 9.4 takes only positive, or only negative, values, then the expansion for K in (8) represents K as an absolutely and uniformly convergent series. This is **Mercer's Theorem** which is of importance, both theoretically and practically.

Exercise 8 In the case when (8) holds, give a quick proof of the Expansion Theorem of section 9.5.

Exercise 9 With the notation of this section, show that, for positive integers m, n and $x, t \in [a,b]$,

(a) $K_{m+n}(x,t) = \int_a^b K_m(x,u)K_n(u,t)\,du,$

(b) $|K_n(x,t)| \leq L^n(b-a)^{n-1}.$

10.4   Neumann series

With the iterated kernels of section 10.3 defined and equation (5) established, we are in a position immediately to re-write the solution of (N), given as (4) in section 10.2 in the following form:

$$
y(x) = f(x) + \sum_{n=1}^{\infty} \lambda^n \int_a^b K_n(x,t)f(t)\,dt, \qquad (x \in [a,b]).
$$

The convergence of the series being uniform for $|\lambda| < [L(b-a)]^{-1}$, where (as before) L is an upper bound on $|K|$ over $[a,b]$, we may interchange summation and integration (by [I] of Chapter 0) to give

(9) $$y(x) = f(x) + \lambda \int_a^b N(x,t;\lambda)f(t)\,dt,$$

for $x \in [a,b]$, where

(10) $$N(x,t;\lambda) = \sum_{n=0}^{\infty} \lambda^n K_{n+1}(x,t),$$

for $x,t \in [a,b]$ and $|\lambda| < [L(b-a)]^{-1}$. The series appearing in (10), and also the series representation of y in (9) (when the substitution is made for $N(x,t;\lambda)$ from (10)), are called *Neumann series*. The function $N(x,t;\lambda)$ is often termed the *reciprocal*, or *resolvent*, *kernel* as the solution of (N), given as (9), may be written as the 'reciprocal integral equation'

$$f(x) = y(x) - \lambda \int_a^b N(x,t;\lambda)f(t)\,dt.$$

Exercise 10 In the Example of section 10.1, show that the kernel $K(x,t) = xt$ gives rise to the resolvent kernel

$$N(x,t;\lambda) = 3xt(3-\lambda)^{-1}$$

when $|\lambda| < 3$, and verify that (9) produces the solution (1).

Exercise 11 Apply the methods of Picard, integral operator powers and Neumann series in turn to find the solution to

$$y(x) = 1 + \lambda \int_0^1 (1 - 3xt)y(t)\,dt, \qquad (x \in [0,1]).$$

Exercise 12 Establish the identities

$$N(x,t;\lambda) = K(x,t) + \lambda \int_a^b K(x,u)N(u,t;\lambda)\,du,$$

$$N(x,t;\lambda) = K(x,t) + \lambda \int_a^b K(u,t)N(x,u;\lambda)\,du,$$

where $x,t \in [a,b]$ and $|\lambda| < [L(b-a)]^{-1}$. [You may find Exercise 9(a) helpful.]

Exercise 13 (Harder) In (N), suppose that the kernel $K = K(x, t)$ is symmetric. Extending the first result in Exercise 7, use the Expansion Theorem of section 9.5 to show that the (symmetric iterated kernels) $K_n = K_n(x, t)$ may be expressed in the form

$$(11) \qquad\qquad K_n(x, t) \;=\; \sum_{i=1}^{\infty} \frac{y_i(x) y_i(t)}{\lambda_i^n}, \qquad (n \geq 2)$$

for x, t in $[a, b]$. For sufficiently small $|\lambda|$, deduce *formally* that the corresponding resolvent kernel is

$$(12) \qquad\quad N(x, t; \lambda) \;=\; K(x, t) + \lambda \sum_{i=1}^{\infty} \frac{y_i(x) y_i(t)}{\lambda_i(\lambda_i - \lambda)}, \qquad (x, t \in [a, b]).$$

[The expansions in (11) and (12) are in fact uniformly convergent in both x and t.]

Note (for those familiar with complex analysis) Formula (12) gives the analytic continuation of the resolvent kernel into the complex (λ-)plane. The resolvent is thus a meromorphic function of λ, and its (simple) poles are the eigenvalues of the corresponding homogeneous equation. The reader might care to ponder the use of evaluating the residues at the eigenvalues.

10.5 A remark on the convergence of iterative methods

When discussing an iterative method, we have given a *sufficient* condition, in terms of an upper bound on $|\lambda|$, for the iteration to converge. The curious reader will have wondered about the best upper bound, about a necessary condition. The answer is a simple one: the series provided by the iterative methods in this set of notes converge when, and only when, $|\lambda| < |\lambda_1|$, where λ_1 is the eigenvalue of the corresponding homogeneous equation of least modulus. With this in mind, the reader is encouraged to review the examples and exercises of this chapter, taking special note of the complete solution given by the Fredholm Alternative in the case of a degenerate kernel.

Exercise 14 Show that the equation (N) has a unique solution, expressible as a power series in λ, when its kernel $K = K(x, t)$ is degenerate and $|\lambda| < |\lambda_1|$, where λ_1 is the eigenvalue of the corresponding homogeneous equation of least modulus.

11 The Calculus of Variations

It is to Queen Dido of Carthage, in the ninth century B.C., that the oldest problem in the Calculus of Variations has been attributed. The ancient Greeks certainly formulated variational problems. Most of the more celebrated analysts of the last three centuries have made substantial contributions to the mathematical theory and, in particular, the names of Bernoulli, Newton, Leibnitz, de l'Hôpital, Euler, Lagrange, Legendre, Dirichlet, Riemann, Jacobi, Weierstrass and Hilbert are attached to important phenomena or results.

Nowadays, the Calculus of Variations not only has numerous applications in rigorous pure analysis and in physical applied mathematics (notably in classical and quantum mechanics and potential theory), but it is also a basic tool in operations research and control theory.

We shall in this chapter use the notation $f \in C^m(A)$ to indicate that f is a real-valued function, defined and m-times continuously differentiable on A, where $A \subseteq \mathbb{R}^n$ and m, n are positive integers.

11.1 The fundamental problem

Suppose that D is an open subset of $\mathbb{R}^3$, that $F \in C^2(D)$, and that a, b, c, d are given real constants with $a < b$. The problem is to find, amongst all $y \in C^1([a,b])$ for which

(1) $y(a) = c, \qquad y(b) = d, \qquad (x, y(x), y'(x)) \in D, \quad$ for all $x \in [a, b]$,

the function y which gives an extreme value (a maximum or a minimum) to the integral

(2) $$I \equiv \int_a^b F(x, y(x), y'(x)) \, dx.$$

We shall show that a C^2-function y which gives such an extreme value must necessarily satisfy Euler's equation

(E) $$\frac{d}{dx}\,F_{y'}\;=\;F_y$$

as well as the given boundary conditions

(B) $$y(a)\;=\;c,\qquad y(b)\;=\;d.$$

A solution of (E), (B) will be called an *extremal* for the problem, though it need not solve the problem. General conditions under which solutions of (E), (B) can also provide extreme values of the integral I are harder to discuss and will not be presented here. However, *ad hoc* arguments can sometimes suffice.

Some physical 'principles' are worded in terms of finding a function y giving a 'stationary value' to such an integral as I. By this will be meant finding an extremal for the problem.

The reader should be clear as to the meaning of the term $F_{y'}$ appearing in Euler's equation (E): one treats $F = F(x, y, z)$ as a function of three independent variables x, y, z; one then differentiates partially with respect to z; finally, one substitutes y' for z wherever the latter appears. Formally,

$$F_{y'}\;\equiv\;\frac{\partial F}{\partial y'}\,(x,y,y')\;=\;\left.\left(\frac{\partial}{\partial z}\,F(x,y,z)\right)\right|_{z=y'}$$

We may use the language of geometry to describe a variational problem. In this language, and with the same conditions on the functions involved, the fundamental problem is expressed as a search for curves, with equation $y = y(x)$, passing through the points (a, c), (b, d) of $\mathbb{R}^2$ and giving a stationary value to the integral I of (2).

There are many extensions of the fundamental problem, even in classical theory. For example, the function y may be subject to a constraint which may take the form

$$J\;\equiv\;\int_a^b G(x,y,y')\;=\;\gamma,$$

where γ is a constant; or, the end-points a, b of the range of integration may vary; or the function F may take one of the forms

$$F(x,y,y',y''),\qquad F(x,y,z,y',z'),\qquad F(x,t,u,u_x,u_t),$$

where $y = y(x)$, $z = z(x)$, $u = u(x, t)$. After presenting some well-known examples to motivate our discussion, we shall proceed to consider these and other cases.

11.2   Some classical examples from mechanics and geometry

In this section, we vary our usual practice in respect of applications because of the historical importance of the examples discussed and because of the motivation which such discussion gives.

Example 1 : the brachistochrone (Greek: $\beta\rho\acute{\alpha}\chi\iota\sigma\tau$os = shortest, $\chi\rho o\upsilon$os = time). This problem deserves mention, not only because it was historically the first variational problem to be formulated mathematically, but because it serves well to illustrate the scope of applicability of the subject to problems of an applied nature.

Let $A = (a,0)$ and $B = (b,d)$, with $a < b$, $d < 0$, denote two points in a vertical plane in real space. The problem is to find, amongst all smooth wires (represented by continuously differentiable functions) that join A to B, the one along which a bead (represented by a mass-point), under the influence of gravity, will slide from A to B in the shortest possible time.

In setting-up our mathematical model, we assume that the motion is frictionless, that the bead starts from rest, and that it is subject only to gravity **g** as external force.

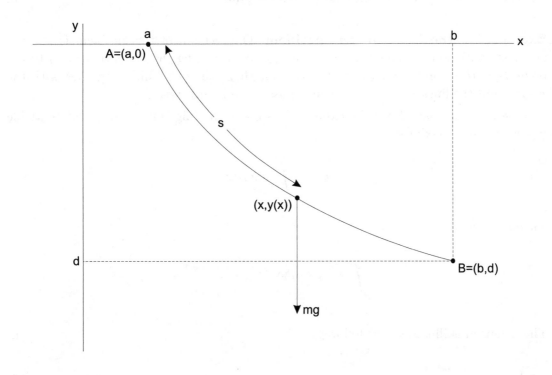

Suppose that $y = y(x)$ gives the equation of the curve representing the wire and that $s = s(t)$ measures the distance the mass-point of mass m has travelled along the curve at time t. The situation is illustrated in the diagram above.

Conservation of energy yields

$$\frac{1}{2} m \left(\frac{ds}{dt} \right)^2 = mgy$$

where $g = |\mathbf{g}|$. As

$$\frac{ds}{dt} = \frac{ds}{dx} \cdot \frac{dx}{dt} \quad \text{and} \quad \left(\frac{ds}{dx} \right)^2 = 1 + \left(\frac{dy}{dx} \right)^2,$$

we can quickly derive (supposing the integral to exist)

$$(3) \qquad\qquad t = \int_a^b \left\{ \frac{1 + (y'(x))^2}{2gy(x)} \right\}^{\frac{1}{2}} dx.$$

The brachistochrone problem reduces thus to that of finding a continuously differentiable function $y = y(x)$ giving a minimum to this integral. $\square$

Example 2 : the isoperimetric problem One version of this problem is to find, amongst continuously differentiable curves $y = y(x)$, joining the point $A = (a, c)$ to the point $B = (b, d)$ and having a fixed given length L, the one which, together with the x−axis and the lines $x = a$, $x = b$, encompasses the largest area I.

The problem may clearly be expressed as one of finding continuously differentiable $y = y(x)$ which maximises

$$(4) \qquad\qquad I = \int_a^b y(x) \, dx$$

when subject to

$$(5) \qquad\qquad \int_a^b \left\{ 1 + (y'(x))^2 \right\} dx = L.$$

The situation is illustrated as follows. $\square$

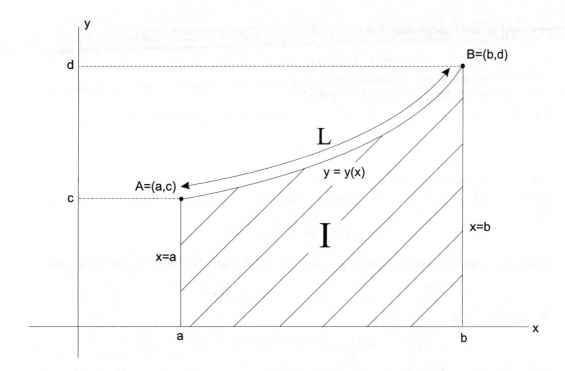

Queen Dido is said to have proposed the classical version of the isoperimetric problem, namely, that of finding the closed (continuously differentiable) curve of given length containing the greatest area. We derive the C^2-extremal, a circle, for this by using polar co-ordinates in section 11.8.

Example 3 : geodesics The problem here is to find the continuously differentiable curve on a smooth surface in $\mathbb{R}^3$ of minimum length and joining given points A, B on the surface.

(i) Suppose that the surface is given by $G(x, y, z) = 0$, where G is continuously differentiable, and the curve is given in the parametric form

$$\mathbf{r} = \mathbf{r}(t) = (x(t), y(t), z(t)),$$

where

$$A = \mathbf{r}(t_0), \qquad B = \mathbf{r}(t_1)$$

and x, y, z are continuously differentiable on $[t_0, t_1]$. Then $\mathbf{r} = \mathbf{r}(t)$ must give a minimum to the integral

(6)
$$I = \int_{t_0}^{t_1} \left(\dot{x}^2 + \dot{y}^2 + \dot{z}^2 \right)^{\frac{1}{2}} dt$$

when subject to the constraint

$$(7) \qquad\qquad G(x(t), y(t), z(t)) \;=\; 0, \qquad (t \in [t_0, t_1]).$$

Dot denotes differentiation with respect to t.

(ii) Suppose, alternatively, that the surface is given in parametric form

$$\mathbf{r} \;=\; \mathbf{r}(u, v) \;=\; (x(u, v), y(u, v), z(u, v)),$$

where $\mathbf{r}$ is continuously differentiable and $\mathbf{r}_u \wedge \mathbf{r}_v$ is never zero. A curve $\mathbf{r} = \mathbf{r}(t)$ on the surface is determined once one can find functions $u = u(t)$, $v = v(t)$; for then we may write

$$\mathbf{r}(t) \;=\; \mathbf{r}(u(t), v(t)).$$

Again, we demand that $u = u(t)$, $v = v(t)$ are continuously differentiable and take

$$A \;=\; \mathbf{r}(t_0) \qquad B \;=\; \mathbf{r}(t_1).$$

By the chain rule,

$$\dot{\mathbf{r}}(t) \;=\; \dot{u}(t).\mathbf{r}_u(u(t), v(t)) + \dot{v}(t).\mathbf{r}_v(u(t), v(t));$$

so, by definition of arc-length, the distance s along the curve from A to B is given by

$$s \;=\; \int_{t_0}^{t_1} (\dot{\mathbf{r}}(t) \cdot \dot{\mathbf{r}}(t))^{\frac{1}{2}} \, dt;$$

that is, by

$$(8) \qquad\qquad s \;=\; \int_{t_0}^{t_1} \left(E\dot{u}^2 + 2F\dot{u}\dot{v} + G\dot{v}^2 \right)^{\frac{1}{2}} \, dt,$$

where $E = \mathbf{r}_u \cdot \mathbf{r}_u$, $F = \mathbf{r}_u \cdot \mathbf{r}_v$, $G = \mathbf{r}_v \cdot \mathbf{r}_v$. The geodesic problem is this time solved by minimising the integral (8). $\square$

Example 4 : Hamilton's Principle and Lagrange's equations Suppose that the 'generalised' co-ordinate function $q_i = q_i(t) \in C^2([t_0, t_1])$ is given as a function of the independent time variable t, for $i = 1, \ldots, n$. The Lagrangian

$$L \;=\; L(t, q_1, \ldots, q_n, \dot{q}_1, \ldots, \dot{q}_n) \in C^2(D),$$

for appropriate open $D \subseteq \mathbb{R}^{2n+1}$, is defined by

$$L \;\equiv\; T - V,$$

where T is the kinetic and V the potential energy of a conservative dynamical system. Suppose further that $q_i(t_0)$, $q_i(t_1)$ are given constants, for $i = 1, \ldots, n$. *Hamilton's Principle* asserts that the actual motion of the system is given by the functions $q_i = q_i(t)$, $i = 1, \ldots, n$, for which the integral

$$(9) \qquad \int_{t_0}^{t_1} L \, dt$$

is stationary.

The problem is solved by functions satisfying *Lagrange's equations*

$$(10) \qquad \frac{d}{dt}\left(\frac{\partial L}{\partial \dot{q}_i}\right) = \frac{\partial L}{\partial q_i}, \qquad (i = 1, \ldots, n). \qquad \square$$

Example 5 : Plateau's problem The problem here is to find the surface S in $\mathbb{R}^3$ of minimum (surface) area, bounded by a given closed curve Γ. Suppose that S is given by the continuously differentiable function $u = u(x, y)$ and that the orthogonal projection of Γ on the (x, y)-plane is the continuously differentiable curve γ with interior Δ. Then S must give rise to a minimum of the surface area integral

$$(11) \qquad \iint_{\Delta} (1 + u_x^2 + u_y^2)^{\frac{1}{2}} \, dx dy. \qquad \square$$

Exercise 1 Show that a surface, generated by revolving $y \in C^1([a, b])$, subject to $y(a) = c$, $y(b) = d$ ($c > 0, d > 0$), about the x−axis, is of minimum surface area when $y = y(x)$ gives a minimum to the integral

$$\int_a^b y(x) \left\{ 1 + (y'(x))^2 \right\}^{\frac{1}{2}} \, dx.$$

Exercise 2 When a heavy uniform string hangs in stable equilibrium in a vertical plane between two fixed points (a, c) and (b, d), its form $y = y(x)$ is such as to minimise its potential energy. Show that if y is continuously differentiable, it must minimise

$$(12) \qquad \int_a^b y(x) \left\{ 1 + (y'(x))^2 \right\}^{\frac{1}{2}} \, dx$$

whilst being subject to

$$(13) \qquad \int_a^b \left\{ 1 + (y'(x))^2 \right\}^{\frac{1}{2}} \, dx = L,$$

where L is the (fixed) length of the string.

11.3   The derivation of Euler's equation for the fundamental problem

Most of the 'hard analysis', as opposed to formal manipulation, required to derive Euler's equation is contained in the following lemma.

Lemma 1 Suppose that $P : [a, b] \to \mathbb{R}$ is continuous and that

$$\int_a^b \eta(x) P(x) \, dx = 0,$$

for every $\eta \in C^2([a, b])$ for which

$$\eta(a) = \eta(b) = 0.$$

Then $P = 0$, identically on $[a, b]$.

Proof Suppose, on the contrary, that $P(x_0) \neq 0$, for some $x_0 \in [a, b]$. With no loss of generality, we may assume that $P(x_0) > 0$. By continuity of P, there are real numbers a', b' for which

$$a < a' < b' < b, \quad \text{and} \quad P(x) > 0 \quad \text{for } x \in [a', b'].$$

Define $\eta : [a, b] \to \mathbb{R}$ by

$$\eta(x) = \begin{cases} \{(x - a')(b' - x)\}^3, & \text{for } x \in [a', b'] \\ 0, & \text{otherwise.} \end{cases}$$

Then $\eta \in C^2([a, b])$, $\eta(a) = \eta(b) = 0$, and η, P are both positive on (a', b'). So

$$\int_a^b \eta(x) P(x) \, dx = \int_{a'}^{b'} \eta(x) P(x) \, dx > 0,$$

contradicting the hypotheses of the lemma.   □

We now proceed directly to showing that, if a C^2-function y solves the fundamental problem described in section 11.1, then y must satisfy Euler's equation (E) of that section. The notation and conditions of 11.1 will continue to apply here.

So, suppose $y \in C^2([a, b])$ satisfies (1) and gives a stationary value to the integral I of (2). Suppose also that $\eta \in C^2([a, b])$ is arbitrary, save that it must satisfy

(14) $$\eta(a) = \eta(b) = 0.$$

Define, for each real number α, a function $y = y_\alpha(x)$ by

$$y_\alpha(x) \;=\; y(x) + \alpha\eta(x), \qquad (x \in [a,b]).$$

Then $y_\alpha \in C^2([a,b])$, $y_\alpha(a) = c$, $y_\alpha(b) = d$, and

$$y_\alpha'(x) \;=\; y'(x) + \alpha\eta'(x), \qquad (x \in [a,b])$$

for each α. As D is open, there exists $\epsilon > 0$ such that

$$(x, y_\alpha(x), y_\alpha'(x)) \in D, \qquad (x \in [a,b])$$

whenever $|\alpha| < \epsilon$. So,

$$I \;=\; I(\alpha) \;=\; \int_a^b F(x, y_\alpha(x), y_\alpha'(x))\, dx,$$

which now varies only with α, is well-defined for $|\alpha| < \epsilon$ and has, as $y_0 = y$, a stationary value at $\alpha = 0$. Because

$$F(x, y_\alpha(x), y_\alpha'(x)) \;=\; F(x, y(x) + \alpha\eta(x), y'(x) + \alpha\eta'(x))$$

is continuous on $[a,b]$ for $|\alpha| < \epsilon$, I is differentiable for $|\alpha| < \epsilon$ and therefore

(15) $$I'(0) \;=\; 0.$$

But, using the chain rule, and evaluating at $\alpha = 0$,

$$I'(0) \;=\; \int_a^b \left\{ F_y(x, y(x), y'(x))\eta(x) + F_{y'}(x, y(x), y'(x))\eta'(x) \right\}\, dx$$

(16) $$= \; \left[\eta(x) F_{y'}(x, y(x), y'(x)) \right]_a^b + \int_a^b \eta \left\{ F_y - \frac{d}{dx} F_{y'} \right\}\, dx,$$

after an integration by parts. The first term of (16) is zero because of condition (14), and

$$P \;\equiv\; F_y - \frac{d}{dx} F_{y'}$$

is continuous because $F \in C^2(D)$. So, (15) combined with Lemma 1 implies that Euler's equation

(E) $$\frac{d}{dx}\{F_{y'}(x, y, y')\} \;=\; F_y(x, y, y')$$

is satisfied by $y = y(x)$ in $[a,b]$.

Exercise 3 Suppose that $P : [a, b] \to \mathbb{R}$ is continuous, that n is a positive integer, and that

$$\int_a^b \eta(x) P(x) \, dx = 0,$$

for every $\eta \in C^n([a, b])$ for which

$$\eta(a) = \eta(b) = 0.$$

Prove that $P = 0$, identically on $[a, b]$.

Exercise 4 Suppose that $A \subseteq A \cup \partial A \subseteq U \subseteq \mathbb{R}^2$, where A is a bounded, connected and simply connected, measurable open set and U is an open set, that $Q : U \to \mathbb{R}$ is continuous, and that

$$\iint_A \zeta(x, y) Q(x, y) \, dx dy = 0,$$

for every $\zeta \in C^2(A)$ for which

$$\zeta = 0 \quad \text{on } \partial A,$$

where ∂A denotes the boundary of A. Prove that $Q = 0$, identically on $A \cup \partial A$.

Note The conditions on the set A, given in Exercise 4, are introduced here and again in section 11.7 in order for the integrals to exist and the usual integration theorems to apply. They are commonly met.

11.4   The special case $F = F(y, y')$

In this special case, where F is not an explicit function of x and $F_x = 0$, one integration of Euler's equation may always be immediately accomplished. We continue to use the notation of section 11.1.

Proposition 1 If F_x is identically zero in D, a first integral of Euler's equation in D is $U = k$, where k is a constant and U is the function defined by

$$U(x, y, y') \equiv F(x, y, y') - y' F_{y'}(x, y, y')$$

for all $(x, y, y') \in D$.

Proof   Using the chain rule and the hypothesis that $F_x \equiv 0$,

$$\frac{dU}{dx} = \left(F_x + y' F_y + y'' F_{y'} \right) - \left(y'' F_{y'} + y' \frac{d}{dx} F_{y'} \right) = y' \left(F_y - \frac{d}{dx} F_{y'} \right) = 0.$$

So, $U = k$, a constant, is a first integral of Euler's equation. $\square$

Example 6 : $F(y, y') = y^n\{1 + (y')^2\}^{\frac{1}{2}}$ In this case,

$$F - y'F_{y'} = y^n\{1 + (y')^2\}^{\frac{1}{2}} - y^n(y')^2\{1 + (y')^2\}^{-\frac{1}{2}}$$

$$= y^n\{1 + (y')^2\}^{-\frac{1}{2}}$$

$$= y^n \cos\psi,$$

where ψ is the angle the tangent to the curve $y = y(x)$ makes with the x-axis. (Note that $y' = \tan\psi$.) By Proposition 1, a first integral of Euler's equation is then

$$(17) \qquad\qquad y^n \cos\psi = K^n,$$

where K is a constant. Differentiating, we obtain, when y is non-zero,

$$y \tan\psi = n\frac{dy}{d\psi}.$$

On using

$$\frac{dy}{dx} = \frac{dy}{d\psi}\Big/\frac{dx}{d\psi} = \tan\psi,$$

this reduces to

$$(18) \qquad\qquad y = n\frac{dx}{d\psi}. \qquad\qquad \square$$

We now present just two amongst a number of important special cases.

(a) $n = -\frac{1}{2}$ This corresponds to the brachistochrone problem of Example 1 above. The first integral (17) is

$$(19) \qquad\qquad y = K\cos^2\psi = \frac{K}{2}(1 + \cos 2\psi).$$

Integrating the corresponding equation (18) gives

$$(20) \qquad\qquad x = L - \frac{K}{2}(2\psi + \sin 2\psi)$$

where L, like K, is a constant. Equations (19), (20) are the parametric equations of (arcs of) *cycloids* with cusps on the x-axis. $\qquad\qquad \square$

(b) $n = 1$ The reader was asked to show in Exercise 1 that this corresponds to the problem of minimising the surface area of a surface generated by revolving a curve about an axis. Equation (17) is now

(21) $$y = K \sec \psi$$

and (18) integrates to

(22) $$x - L = K \log(\sec \psi + \tan \psi)$$

where K, L are constants.

Equations (21), (22) give an extremal in parametric form. To derive a solution in the form $y = y(x)$, we proceed as follows:

$$e^{(x-L)/K} = \sec \psi + \tan \psi = (\sec \psi - \tan \psi)^{-1},$$

as $\sec^2 \psi = 1 + \tan^2 \psi$, and therefore

$$\cosh\left(\frac{x-L}{K}\right) = \frac{1}{2}\left(e^{(x-L)/K} + e^{-(x-L)/K}\right) = \sec \psi$$

when, additionally, K is non-zero. Using (21), we obtain the extremal

(23) $$\frac{y}{K} = \cosh\left(\frac{x-L}{K}\right),$$

which satisfies Euler's equation whenever $K \neq 0$. This curve is a *catenary*.   □

Note The techniques provided by Proposition 1, for the derivation of an extremal from a first integral or in parametric form, can be very useful but do not always provide the best method of solving a particular Euler equation. The reader should always check first to see, in any particular case, if Euler's equation can be solved directly. An opportunity for comparing methods is provided by Exercise 5 below.

Exercise 5 Derive the solution (23) of Example 6(b) directly from the first integral

$$y\{1 + (y')^2\}^{\frac{1}{2}} - y'\frac{\partial}{\partial y'}\left(y\{1 + (y')^2\}^{\frac{1}{2}}\right) = k$$

obtained from Proposition 1.

Exercise 6 Find extremals corresponding to $F(y, y') = y^n\{1 + (y')^2\}^{\frac{1}{2}}$ when $n = \frac{1}{2}$ and $n = -1$.

Exercise 7 Find extremals corresponding to the following problems:

(a) $\int_0^1 (y^2 + y' + (y')^2)\, dx$,   subject to $y(0) = 0$, $y(1) = 1$,

(b) $\int_0^1 (y')^2\, dx + \{y(1)\}^2$,   subject to $y(0) = 1$.

[HINT for (b): Re-write in terms of an integral minus a constant independent of y.]

Exercise 8 Suppose that $F(y, y') = y^n\{(1 + (y')^2)^{\frac{1}{2}}\}$, that y' is never zero and that $n > \frac{1}{2}$. Use the first integral

$$y^{2n} = K^{2n}(1 + (y')^2)$$

of Euler's equation, where K is a positive constant, to show that, if

$$\frac{y}{K} = g\left(\frac{x}{K}\right)$$

is an extremal, then g satisfies the differential equation $g'' = ng^{2n-1}$. If $g(x) > L > 0$, for some constant L and all $x \geq 0$, show that there exists precisely one $\alpha > 0$ such that the line given by $y = g'(\alpha).x$ touches the extremal whatever the value of $K > 0$.

11.5   When F contains higher derivatives of y

In this section, we consider the problem of finding the function $y \in C^4([a, b])$ which satisfies

(24) $y(a) = c$, $y(b) = d$, $y'(a) = e$, $y'(b) = f$,

where a, b, c, d, e, f are constants, and which gives a stationary value to the integral

(25) $I = \int_a^b F(x, y, y', y'')\, dx$,

where $F \in C^3(D)$ for an appropriate $D \subseteq \mathbb{R}^4$.

Here, and subsequently, we shall not fill in all the details, but rather pick out the essential differences which modify the argument of section 11.3. The reader should have no difficult in completing the discussion.

Suppose that $y \in C^4([a, b])$ satisfies (24) and gives a stationary value to (25), and that $\eta \in C^4([a, b])$ is arbitrary, save that it must satisfy

(26) $\eta(a) = \eta(b) = \eta'(a) = \eta'(b) = 0$,

with

$$y_\alpha(x) \;=\; y(x) + \alpha\eta(x)$$

and for small $|\alpha|$, we may differentiate

$$I \;=\; I(\alpha) \;=\; \int_a^b F(x, y_\alpha(x), y_\alpha'(x), y_\alpha''(x))\, dx$$

and set $\alpha = 0$ to obtain

$$I'(0) \;=\; \int_a^b (\eta F_y + \eta' F_{y'} + \eta'' F_{y''})\, dx$$

$$= \;\left[\eta F_{y'} + \eta' F_{y''}\right]_a^b + \int_a^b \{\eta(F_y - \frac{d}{dx} F_{y'}) - \eta'\frac{d}{dx} F_{y''}\}\, dx$$

$$= \;\left[\eta F_{y'} + \eta' F_{y''} - \eta\frac{d}{dx} F_{y''}\right]_a^b + \int_a^b \eta \left\{F_y - \frac{d}{dx} F_{y'} + \frac{d^2}{dx^2} F_{y''}\right\}\, dx.$$

As $I'(0) = 0$ and η satisfies (26),

$$\int_a^b \eta \left\{F_y - \frac{d}{dx} F_{y'} + \frac{d^2}{dx^2} F_{y''}\right\}\, dx \;=\; 0,$$

where η is an otherwise arbitrary $4-$times continuously differentiable function. Since $F \in C^3(D)$, every term in the integrand of the last equation is continuous, and we may deduce that y satisfies the Euler equation

(27) $$\frac{d^2}{dx^2} F_{y''} - \frac{d}{dx} F_{y'} + F_y \;=\; 0.$$

The reader is asked to provide and to prove the analogous result to Lemma 1 of section 11.3, which establishes the last step.

Extremals for this problem are functions $y = y(x)$ solving (27) together with (24).

Exercise 9 Formulate the problem corresponding to

$$F \;=\; F(x, y, y', \ldots, y^{(n)})$$

and write down the appropriate Euler equation.

Exercise 10 State and prove a lemma, analogous to Lemma 1 of section 11.3, which provides the last step in the derivation of (27) above.

Exercise 11 Show that the Euler equation (27) has the first integral

$$F - y' \left(F_{y'} - \frac{d}{dx} F_{y''} \right) - y'' F_{y''} = k,$$

where k is a constant, whenever $F = F(y, y', y'')$ (that is, whenever F_x is identically zero).

Exercise 12 Find extremals for the problems given by the following functions $F = F(x, y, y', y'')$ and boundary conditions:

(a) $F = (y')^2 + (y'')^2$, $y(0) = y'(0) = 0$, $y(1) = y'(1) = 1$,

(b) $F = y^2 - 2(y')^2 + (y'')^2$, $y(0) = y'(\pi/2) = 0$, $y'(0) = y(\pi/2) = 1$.

11.6   When F contains more dependent functions

Consider the problem of finding the functions $y, z \in C^2([a, b])$ which satisfy

(28) $y(a) = c$, $y(b) = d$, $z(a) = e$, $z(b) = f$,

where a, b, c, d, e, f are constants, and which gives a stationary value to the integral

(29) $I = \int_a^b F(x, y, z, y', z') \, dx,$

where $F \in C^2(D)$ for an appropriate open $D \subseteq \mathbb{R}^5$.

We define

$$y_\alpha(x) = y(x) + \alpha \eta(x), \qquad z_\alpha(x) = z(x) + \alpha \zeta(x),$$

for small $|\alpha|$, where the otherwise arbitrary functions $\eta, \zeta \in C^2([a, b])$ are chosen to satisfy

(30) $\eta(a) = \eta(b) = \zeta(a) = \zeta(b) = 0.$

Integration by parts yields

$$I'(0) = \int_a^b (\eta F_y + \zeta F_z + \eta' F_{y'} + \zeta' F_{z'}) \, dx$$

$$= \left[\eta F_{y'} + \zeta F_{z'} \right]_a^b + \int_a^b (\eta P + \zeta Q) \, dx$$

$$= \int_a^b (\eta P + \zeta Q) \, dx,$$

using (30), where

$$P \equiv F_y - \frac{d}{dx} F_{y'}, \qquad Q \equiv F_z - \frac{d}{dx} F_{z'}.$$

Choosing $\zeta = 0$, identically on $[a, b]$, which satisfies (30), $I'(0) = 0$ gives (using Lemma 1 and the continuity of P) the Euler equation

$$(31) \qquad\qquad \frac{d}{dx} F_{y'} = F_y.$$

Similarly, $\eta = 0$, identically on $[a, b]$, gives

$$(32) \qquad\qquad \frac{d}{dx} F_{z'} = F_z.$$

Thus, we get an Euler equation for each dependent function. Similarly, n dependent functions give rise to n Euler equations.

Example 7 : geodesics on $\mathbf{r} = \mathbf{r}(u, v)$ We showed in Example 3(ii) of section 11.2 that a curve $\mathbf{r} = \mathbf{r}(t)$ on the surface $\mathbf{r} = \mathbf{r}(u, v)$, given by $u = u(t)$, $v = v(t)$, is a geodesic from $\mathbf{r}(t_0)$ to $\mathbf{r}(t_1)$ if and only if it minimises the integral

$$s = \int_{t_0}^{t_1} \sqrt{2T} \, dt,$$

where

$$T \;=\; T(u,v,\dot{u},\dot{v}) \;=\; \frac{1}{2}(\dot{\mathbf{r}}\boldsymbol{.}\dot{\mathbf{r}}) \;=\; \frac{1}{2}(E\dot{u}^2 + 2F\dot{u}\dot{v} + G\dot{v}^2)$$

and

$$E \;=\; \mathbf{r}_u\boldsymbol{.}\mathbf{r}_u, \qquad F \;=\; \mathbf{r}_u\boldsymbol{.}\mathbf{r}_v, \qquad G \;=\; \mathbf{r}_v\boldsymbol{.}\mathbf{r}_v.$$

We assume from now on that $\mathbf{r} = \mathbf{r}(u,v) \in C^3(A)$ and $u,v \in C^3(B)$, where A, B are open sets in, respectively, $\mathbb{R}^2$ and $\mathbb{R}$; so that, $T \in C^2(D)$, where D is open in $\mathbb{R}^4$ (T does not contain the independent variable t explicitly). We also assume that

$$\dot{\mathbf{r}}(t), \quad \ddot{\mathbf{r}}(t), \quad \mathbf{r}_u(u,v), \quad \mathbf{r}_v(u,v)$$

are each never zero in appropriate subsets of $\mathbb{R}$ and $\mathbb{R}^2$. In particular, this ensures that T is never zero and that $\mathbf{r} = \mathbf{r}(u,v)$ has a well-defined tangent plane and normal at each point.

In these circumstances, the Euler equations corresponding to (31) and (32) are

$$(33) \qquad \frac{d}{dt}\left(\frac{1}{\sqrt{2T}}\,T_{\dot{u}}\right) \;=\; \frac{1}{\sqrt{2T}}\,T_u, \qquad \frac{d}{dt}\left(\frac{1}{\sqrt{2T}}\,T_{\dot{v}}\right) \;=\; \frac{1}{\sqrt{2T}}\,T_v.$$

Now choose the parameter t to be arc-length s along the geodesic. Letting dash denote differentiation with respect to s, certainly $\mathbf{r}'(s) = 1$ along the geodesic and hence $T = \frac{1}{2}$, a constant. The Euler equations (33) then reduce to

$$\frac{d}{ds}\,T_{u'} \;=\; T_u, \qquad \frac{d}{ds}\,T_{v'} \;=\; T_v,$$

which may alternatively be written

$$(34) \qquad \begin{cases} \dfrac{d}{ds}\,(Eu' + Fv') \;=\; \dfrac{1}{2}(E_u u'^2 + 2F_u u'v' + G_u v'^2) \\[4mm] \dfrac{d}{ds}\,(Fu' + Gv') \;=\; \dfrac{1}{2}(E_v u'^2 + 2F_v u'v' + G_v v'^2) \end{cases} \qquad \square$$

Example 8 : Hamilton's Principle and Lagrange's equations   In Example 4 of section 11.2, we described *Hamilton's Principle* which asserts that the actual motion of a conservative mechanical system with Lagrangian

$$L = L(t, q_1, \ldots, q_n, \dot{q}_1, \ldots, \dot{q}_n) \in C^2(D),$$

where D is open in $\mathbb{R}^{2n+1}$ and $q_i = q_i(t) \in C^2([t_0, t_1])$ are co-ordinate functions $(i = 1, \ldots, n)$, arises from a stationary value of the integral

$$\int_{t_0}^{t_1} L \, dt.$$

Direct application of the theory of this section gives the n Euler equations

$$\frac{d}{dt}\left(\frac{\partial L}{\partial \dot{q}_i}\right) = \frac{\partial L}{\partial q_i}, \qquad (i = 1, \ldots, n).$$

In this context, the Euler equations are known as *Lagrange's equations*.   □

Exercise 13 Find extremals $y = y(x)$, $z = z(x)$ corresponding to the problem of finding a stationary value of the integral

$$\int_0^{\frac{\pi}{2}} \left((y')^2 + (z')^2 + 2yz\right) dx$$

when subject to $y(0) = z(0) = 0$, $y(\pi/2) = z(\pi/2) = 1$.

Exercise 14 Use (34) to show that the geodesics on the sphere

$$\mathbf{r} = \mathbf{r}(\theta, \phi) = a(\sin\theta\cos\phi, \sin\theta\sin\phi, \cos\theta), \qquad (0 \le \theta \le \pi,\ 0 \le \phi < 2\pi)$$

of radius a are arcs of great circles. State clearly any assumptions that you make.

Exercise 15 Show that the problem of finding geodesics on the cylinder $x^2 + y^2 = a^2$ through the points $(a, 0, 0)$ and $(a\cos\alpha, a\sin\alpha, b)$, where a, b, α are constants and $0 < \alpha < 2\pi$, gives rise to the extremals

$$x = a\cos t, \qquad y = a\sin t, \qquad z = bt/(\alpha + 2n\pi),$$

for all integers n such that $\alpha + 2n\pi \ne 0$.

Exercise 16 Show that the geodesic equations (34) may be re-written

$$\mathbf{r}_u \cdot \mathbf{r}'' = \mathbf{r}_v \cdot \mathbf{r}'' = 0.$$

(This shows that the direction $\mathbf{r}''$ of the principal normal at a point of a geodesic is parallel to the normal direction $\mathbf{r}_u \wedge \mathbf{r}_v$ (at the same point) of the surface on which it lies.)

Exercise 17 (a) Define momenta $p_i = p_i(t)$, $i = 1, \ldots, n$, and the Hamiltonian

$$H = H(t, q_1, \ldots, q_n, p_1, \ldots, p_n)$$

by

$$p_i = \frac{\partial L}{\partial \dot{q}_i}, \qquad (i = 1, \ldots, n)$$

and

$$H = \sum_{i=1}^{n} p_i \dot{q}_i - L,$$

where L is the Lagrangian of Examples 4 and 8. Show that Lagrange's equations are equivalent to *Hamilton's equations*:

$$\dot{q}_i = \frac{\partial H}{\partial p_i}, \qquad \dot{p}_i = -\frac{\partial H}{\partial q_i}, \qquad (i = 1, \ldots, n).$$

Show also that Hamilton's equations arise from a search for functions giving a stationary value to the integral

$$\int_{t_0}^{t_1} \left(\sum_{i=1}^{n} p_i \dot{q}_i - H \right) dt.$$

(b) Show that the function $F = F(t, q_1, \ldots, q_n)$, defined as

$$\int_{t_0}^{t} \left(\sum_{i=1}^{n} p_i \dot{q}_i - H(u, q_1, \ldots, q_n, p_1, \ldots, p_n) \right) du,$$

where H is the Hamiltonian in part (a), satisfies the *Hamilton–Jacobi equation*

$$\frac{\partial F}{\partial t} + H\left(t, q_1, \ldots, q_n, \frac{\partial F}{\partial q_1}, \ldots, \frac{\partial F}{\partial q_n} \right) = 0.$$

(c) Show that the Hamilton–Jacobi equation for the one-dimensional harmonic oscillator, where the Lagrangian L is given by

$$L = \frac{1}{2} (m \dot{q}^2 - k q^2)$$

and k, m are positive constants, is

$$\frac{\partial F}{\partial t} + \frac{1}{2m} \left(\frac{\partial F}{\partial q} \right)^2 + \frac{k q^2}{2} = 0.$$

11.7   When F contains more independent variables

Consider the problem of finding a function $u = u(x, y) \in C^2(D)$, where D is an open subset of $\mathbb{R}^2$, which gives a stationary value to the integral

$$(35) \qquad\qquad I = \iint_A F(x, y, u, u_x, u_y)\, dx dy$$

and satisfies

$$(36) \qquad\qquad u = f \quad \text{on } \partial A,$$

where ∂A denotes the continuously differentiable simple closed curve which bounds the connected and simply connected open set A, and $A \cup \partial A \subseteq D$. Of course, we also need $F \in C^2(E)$, where E is an appropriate open subset of $\mathbb{R}^5$. (Concerning these conditions, see the note at the end of section 11.3.) We define

$$u_\alpha(x, y) = u(x, y) + \alpha \eta(x, y),$$

for small $|\alpha|$, where the otherwise arbitrary function $\eta \in C^2(D)$ is chosen to satisfy

$$\eta = 0 \quad \text{on } \partial A.$$

Then

$$I = I(\alpha) = \iint_A F\left(x, y, u_\alpha, \frac{\partial}{\partial x} u_\alpha, \frac{\partial}{\partial y} u_\alpha\right) dx dy.$$

Instead of using integration by parts, which was appropriate for a single integral, we apply *Green's Theorem in the Plane*[1] as follows:

$$I'(0) = \iint_A (F_u \eta + F_{u_x} \eta_x + F_{u_y} \eta_y)\, dx dy$$

$$= \iint_A \eta \left\{ F_u - \frac{\partial}{\partial x} F_{u_x} - \frac{\partial}{\partial y} F_{u_y} \right\} dx dy + \iint_A \left\{ \frac{\partial}{\partial x}(\eta F_{u_x}) + \frac{\partial}{\partial y}(\eta F_{u_y}) \right\} dx dy$$

$$= \iint_A \eta \left\{ F_u - \frac{\partial}{\partial x} F_{u_x} - \frac{\partial}{\partial y} F_{u_y} \right\} dx dy + \int_{\partial A} \eta \left\{ -F_{u_y}\, dx + F_{u_x}\, dy \right\}.$$

---

[1]This theorem states that if $P = P(x, y)$, $Q = Q(x, y) \in C^1(D)$ then, with A as given above,

$$\iint_A \left(\frac{\partial Q}{\partial x} - \frac{\partial P}{\partial y} \right) dx dy = \int_{\partial A} (P\, dx + Q\, dy)$$

As $I'(0) = 0$, $\eta = 0$ on ∂A, and $F \in C^2(E)$, we can use Exercise 4 to deduce that u must satisfy the Euler equation

$$(37) \qquad \frac{\partial}{\partial x} F_{u_x} + \frac{\partial}{\partial y} F_{u_y} = F_u.$$

The corresponding result for three independent variables is left as an exercise for the reader.

Example 9 : Dirichlet's integral and Laplace's equation   With the same notation and conditions as above, consider the problem of finding the function u which gives a stationary value to *Dirichlet's integral*

$$\iint_A (\operatorname{grad} u)^2 \, dxdy = \iint_A (u_x^2 + u_y^2) \, dxdy,$$

when u is equal to a given continuous function on the boundary ∂A of A. From (37), the corresponding Euler equation is

$$\frac{\partial}{\partial x} (2u_x) + \frac{\partial}{\partial y} (2u_y) = 0;$$

that is, *Laplace's equation*

$$\nabla^2 u \equiv u_{xx} + u_{yy} = 0. \qquad \square$$

Example 10 : the wave equation   For a non-homogeneous string of length a, with $p(x)$ denoting modulus of elasticity multiplied by the cross-sectional area at x and $\rho(x)$ denoting the mass per unit length at x, the kinetic energy T and the potential energy V are given, for a transverse displacement $u = u(x,t)$, by

$$T = \tfrac{1}{2} \int_0^a \rho u_t^2 \, dx, \qquad V = \tfrac{1}{2} \int_0^a p u_x^2 \, dx.$$

From Hamilton's Principle (Examples 4 and 8), the actual motion of the string corresponds to a function $u = u(x,t)$ which gives a stationary value to the integral

$$\int_{t_0}^{t_1} L \, dt = \tfrac{1}{2} \int_{t_0}^{t_1} \int_0^a (\rho u_t^2 - p u_x^2) \, dxdt,$$

where $L \equiv T - V$ is the Lagrangian of the system. In this case the Euler equation (37) reduces to the *wave equation*

$$(p u_x)_x = \rho u_{tt}.$$

The mathematical model of the string provides sufficiently strong differentiability conditions for the above analysis to be valid.   □

Exercise 18 Find an extremal corresponding to

$$\iint_A \left\{ (u_x)^2 + (u_y)^2 \right\} \, dxdy,$$

given that $u(\cos\theta, \sin\theta) = \cos\theta$ $(0 \le \theta < 2\pi)$ and A is the open disc $\{(x,y) : x^2 + y^2 < 1\}$.

Exercise 19 Show that the Euler equation for Plateau's problem (Example 5), where

$$F(x, y, u, u_x, u_y) \;=\; (1 + u_x^2 + u_y^2)^{\frac{1}{2}},$$

can be written

$$(1 + q^2)r - 2pqs + (1 + p^2)t \;=\; 0,$$

where $p = u_x$, $q = u_y$, $r = u_{xx}$, $s = u_{xy}$, $t = u_{yy}$ satisfy suitable differentiability conditions.

Exercise 20 Use the *Divergence Theorem*, which states that, for continuously differentiable $P = P(x, y, z)$, $Q = Q(x, y, z)$, $R = R(x, y, z)$,

$$\iiint_V \left(\frac{\partial P}{\partial x} + \frac{\partial Q}{\partial y} + \frac{\partial R}{\partial z} \right) dxdydz \;=\; \iint_{\partial V} (P, Q, R) \cdot \mathbf{n} \, dS,$$

where $\mathbf{n}$ is the outward-drawn unit normal to the simple closed, continuously differentiable surface ∂V bounding the volume V, to derive the Euler equation

(37′) $$\frac{\partial}{\partial x} F_{u_x} + \frac{\partial}{\partial y} F_{u_y} + \frac{\partial}{\partial z} F_{u_z} \;=\; F_u,$$

for the case of three independent variables x, y, z, where $u = u(x, y, z)$ gives a stationary value to the integral

$$\iiint_V F(x, y, z, u, u_x, u_y, u_z) \, dxdydz$$

amongst functions u equal to a given fixed continuous function on the boundary ∂V of V.

Exercise 21 Find an extremal corresponding to

$$\iiint_V \left\{ (u_x)^2 + (u_y)^2 + (u_z)^2 + ku^2 \right\} \, dxdydz,$$

given that $u = 1$ on $x^2 + y^2 + z^2 = 1$, k is a positive constant and

$$V \;=\; \{(x, y, z) : x^2 + y^2 + z^2 < 1\}.$$

[HINT: Try to find a spherically symmetric solution; that is, a solution in the form $u = u(r)$, where $r^2 = x^2 + y^2 + z^2$.]

11.8 Integral constraints

Consider the problem of finding a function $y = y(x) \in C^2([a, b])$, which gives a stationary value to the integral

$$(38) \qquad I \; = \; \int_a^b F(x, y, y') \, dx,$$

while satisfying

$$(39) \qquad y(a) \; = \; c, \qquad y(b) \; = \; d,$$

and also the 'integral constraint'

$$(40) \qquad J \; \equiv \; \int_a^b G(x, y, y') \, dx \; = \; \gamma,$$

where a, b, c, d and γ are constants and $F, G \in C^2(D)$ for some appropriate open $D \subseteq \mathbb{R}^3$.

This time, it is necessary to introduce two arbitrary functions $\eta_1, \eta_2 \in C^2([a, b])$, satisfying

$$\eta_1(a) \; = \; \eta_1(b) \; = \; \eta_2(a) \; = \; \eta_2(b) \; = \; 0,$$

and to define

$$y_{\alpha_1, \alpha_2}(x) \; = \; y(x) + \alpha_1 \eta_1(x) + \alpha_2 \eta_2(x),$$

for small $|\alpha_1|$, $|\alpha_2|$. If y gives a stationary value to

$$I \; = \; I(\alpha_1, \alpha_2) \quad \text{when subject to} \quad J \; = \; J(\alpha_1, \alpha_2) \; = \; \gamma,$$

there must be a *Lagrange multiplier*[2] λ (a non-zero constant) such that

$$(41) \qquad \frac{\partial I}{\partial \alpha_1} = \lambda \frac{\partial J}{\partial \alpha_1}, \qquad \frac{\partial I}{\partial \alpha_2} = \lambda \frac{\partial J}{\partial \alpha_2}, \qquad \text{when } \alpha_1 = \alpha_2 = 0.$$

(We assume that grad(J) is non-zero when evaluated at $\alpha_1 = \alpha_2 = 0$.) But, equations (41) are equivalent to

$$(42) \qquad \frac{\partial K}{\partial \alpha_i} = 0 \qquad \text{at } \alpha_1 = \alpha_2 = 0, \qquad (i = 1, 2)$$

---

[2]See, for example, T.M. Apostol, *Mathematical Analysis*, 2nd Edition, page 381.

for the function $K \equiv I - \lambda J$.

So, our problem is equivalent to the (unconstrained) fundamental problem for the integral

$$\int_a^b (F - \lambda G),$$

and has Euler equation

(43) $$\frac{d}{dx}(F_{y'} - \lambda G_{y'}) \;=\; F_y - \lambda G_y .$$

Notes

1. The analysis can clearly be extended to the cases where I, J are both double, or both triple, integrals giving Euler equations of the form (37), or (37'), for $F - \lambda G$.

2. The Euler equation here possesses the additional unknown λ, but there is the additional equation $J = \gamma$ to help determine it.

3. One function η, with $y_\alpha(x) = y(x) + \alpha\eta(x)$, does not suffice, because

$$J \;=\; \int_a^b G(x, y(x) + \alpha\eta(x), y'(x) + \alpha\eta'(x))\, dx \;=\; \gamma$$

would then fix α as a function of γ, and we should gain no useful result from differentiating.

4. The reader might ponder the need for both equations (42) in deriving the single equation (43).

5. The problem of this section is equivalent to finding a function y which gives a stationary value to the quotient

$$\lambda \;\equiv\; \frac{I}{J} \;=\; \frac{\int_a^b F(x, y, y')\, dx}{\int_a^b G(x, y, y')\, dx},$$

subject to $y(a) = c$ and $y(b) = d$, when J is non-zero. To see this, we again put

$$y_\alpha(x) \;=\; y(x) + \alpha\eta(x),$$

for small $|\alpha|$ and with $\eta \in C^2([a, b])$ satisfying

$$\eta(a) \;=\; \eta(b) \;=\; 0.$$

Then,

$$\lambda = \lambda(\alpha) = \frac{I(\alpha)}{J(\alpha)},$$

and

$$\lambda' = \frac{I'J - IJ'}{J^2} = \frac{1}{J}(I' - \lambda J')$$

is zero at $\alpha = 0$, provided that $\lambda = \lambda(0)$ satisfies

$$I'(0) - \lambda J'(0) = 0.$$

As above, this gives rise to the Euler equation (43).

Example 11 : the isoperimetric problem In Example 2, we modelled this problem in Cartesian co-ordinates. Rather than solve the problem in this form, we shall use polar co-ordinates.

Suppose that the extremal curve $r = r(\theta)$ is of length $L > 0$. It is clear that the area it encloses, which is a maximum amongst all areas enclosed by curves of length L, must be convex. In particular, for every point on the curve, the tangent at that point must lie wholly on one side of the curve.

Take one tangent as polar axis, with the pole at the point of contact and the curve in the upper half-plane. We search for twice continuously differentiable $r = r(\theta)$ giving a stationary value to the area integral

$$I \equiv \tfrac{1}{2} \int_0^\pi r^2 \, d\theta$$

when subject to the constraint

$$J \equiv \int_0^\pi (r^2 + \dot{r}^2)^{\frac{1}{2}} \, d\theta = L,$$

where dot denotes differentiation with respect to θ. According to the theory of this section and Proposition 1 of section 11.4, a first integral of the Euler equation for this problem is

$$\left(\frac{1}{2} r^2 - \lambda(r^2 + \dot{r}^2)^{\frac{1}{2}} \right) - \dot{r} \frac{\partial}{\partial \dot{r}} \left(\frac{1}{2} r^2 - \lambda(r^2 + \dot{r}^2)^{\frac{1}{2}} \right) = k, \quad (k \text{ constant}).$$

This quickly reduces to

$$\frac{1}{2}r^2 - \lambda r^2 (r^2 + \dot{r}^2)^{-\frac{1}{2}} = k.$$

As $r = 0$ when $\theta = 0$, we must have $k = 0$. As $r = 0$, identically on $[0, \pi]$, encloses zero area and does not satisfy $J = L$, we must have

$$r^2 + \dot{r}^2 = 4\lambda^2, \qquad (\theta \in [0, \pi]).$$

The solution of this equation, satisfying $r(0) = 0$, is

$$r = 2\lambda \sin \theta,$$

which represents a circle. The constraint $J = L$ then shows that $\lambda = L/2\pi$, as would be expected.   □

Exercise 22 Show that extremals for the problem of the heavy uniform string, hanging in stable equilibrium under gravity (see Exercise 2), are catenaries.
[HINT: Use Example 6(b) of section 11.4.]

Exercise 23 Find an extremal corresponding to the area integral

$$\int_{-1}^{1} y \, dx$$

when subject to $y(-1) = y(1) = 0$ and

$$\int_{1}^{1} \{y^2 + (y')^2\} \, dx = 1.$$

Exercise 24 Find the extremal curve $y = y(x)$ corresponding to the problem of minimising the length of the curve in the upper half plane $\{(x, y) \in \mathbb{R}^2 : y \geq 0\}$ which joins the point $(-1, 0)$ to the point $(1, 0)$ whilst enclosing, together with the x-axis, a given fixed area A.

Exercise 25 Find extremals corresponding to

$$\iiint_V \{(u_x)^2 + (u_y)^2 + (u_z)^2\} \, dx dy dz,$$

where $V = \{(x, y, z) : x^2 + y^2 + z^2 \leq 1\}$,

(a)   when subject to $u = 1$ on $x^2 + y^2 + z^2 = 1$ and

$$\iiint_V u \, dx dy dz = 4\pi,$$

(b)   when subject to $u = 0$ on $x^2 + y^2 + z^2 = 1$ and

$$\iiint_V u^2 \, dx dy dz = 1.$$

11.9   Non-integral constraints

A variational problem, where a constraint is not given in integral form, can be difficult to solve. We consider here only the problem of finding a curve

$$\mathbf{r} \;=\; \mathbf{r}(t) \;=\; (x(t), y(t), z(t))$$

on the surface

(44) $G(x, y, z) \;=\; 0,$

which passes through the points $\mathbf{r}(t_0)$, $\mathbf{r}(t_1)$ and gives a stationary value to the integral

(45) $\displaystyle\int_{t_0}^{t_1} F(x, y, z, \dot{x}, \dot{y}, \dot{z}) \, dt,$

where dot denotes differentiation with respect to t. We assume that $x, y, z \in C^2([t_0, t_1])$, that $G \in C^1(D)$, and that $F \in C^2(E)$, for appropriate open $D \subseteq \mathbb{R}^3$, $E \subseteq \mathbb{R}^6$. We also suppose that $\mathrm{grad}(G) = (G_x, G_y, G_z)$ is never zero along an extremal. Note that F is not an explicit function of the variable t.

When (with no loss of generality) G_z is non-zero, we use the Implicit Function Theorem of the differential calculus[3] to solve (44) in a neighbourhood of an extremal to give

(46) $z \;=\; g(x, y) \in C^1(N);$

for an appropriate open $N \subseteq \mathbb{R}^2$, so that

(47) $g_x \;=\; -G_x/G_z, \qquad g_y \;=\; -G_y/G_z.$

---

[3]See, for example, T.M. Apostol, *Mathematical Analysis*, 2nd Edition, page 374.

Note that

(48) $$\dot{z} \;=\; g_x \dot{x} + g_y \dot{y}$$

and hence that

(49) $$\frac{\partial}{\partial \dot{x}} \, \dot{z} \;=\; g_x, \qquad \frac{\partial}{\partial \dot{y}} \, \dot{z} \;=\; g_y.$$

Further, *assuming* $g \in C^2(N)$,

(50) $$\frac{d}{dt} \, g_x \;=\; g_{xx} \dot{x} + g_{xy} \dot{y} \;=\; \frac{\partial}{\partial x} \left(g_x \dot{x} + g_y \dot{y} \right) \;=\; \frac{\partial \dot{z}}{\partial x}$$

and, similarly,

(51) $$\frac{d}{dt} \, g_y \;=\; \frac{\partial \dot{z}}{\partial y} .$$

Using the substitution (46), in order to dispense with the need for the constraint, the problem reduces to a search for curves $x = x(t)$, $y = y(t)$ passing through $(x(t_0), y(t_0))$ and $(x(t_1), y(t_1))$, which give a stationary value to the integral

$$\int_{t_0}^{t_1} H(x, y, \dot{x}, \dot{y}) \, dt$$

where H is given as a C^2-function by

$$H(x, y, \dot{x}, \dot{y}) \;\equiv\; F(x, y, g(x, y), \dot{x}, \dot{y}, \dot{x} g_x + \dot{y} g_y)$$

wherever F, g are defined. The corresponding Euler equations, given by the theory of two dependent functions in section 11.6, are

(52) $$\frac{d}{dt} \, H_{\dot{x}} \;=\; H_x, \qquad \frac{d}{dt} \, H_{\dot{y}} \;=\; H_y .$$

However, using (50), and then (47),

$$\frac{d}{dt} \, H_{\dot{x}} - H_x \;=\; \frac{d}{dt} \left(F_{\dot{x}} + F_{\dot{z}} g_x \right) - \left(F_x + F_z g_x + F_{\dot{z}} \frac{d}{dt} g_x \right)$$

$$=\; \left(\frac{d}{dt} \, F_{\dot{x}} - F_x \right) - \frac{G_x}{G_z} \left(\frac{d}{dt} \, F_{\dot{z}} - F_z \right) .$$

Similarly,

$$\frac{d}{dt} H_{\dot{y}} - H_y = \left(\frac{d}{dt} F_{\dot{y}} - F_y \right) - \frac{G_y}{G_z} \left(\frac{d}{dt} F_{\dot{z}} - F_z \right).$$

If λ denotes the *function*

$$\lambda \equiv \frac{1}{G_z} \left(\frac{d}{dt} F_{\dot{z}} - F_z \right),$$

equations (52) become

$$\frac{d}{dt} F_{\dot{x}} - F_x = \lambda G_x, \qquad \frac{d}{dt} F_{\dot{y}} - F_y = \lambda G_y,$$

where

$$\frac{d}{dt} F_{\dot{z}} - F_z = \lambda G_z.$$

These three equations together are known as the Euler equations in this case.

Example 12 : geodesics   We return to Example 3(i) of section 11.2, where geodesics $\mathbf{r} = \mathbf{r}(t)$ through

$$A = \mathbf{r}(t_0), \qquad B = \mathbf{r}(t_1)$$

gave rise to a minimum of the integral

$$\int_{t_0}^{t_1} (\dot{x}^2 + \dot{y}^2 + \dot{z}^2)^{\frac{1}{2}} \, dt$$

when subject to the constraint

$$G(x(t), y(t), z(t)) = 0, \qquad (t \in [t_0, t_1]).$$

With the differentiability conditions of this section satisfied, extremals must satisfy the Euler equations

(53) $\qquad \dfrac{d}{dt} \left(\dfrac{\dot{x}}{K} \right) = \lambda G_x, \qquad \dfrac{d}{dt} \left(\dfrac{\dot{y}}{K} \right) = \lambda G_y, \qquad \dfrac{d}{dt} \left(\dfrac{\dot{z}}{K} \right) = \lambda G_z,$

where $K = (\dot{x}^2 + \dot{y}^2 + \dot{z}^2)^{\frac{1}{2}}$. Now choose t to be arc-length s, so that $K = 1$. The Euler equations then become

(54) $\qquad\qquad x'' = \lambda G_x, \qquad y'' = \lambda G_y, \qquad z'' = \lambda G_z,$

where dash denotes differentiation with respect to s. $\qquad\qquad\qquad\qquad\qquad$ $\square$

Exercise 26 Suppose, in the notation of Example 12, that

$$G(x, y, z) \equiv x^2 + y^2 + z^2 - a^2,$$

where a is a positive constant; so that, the geodesic

$$\mathbf{r} = \mathbf{r}(s) = (x(s), y(s), z(s))$$

lies on a sphere where $\mathrm{grad}(G)$ is never zero. Show that the geodesic must satisfy

$$xx'' + yy'' + zz'' = -1,$$

and hence, that the Euler equations (54) become

$$a^2 x'' + x = a^2 y'' + y = a^2 z'' + z = 0.$$

Deduce that an extremal must lie on a plane through the origin and therefore must be an arc of a great circle.

11.10   Varying boundary conditions

So far, we have only considered problems where the boundary (end-points, curves, surfaces) has been fixed. Varying the boundary will allow us to consider such elementary problems as that of finding the minimal distance from a point to a curve.

We start by searching for a function $y = y(x) \in C^2([a, b])$, where a, b, c are real constants, satisfying only

$$(55) \qquad\qquad\qquad\qquad y(a) = c$$

and giving a minimum (or maximum) to the integral

$$(56) \qquad\qquad\qquad\qquad I = \int_a^b F(x, y, y')\, dx.$$

Geometrically, we are looking for a curve through the point (a, c). But the point at which it meets the line $x = b$ is not given.

Suppose now that $y = y(x)$ actually gives such a minimum and that $F \in C^2(D)$ for some open $D \subseteq \mathbb{R}^3$ for which

$$(x, y(x), y'(x)) \in D, \qquad (x \in [a, b]).$$

Again, for small $|\alpha|$, we introduce the function $y = y_\alpha(x)$:

$$y_\alpha(x) = y(x) + \alpha\eta(x), \qquad (x \in [a, b])$$

where $\eta \in C^2([a, b])$ is arbitrary, save that it must satisfy

$$\eta(a) = 0.$$

We may visualise the situation as in the following diagram.

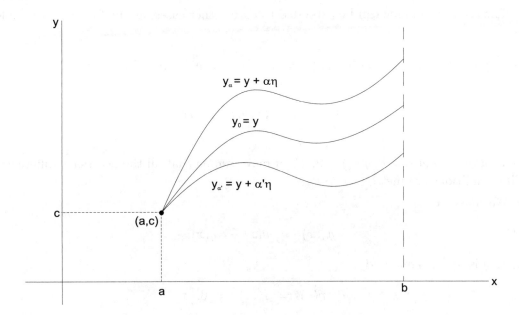

Substituting in the integral I and integrating $I'(0)$ by parts (as with the fundamental problem in section 11.3) give

(57) $$I'(0) \; = \; \eta(b)F_{y'}(b, y(b), y'(b)) + \int_a^b \eta(F_y - \frac{d}{dx} F_{y'}) \, dx,$$

as we only know that $\eta(a) = 0$. Now, y gives a minimum to the integral I amongst curves satisfying (55); so, it must *a fortiori* give a minimum to I amongst curves passing through $(b, y(b))$ (corresponding to $\eta(b) = 0$). In that Fundamental Problem, y satisfies Euler's equation

(58) $$\frac{d}{dx} F_{y'} \; = \; F_y,$$

which an extremal must also, therefore, satisfy here. But then, $I'(0) = 0$, coupled with (57), yields

$$\eta(b)F_{y'}(b, y(b), y'(b)) \; = \; 0.$$

As η may clearly be chosen so that $\eta(b)$ is non-zero, we derive the boundary condition

(59) $$F_{y'} \; = \; 0 \quad \text{at } x = b.$$

This replaces the condition $y(b) = d$ in the Fundamental Problem. An extremal here will thus be a solution of (58) satisfying (55) and (59).

The above technique can be extended to cover other cases, as the following examples show.

Example 13 Find the conditions to be satisfied by an extremal corresponding to

$$(60) \qquad\qquad I \;=\; \int_a^b F(x, y, y', y'')\, dx$$

when subject to $y(a) = c$, $y'(a) = d$. (You may assume that all the necessary differentiability conditions are met.)

We consider

$$(61) \qquad\qquad y_\alpha(x) \;=\; y(x) + \alpha \eta(x),$$

where y is an extremal and

$$(62) \qquad\qquad \eta(a) \;=\; \eta'(a) \;=\; 0.$$

As in the above argument, y must satisfy the Euler equation (27):

$$\frac{d^2}{dx^2} F_{y''} - \frac{d}{dx} F_{y'} + F_y \;=\; 0.$$

Substituting from (61) into $I = I(\alpha)$, putting $I'(0) = 0$ and integrating by parts as in section 11.5, we see that

$$\eta F_{y'} + \eta' F_{y''} - \eta \frac{d}{dx} F_{y''} \;=\; 0 \quad \text{at } x = b.$$

(We have used Euler's equation and (62).)

Choosing η in turn to satisfy $\eta(b) \neq 0$, $\eta'(b) = 0$, and then $\eta(b) = 0$, $\eta'(b) \neq 0$, we see that

$$F_{y'} - \frac{d}{dx} F_{y''} \;=\; 0 \quad \text{at } x = b$$

$$F_{y''} \;=\; 0 \quad \text{at } x = b$$

are the two additional boundary conditions which an extremal y (satisfying Euler's equation and $y(a) = c$, $y'(a) = d$) must also satisfy. $\qquad\square$

Example 14 Show that an extremal corresponding to

$$\int_a^b F(x, y, y')\, dx,$$

subject to

$$y(a) \;=\; c$$

and to

$$\int_a^b G(x, y, y')\, dx \;=\; \gamma,$$

where a, b, c, d and γ are constants, must satisfy

(63) $$F_{y'} - \lambda G_{y'} \;=\; 0 \quad \text{at } x = b.$$

for some constant λ. (Again, you may assume that all the necessary differentiability conditions are met.)

The analysis of section 11.8 shows that the integral constraint problem

$$\int_a^b F \quad \text{subject to} \quad \int_a^b G \;=\; \gamma$$

can be reduced to the unconstrained problem for

$$\int_a^b (F - \lambda G),$$

where λ is a constant. Applying the work of this section to the latter gives (63) as the condition corresponding to (59). $\qquad\square$

We conclude this chapter by finding the condition an extremal curve $y = y(x)$ corresponding to (56), subject to (55), must satisfy in order that $y(b)$ should lie on the (given) continuously differentiable curve $y = f(x)$. On this occasion, it is necessary to allow the upper limit b of integration to vary.

Again making the substitution

$$y_\alpha(x) \;=\; y(x) + \alpha\eta(x)$$

for an extremal $y = y(x)$, small $|\alpha|$, and $\eta(a) = 0$, we see that

$$I \;=\; I(\alpha) \;=\; \int_a^{b(\alpha)} F(x, y + \alpha\eta, y' + \alpha\eta')\, dx.$$

Assuming that b is a continuously differentiable function of α, in addition to our earlier differentiability assumptions, the condition $I'(0) = 0$ leads quickly to

(64) $$b'.F(b, y(b), y'(b)) + \eta(b).F_{y'}(b, y(b), y'(b)) \;=\; 0$$

at $\alpha = 0$. But, for small α, $y = y_\alpha(x)$ meets $y = f(x)$ when $x = b(\alpha)$, that is, when

$$y(b(\alpha)) + \alpha\eta(b(\alpha)) \;=\; f(b(\alpha)).$$

So, differentiating with respect to α and putting $\alpha = 0$,

$$y'(b(0)).b'(0) + \eta(b(0)) \;=\; f'(b(0)).b'(0),$$

and hence

(65) $$b'(f'(b) - y'(b)) \;=\; \eta(b)$$

at $\alpha = 0$. From (64), (65), we obtain

$$\eta(b).\{F(b, y(b), y'(b)) + (f'(b) - y'(b)).F_{y'}(b, y(b), y'(b))\} \;=\; 0$$

at $\alpha = 0$. As we may choose $\eta(b)$ to be non-zero, the extremal must satisfy the second boundary condition

(66) $$F + (f' - y')F_{y'} \;=\; 0 \quad \text{at } x = b.$$

This condition is the so-called *transversal condition*.

Example 15 Show that there are two extremals corresponding to

$$\int_1^b y^2(1 - y')^2 \, dx$$

satisfying $y(1) = 1$ and $y(b) = 3$ and that they occur when $b = -1$ and $b = 3$.

Using Proposition 1 of section 11.4, a first integral of the Euler equation in this case is

(67) $$y^2(1 - y')^2 - y'\frac{\partial}{\partial y'}\{y^2(1 - y')^2\} \;=\; k,$$

where k is a constant. As $f(x) = 3$, for all x, condition (66) implies that $k = 0$. When y is non-zero, (67) leads quickly, on integration and making use of $y(1) = 1$, to the extremals

$$y \;=\; -x + 2, \qquad y \;=\; x.$$

The value $y = 3$ corresponds to $x = -1$, respectively $x = 3$. (Note that $y = 0$, identically on $[1, b]$, does not satisfy the boundary conditions.) □

The reader may already have determined the relevant Euler equation and its solution for some of the following in attempting earlier exercises in this chapter.

Exercise 27 Find extremals corresponding to

(a) $$\int_0^1 \{y^2 + y' + (y')^2\} \, dx, \qquad\qquad y(0) = 0,$$

(b) $$\int_0^1 \{(y')^2 + (y'')^2\} \, dx, \qquad\qquad y(0) = y'(0) = 0, \qquad y(1) = 1,$$

(c) $$\int_0^{\frac{\pi}{2}} \{y^2 - 2(y')^2 + (y'')^2\} \, dx, \qquad\qquad y(0) = 1, \qquad y'(0) = 0.$$

Exercise 28 Find the curve $y = f(x)$ $(0 \le x \le 1)$, having length $\frac{\pi}{2}$ and where $f(0) = 0$, which maximises the area of the set

$$\{(x, y) : 0 \le x \le 1, 0 \le y \le f(x)\}.$$

State clearly any assumptions you need to make.

Exercise 29 Stating clearly any assumptions you make, find the shortest distance from the point $(x, y) = (-5, 0)$ to the parabola $y = 1 + x^2$.

Exercise 30 Find the extremals corresponding to

$$\int_0^X \{(y')^2 + 4y\} \, dx,$$

given that $y(0) = 1$ and $y(X) = X^2$.

Exercise 31 Consider the problem of finding extremals

$$\mathbf{r} = \mathbf{r}(t) = (x(t), y(t), z(t))$$

corresponding to

$$\int_0^T F(x, y, z, \dot{x}, \dot{y}, \dot{z}) \, dt,$$

where T is a constant, $\mathbf{r}(0)$ is specified, and $\mathbf{r}(T)$ lies on the curve

$$\mathbf{r} = \mathbf{R}(s) = (X(s), Y(s), Z(s)).$$

Show that the extremal satisfies

$$X'F_{\dot{x}} + Y'F_{\dot{y}} + Z'F_{\dot{z}} \; = \; 0$$

at $t = T$. In the particular case

$$F \; = \; \dot{x}^2 + \dot{y}^2 + \dot{z}^2 - 2gz, \qquad \mathbf{r}(0) \; = \; \mathbf{0},$$

with

$$\mathbf{R}(s) \; = \; s(\cos s, \sin s, 1),$$

show that that extremal meets $\mathbf{r} = \mathbf{R}(s)$ when $s = 0$ and $s = gT^2/4$.

Exercise 32 An elastic beam has vertical displacement $y(x)$ for $0 \le x \le L$. The displacement minimises the gravitational energy

$$\int_0^L \{ \tfrac{1}{2} D(y'')^2 + \rho g y \}$$

where D, ρ and g are positive constants. Consider the following two cases.

(a) The ends of the beam are supported; that is, $y(0) = y(L) = 0$. Show that an extremal for the problem is

$$y(x) \; = \; -\frac{\rho g}{24D}\, x(L - x)(L^2 - x(L - x)).$$

(b) The end at $x = 0$ is clamped; that is, $y(0) = y'(0) = 0$ (the case of an elastic springboard). Determine an extremal for the problem.

Exercise 33 It is required to find a C^2-function ϕ that makes the functional J stationary, where

$$J[\phi] \; = \; \iint_D |\nabla\phi - \mathbf{v}(x, y)|^2 \, dxdy$$

$$= \; \iint_D \left(\left(\frac{\partial\phi}{\partial x} - v_1(x, y) \right)^2 + \left(\frac{\partial\phi}{\partial y} - v_2(x, y) \right)^2 \right) dxdy.$$

In this, $\mathbf{v} = (v_1, v_2)$ is a continuously differentiable vector field in the plane, and D is a bounded open region of the plane with smooth boundary ∂D and unit outward normal $\mathbf{n}$.

Show that if ϕ makes J stationary then it must satisfy

$$\nabla^2\phi \; = \; \operatorname{div} \mathbf{v} \;\; \text{in } D \qquad \text{and} \qquad \operatorname{grad}\phi \,.\, \mathbf{n} \; = \; \mathbf{v} \,.\, \mathbf{n} \;\; \text{on } \partial D.$$

If D is the ellipse $x^2/a^2 + y^2/b^2 < 1$ and the functions $v_1 = v_1(x, y)$ and $v_2 = v_2(x, y)$ are given by

$$v_1(x, y) \; = \; y, \qquad v_2(x, y) \; = \; -x,$$

show that the conditions on ϕ can be satisfied by $\phi(x, y) = \alpha xy$ for a suitable constant α which should be determined. Hence, show that the stationary value of J is $\pi a^3 b^3/(a^2 + b^2)$.

12 The Sturm–Liouville Equation

This chapter concerns the homogeneous equation

$$\text{(SL)} \qquad Ly + \lambda\rho(x)y \;\equiv\; (p(x)y')' + q(x)y + \lambda\rho(x)y \;=\; 0, \qquad (x \in [a,b])$$

where $p : [a,b] \to \mathbb{R}$ is continuously differentiable, $q : [a,b] \to \mathbb{R}$ and $\rho : [a,b] \to \mathbb{R}$ are continuous, $p(x) > 0$ and $\rho(x) > 0$, for all x in $[a,b]$, and λ is a real constant.

The equation generalises (H) of section 4.2 and L denotes the operator that was defined in that section. Further, we shall impose the same homogeneous boundary conditions, namely:

$$(\alpha) \qquad\qquad\qquad\qquad A_1 y(a) + B_1 y'(a) \;=\; 0,$$

$$(\beta) \qquad\qquad\qquad\qquad A_2 y(b) + B_2 y'(b) \;=\; 0,$$

where A_1, A_2, B_1, B_2 are constants (A_1, B_1 not both zero and A_2, B_2 not both zero).

A value of the constant λ, for which (SL), together with (α) and (β), has a solution $y = y(x)$ which is not identically zero on $[a,b]$, is called an *eigenvalue of* (SL), (α), (β), and such a non-zero solution $y = y(x)$ is called an *eigenfunction of* (SL), (α), (β) ('corresponding to the eigenvalue λ').

Our discussion will bring together much of the work in earlier chapters, from Green's functions, the correspondence between integral and differential equations and the existence-uniqueness theorems for the latter, to the use of the Calculus of Variations, of the Fredholm Alternative Theorem and of the Expansion Theorem for symmetric kernels.

We should like to stress the importance of the material in this chapter for applications, especially in physics. The Note following Exercise 5 below gives an important example. Others can be found via the list of references in the Bibliography.

12.1   Some elementary results on eigenfunctions and eigenvalues

Suppose that λ_1, λ_2 are eigenvalues of (SL), (α), (β) corresponding, respectively, to eigenfunctions y_1, y_2. It is easily verified that

$$(1) \qquad\qquad (\lambda_1 - \lambda_2)\rho y_1 y_2 \;=\; -y_2 L y_1 + y_1 L y_2$$

$$= \; (p(y_1 y_2' - y_2 y_1'))'$$

on $[a, b]$, and hence, on integrating from a to b and applying the boundary conditions (α), (β), that

$$(2) \qquad\qquad (\lambda_1 - \lambda_2) \int_a^b \rho(x) y_1(x) y_2(x)\,dx \;=\; 0.$$

So, when λ_1, λ_2 are distinct, the functions $z_1 \equiv y_1\sqrt{\rho}$ and $z_2 \equiv y_2\sqrt{\rho}$ are orthogonal.

Direct differentiation shows that the function $z = y\sqrt{\rho}$ satisfies the differential equation

$$(3) \qquad\qquad (p_1(x)z')' + q_1(x)z + \lambda z \;=\; 0, \qquad (x \in [a, b])$$

once one defines

$$p_1 \;=\; \frac{p}{\rho}, \qquad q_1 \;=\; \frac{1}{\sqrt{\rho}}\frac{d}{dx}\left(p\frac{d}{dx}\left(\frac{1}{\sqrt{\rho}}\right)\right) + \frac{q}{\rho}$$

on $[a, b]$.

Suppose that (y_n) is a sequence of eigenfunctions of (SL), (α), (β), corresponding to distinct eigenvalues. We can ensure that the orthogonal sequence (z_n), where $z_n = y_n\sqrt{\rho}$ for each n, is in fact orthonormal as in section 9.3. This corresponds to imposing the condition

$$(4) \qquad\qquad \int_a^b \rho y_n^2 \;=\; 1$$

on y_n, for each n.

There are, as we shall show in the next section, a countable number of eigenvalues. To each such eigenvalue, there corresponds, however, essentially only one eigenfunction, as the following result shows.

Proposition 1 Every eigenvalue λ of (SL), (α), (β) is simple (that is, there do not exist two linearly independent eigenfunctions of (SL), (α), (β) corresponding to λ).

Proof Suppose that the linearly independent eigenfunctions y_1, y_2 both correspond to λ. Then, every solution y of (SL) corresponding to λ can be expressed in the form

$$y = c_1 y_1 + c_2 y_2,$$

where c_1, c_2 are constants. Hence, every solution of (SL) corresponding to λ must satisfy (α), (β) and be an eigenfunction. However, Theorem 4 of Chapter 2 tells us that we may find a solution y to (SL) together with any prescribed conditions $y(x_0) = c$, $y'(x_0) = d$ at a point $x_0 \in [a, b]$, even if these conditions were incompatible with (α) or (β). We have reached a palpable contradiction. $\qquad\square$

Exercise 1 Establish the validity of equations (1), (2), (3) above.

Exercise 2 Suppose that λ is an eigenvalue of (SL), (α), (β), with corresponding eigenfunction y, where for convenience we adopt the normalisation

$$\int_a^b \rho y^2 = 1.$$

Show that

$$\lambda = -\Big[pyy'\Big]_a^b + \int_a^b (p(y')^2 - qy^2),$$

and hence, that λ is always positive when both

(5) $$B_1 \neq 0 \neq B_2, \qquad \frac{A_1}{B_1} \leq 0, \qquad \frac{A_2}{B_2} \geq 0,$$

and

(6) $$q(x) \leq 0, \quad \text{for every } x \text{ in } [a, b].$$

Exercise 3 In place of (α), (β), impose the boundary conditions

(γ) $$y(a) = y(b),$$

(δ) $$p(a)y'(a) = p(b)y'(b).$$

If y_1, y_2 are eigenfunctions of (SL), (γ), (δ), corresponding, respectively, to eigenvalues λ_1, λ_2, prove that

(i) y_1, y_2 are orthogonal

(ii) λ_1, λ_2 are both positive, provided condition (6) of Exercise 2 is also imposed.

Exercise 4 Show that, when $q = 0$ on $[a, b]$, the substitutions

$$t(x) \;=\; \int_a^x \frac{du}{p(u)}, \qquad \sigma(x) \;=\; \rho(x)p(x), \qquad (x \in [a, b])$$

reduce (SL) to

$$\ddot{y} + \lambda \sigma y \;=\; 0, \qquad \left(t \in \left[0, \int_a^b p^{-1}\right]\right)$$

where dot denotes differentiation with respect to t.

Exercise 5 Suppose that $u = u(x, t)$ satisfies the partial differential equation

(7) $$(p(x)u_x)_x \;=\; \rho(x)u_{tt}, \qquad (x \in [a, b], \, t \geq 0)$$

where $p : [a, b] \to \mathbb{R}$ is continuously differentiable and $\rho : [a, b] \to \mathbb{R}$ is continuous. Show that the substitution

$$u(x, t) \;=\; y(x)z(t)$$

leads to y satisfying (SL) with $q \equiv 0$, whenever λ, z satisfy

$$\ddot{z} + \lambda z \;=\; 0$$

and dot denotes differentiation with respect to t.

Note Exercise 5 shows how a Sturm–Liouville equation arises from the equation of motion of a non-homogeneous string, where $p(x)$ denotes the modulus of elasticity multiplied by the cross-sectional area at x, and $\rho(x)$ denotes the mass per unit length at x. The boundary conditions (α), (β) correspond to

(i) fixed ends, if $B_1 = B_2 = 0$,

(ii) free ends, if $A_1 = A_2 = 0$,

(iii) elastically attached ends, if condition (5) of Exercise 2 is met.

The conditions (γ), (δ) of Exercise 3 correspond to periodic solutions when $p(a) = p(b)$. If $\lambda = \mu^2$ is taken positive and $y = y(x)$ is a solution of (SL), together with its relevant boundary conditions, a solution of (7) may be written

$$u(x, t) \;=\; y(x)(A \sin \mu t + B \cos \mu t),$$

where A, B are constants. If the physically consistent conditions (5), (6) are imposed, Exercise 2 shows that λ must be positive.

12.2   The Sturm–Liouville Theorem

In this section, we use the theory of Green's functions (section 4.2), the Fredholm Alternative Theorem (section 8.3), and the Hilbert–Schmidt Expansion Theorem (section 9.5) to analyse the non-homogeneous equation

(SLN) $$Ly + \lambda\rho(x)y \; = \; f(x), \qquad (x \in [a, b])$$

and its homogeneous counterpart, the Sturm–Liouville equation

(SL) $$Ly + \lambda\rho(x)y \; = \; 0, \qquad (x \in [a, b])$$

both taken together with the homogeneous boundary conditions

(α) $$A_1 y(a) + B_1 y'(a) \; = \; 0,$$

(β) $$A_2 y(b) + B_2 y'(b) \; = \; 0.$$

To the notation defined and the conditions imposed at the start of this chapter, we add that $f : [a, b] \to \mathbb{R}$ must be continuous. We also insist, as in Theorem 1 of section 4.2, that $Ly = 0$, *taken together with both* (α) *and* (β), *has only the trivial solution* $y = 0$, *identically on* $[a, b]$. This permits, by Lemma 2 of that section, the existence of linearly independent solutions u solving $Ly = 0$ with (α) and v solving $Ly = 0$ with (β). In turn, this allows the definition of a Green's function

$$G(x, t) \; = \; \begin{cases} \dfrac{u(x)v(t)}{A}, & \text{for } a \le x \le t \le b, \\[3mm] \dfrac{u(t)v(x)}{A}, & \text{for } a \le t \le x \le b, \end{cases}$$

where A is the non-zero constant given by $pW(u, v) = A$, which is, by Lemma 1 of section 4.2, valid on $[a, b]$. Note the important fact that G *is symmetric*.

Replacing f in the Green's function theorem by $f - \lambda\rho y$, that theorem allows us to re-write (SLN) with (α), (β) in the equivalent form

(N) $$y(x) \; = \; F(x) - \lambda \int_a^b G(x, t)\rho(t)y(t)\, dt, \qquad (x \in [a, b])$$

where

$$F(x) \; = \; \int_a^b G(x, t)f(t)\, dt, \qquad (x \in [a, b])$$

and to re-write (SL) with (α), (β) as

$$\text{(H)} \qquad\qquad y(x) \;=\; -\lambda \int_a^b G(x,t)\rho(t)y(t)\,dt, \qquad (x \in [a,b])$$

corresponding to the homogeneous case when $f = 0$ on $[a,b]$. The substitutions

$$z(x) \;=\; y(x)\sqrt{\rho(x)},$$

$$g(x) \;=\; F(x)\sqrt{\rho(x)},$$

$$K(x,t) \;=\; -G(x,t)\sqrt{\rho(x)\rho(t)},$$

for x, t in $[a,b]$, can now be used to reduce (N) and (H), respectively, to

$$\text{(N}_1\text{)} \qquad\qquad z(x) \;=\; g(x) + \lambda \int_a^b K(x,t)z(t)\,dt, \qquad (x \in [a,b])$$

and

$$\text{(H}_1\text{)} \qquad\qquad z(x) \;=\; \lambda \int_a^b K(x,t)z(t)\,dt, \qquad (x \in [a,b]).$$

Note that, as G is symmetric, so is K.

Theorem 1 (The Sturm–Liouville Theorem)   For each (fixed) real number λ, exactly one of the following two statements is true:

(SL1) The equation (SLN), taken together with the boundary conditions (α), (β), possesses a unique twice continuously differentiable solution and, in particular, the only such solution of (SL) with (α), (β) is the trivial solution.

(SL2) The Sturm–Liouville equation (SL), taken together with (α), (β), possesses a non-trivial solution and (SLN), with (α), (β), possesses a solution if and only if

$$\int_a^b yf \;=\; 0,$$

for every eigenfunction y of (SL), (α), (β) corresponding to λ.

Further, (SL), (α), (β) has infinite sequences (λ_n) of eigenvalues and (y_n) of corresponding eigenfunctions satisfying the 'weighted orthonormality relations'

$$(8) \qquad \int_a^b \rho y_n^2 = 1, \qquad \int_a^b \rho y_m y_n = 0 \qquad (m \neq n)$$

and allowing an arbitrary twice continuously differentiable function $h : [a, b] \rightarrow \mathbb{R}$, satisfying ($\alpha$), ($\beta$), to be expanded in the uniformly convergent series

$$(9) \qquad h = \sum_{n=1}^\infty c_n y_n, \quad \text{where } c_n = \int_a^b \rho h y_n, \qquad (n \geq 1)$$

on $[a, b]$.

Proof It is readily verified that y satisfies (N), respectively (H), if and only if z satisfies (N$_1$), respectively (H$_1$).

Note that, if z is an eigenfunction of (H$_1$) = (H$_1^T$) corresponding to λ (which must therefore be non-zero), then, by Fubini's Theorem ([F] of Chapter 0) and the symmetry of G,

$$\int_a^b z(x)g(x)\,dx = \int_a^b y(x)\rho(x)\left\{ \int_a^b G(x,t)f(t)\,dt \right\} dx$$

$$= \int_a^b f(t)\left\{ \int_a^b G(t,x)\rho(x)y(x)\,dx \right\} dt.$$

Hence,

$$\int_a^b z(x)g(x)\,dx = -\frac{1}{\lambda} \int_a^b y(t)f(t)\,dt.$$

The alternatives (SL1), (SL2) are therefore just the Fredholm alternatives given by Theorem 1 of section 8.3, once one replaces (SLN) and (SL), both taken with (α), (β), by the integral equations (N$_1$) and (H$_1$) above.

Further, as the kernel K is symmetric, we may apply the Hilbert–Schmidt theory of Chapter 9 to (H$_1$). The reader is asked to show that K cannot be degenerate, as an exercise. It follows from Proposition 4 of section 9.4 that (H$_1$), and hence (SL), (α), (β) must have an infinite sequence (λ_n) of distinct eigenvalues. Suppose that y_n is an eigenfunction of (SL), (α), (β) corresponding to λ_n. (Remember that, by Proposition 1 of the last section, each λ_n is in fact simple.) Then, the discussion of section 12.1 shows that the weighted orthonormality relations (8) are satisfied.

As h is twice continuously differentiable, we may define a continuous function $k : [a, b] \to \mathbb{R}$ by $Lh = -k\sqrt{\rho}$. But then, as h satisfies (α), (β), the Green's function theorem gives

$$h(x) \; = \; -\int_a^b G(x, t)k(t)\sqrt{\rho(t)}\, dt, \qquad (x \in [a, b]).$$

Hence,

$$h(x)\sqrt{\rho(x)} \; = \; \int_a^b K(x, t)k(t)\, dt, \qquad (x \in [a, b]).$$

Putting $z_n = y_n\sqrt{\rho}$ for each n, we now invoke the Expansion Theorem of section 9.5 to derive the uniformly convergent expansion

$$h\sqrt{\rho} \; = \; \sum_{n=1}^{\infty} c_n z_n, \quad \text{where } c_n \; = \; \int_a^b h\sqrt{\rho}z_n, \qquad (n \geq 1).$$

But, this is precisely the expansion (9). □

Exercise 6 Show that the kernel K, defined in the above proof, cannot be degenerate.

Exercise 7 If (λ_n) is an infinite sequence of distinct eigenvalues of (SL), (α), (β) and if conditions (5), (6) of Exercise 2 of section 12.1 are satisfied, show that $\lambda_n \to \infty$ as $n \to \infty$.

Exercise 8 With the notation of Theorem 1, prove that a solution y of (SLN), (α), (β) may be written

$$y \; = \; \sum_{n=1}^{\infty} d_n y_n$$

where

$$d_n \; = \; \frac{1}{\lambda - \lambda_n}\int_a^b fy_n, \qquad (n \geq 1)$$

whenever λ is not one of the eigenvalues λ_n.

[HINT: Multiply (SLN) through by y_n and integrate by parts.]

Exercise 9 For each of the following Sturm–Liouville equations

$$(py')' + qy + \lambda\rho y \; = \; 0, \qquad (x \in [1, 2])$$

subject to boundary conditions

$$y(1) \; = \; y(2) \; = \; 0,$$

find the eigenvalues $\lambda_1, \lambda_2, \ldots$ and the corresponding eigenfunctions $y_1, y_2, \ldots$ and use Exercise 8 to determine a solution of the non-homogeneous equation

$$(py')' + qy + \lambda \rho y \;=\; f, \qquad (x \in [1, 2])$$

when the positive constant λ is not an eigenvalue, f is continuous and

(a) $p(x) = 1/x$, $q = 0$ and $\rho(x) = 1/x^3$,   or

(b) $p(x) = x^3$ and $q(x) = \rho(x) = x$.

12.3   Derivation from a variational principle

Before proceeding to the Sturm–Liouville equation, we give an extension of the fundamental result of the Calculus of Variations to take care of a wider class of boundary conditions.

Proposition 2 Suppose that $f = f(x, y)$ and $g = g(x, y)$ are continuously differentiable. Then, a twice continuously differentiable extremal $y = u(x)$ for

$$I \;\equiv\; \int_a^b F(x, y(x), y'(x))\, dx - f(a, y(a)) + g(b, y(b)),$$

where F is also twice continuously differentiable, must satisfy Euler's equation

$$\frac{d}{dx} F_{y'} - F_y \;=\; 0,$$

together with the boundary conditions

$$F_{y'}(a, y(a), y'(a)) + f_y(a, y(a)) \;=\; 0,$$

$$F_{y'}(b, y(b), y'(b)) + g_y(b, y(b)) \;=\; 0.$$

Proof   Writing $y = u + \alpha\eta$, where $\eta : [a, b] \to \mathbb{R}$ is twice continuously differentiable,

$$I \;=\; I(\alpha)$$

$$=\; \int_a^b F(x, u(x) + \alpha\eta(x), u'(x) + \alpha\eta'(x))\, dx - f(a, u(a) + \alpha\eta(a)) + g(b, u(b) + \alpha\eta(b)).$$

As the extremal $y = u$ corresponds to $I'(0) = 0$, it must satisfy, on integrating by parts,

$$(10) \qquad \int_a^b \eta\{F_y - \frac{d}{dx}F_{y'}\} + \left[\eta F_{y'}\right]_a^b - \eta(a)f_y(a, u(a)) + \eta(b)g_y(b, u(b)) \;=\; 0.$$

As with the fundamental theory in Chapter 11, if we choose $\eta(a) = \eta(b) = 0$, $y = u$ must satisfy the Euler equation

$$\frac{d}{dx}F_{y'} - F_y \;=\; 0.$$

The equation (10) then reduces to

$$\eta(a)\{F_{y'}(a, u(a), u'(a)) + f_y(a, u(a))\} \;=\; \eta(b)\{F_{y'}(b, u(b), u'(b)) + g_y(b, u(b))\}.$$

Choosing η in turn to satisfy $\eta(a) \neq 0$, $\eta(b) = 0$ and then $\eta(a) = 0$, $\eta(b) \neq 0$, the required boundary conditions are met. $\qquad\square$

We now assume that the constants B_1, B_2 appearing in the homogeneous boundary conditions (α), (β) are both non-zero. The variational principle we shall invoke is that a stationary value be given to the integral

$$I_1 \;\equiv\; \int_a^b \{p(y')^2 - qy^2\} - \frac{A_1}{B_1}p(a)(y(a))^2 + \frac{A_2}{B_2}p(b)(y(b))^2,$$

when subject to the normalisation condition

$$J_1 \;\equiv\; \int_a^b \rho y^2 \;=\; 1.$$

The integral constraint theory (section 11.8) shows that this principle is equivalent to that whereby there is a constant λ for which $I \equiv I_1 - \lambda J_1$ achieves a stationary value. With

$$F \;\equiv\; p(y')^2 - qy^2 - \lambda\rho y^2$$

and

$$f \;=\; \frac{A_1}{B_1}py^2, \qquad g \;=\; \frac{A_2}{B_2}py^2,$$

Proposition 2 shows that an extremal must satisfy (SL), together with the boundary conditions (α), (β).

We may, of course, achieve the same result by applying variational methods to determine the minimum λ of the quotient I_1/J_1 (see section 11.8, Note 5).

Exercise 10 In the case $B_1 = B_2 = 0$, show that (SL), (α), (β) follow from a variational principle.

Exercise 11 Use the results of this section to find an extremal corresponding to the problem in Exercise 7(b) of Chapter 11.

Exercise 12 Give a variational principle which gives rise to the following differential equation and boundary conditions:

$$(s(x)y'')'' + (p(x)y')' + (q(x) - \lambda\rho(x))y = 0, \qquad (x \in [a, b])$$

subject to

$$y(a) = y_0, \qquad y'(a) = y_1, \qquad (y_0, y_1 \text{ constants})$$

and

$$(sy'')' + py' = sy'' = 0 \quad \text{at } x = b.$$

State carefully (non-trivial) sufficient conditions for your result to hold.

12.4 Some singular equations

A number of famous functions, which arise classically in models of physical situations, derive naturally from equations of Sturm–Liouville type. Frequently, however, these equations are *singular* insofar as the function p may vanish at certain points, often end-points, of the domain in which a solution is required, or the boundary conditions may not be homogeneous or periodic there. We give here, in the text and in the exercises, some examples of the most well-known equations and functions. Amongst these, the reader will already have become acquainted with Legendre polynomials and Bessel functions in Chapter 7. Complete proofs are not provided here, but further details will be found in the context of series solutions in the next chapter.

Legendre polynomials For the Sturm–Liouville equation

$$(11) \qquad ((1 - x^2)y')' + \lambda y = 0, \qquad (x \in [-1, 1])$$

$p(x) = 1 - x^2$ has zeros in the domain $[-1, 1]$. Appropriate boundary conditions are that the solution be finite at $x = \pm 1$. Polynomial eigenfunctions $y = P_n(x)$, given by Rodrigues' Formula,

$$(12) \qquad P_n(x) = \frac{1}{2^n n!} \frac{d^n}{dx^n} (x^2 - 1)^n, \qquad (n = 0, 1, 2 \ldots)$$

correspond to eigenvalues $\lambda = n(n+1)$. These Legendre polynomials already form an orthogonal sequence (as $\rho = 1$).

Bessel functions For the equation

(13) $$x^2 y'' + xy' - n^2 y + \lambda x^2 y \;=\; 0, \qquad (x \in [0,1])$$

appropriate boundary conditions are

$$y(0) \text{ finite}, \qquad y(1) \;=\; 0.$$

The equation may be re-written, when $x \neq 0$, in Sturm–Liouville form:

(13′) $$(xy')' - \frac{n^2}{x} y + \lambda xy \;=\; 0,$$

where $p(x) = x$ is zero at the origin. Eigenfunctions are the Bessel functions $y = J_n(\sqrt{\lambda}x)$ corresponding to eigenvalues λ determined from the boundary condition at $x = 1$, where

$$J_n(t) \;=\; \frac{t^n}{2^n n!} \left\{ 1 - \frac{t^2}{2(2n+2)} + \frac{t^4}{2.4(2n+2)(2n+4)} - \cdots \right\}$$

is a series convergent for all values of t (see Chapter 13). As $\rho(x) = x$, the associated orthogonal functions $z = z_n(x)$ are given by

$$z_n(x) \;=\; \sqrt{x} J_n(\sqrt{\lambda}x), \qquad (x \in [0,1])$$

and satisfy

$$x^2 z'' - (n^2 - \tfrac{1}{4})z + \lambda x^2 z \;=\; 0, \qquad (x \in [0,1])$$

a particular case of equation (3) of section 12.1.

Hermite polynomials The differential equation

(14) $$y'' - 2xy' + \lambda y \;=\; 0, \qquad (-\infty < x < \infty)$$

may be re-written

(14′) $$(e^{-x^2} y')' + \lambda e^{-x^2} y \;=\; 0,$$

where $p(x) = e^{-x^2}$ tends to zero as $x \to \pm\infty$. Appropriate boundary conditions this time are

$$y(x) \;=\; O(x^k), \text{ for some positive integer } k, \text{ as } x \to \pm\infty.$$

Polynomial eigenfunctions $y = H_n(x)$ are given by

$$(15) \qquad H_n(x) \; = \; (-1)^n e^{x^2} \frac{d^n}{dx^n} (e^{-x^2}), \qquad (-\infty < x < \infty)$$

and correspond to eigenvalues $\lambda = 2n$, where n is a non-negative integer. As $\rho(x) = e^{-x^2}$, the associated orthogonal functions $z = z_n(x)$ are given by

$$z_n(x) \; = \; H_n(x) e^{-x^2/2}$$

and satisfy

$$z'' + (1 - x^2)z + \lambda z \; = \; 0.$$

Laguerre polynomials The differential equation

$$(16) \qquad xy'' + (1 - x)y' + \lambda y \; = \; 0, \qquad (0 \leq x < \infty)$$

has Sturm–Liouville form

$$(16') \qquad (xe^{-x}y')' + \lambda e^{-x}y \; = \; 0$$

and, since

$$p(x) \; = \; xe^{-x} \to 0 \text{ as } x \to 0 \text{ and as } x \to \infty,$$

appropriate boundary conditions turn out to be that y is finite at $x = 0$ and

$$y(x) \; = \; O(x^k), \text{ for some positive integer } k, \text{ as } x \to \infty.$$

Polynomial eigenfunctions $y = L_n(x)$ are given by

$$(17) \qquad L_n(x) \; = \; e^x \frac{d^n}{dx^n} (x^n e^{-x}), \qquad (0 \leq x < \infty)$$

and correspond to eigenvalues $\lambda = n$, where n is a positive integer. (This is discussed again, in the context of the Laplace transform, in Examples 7 and 10 in Chapter 14.) This time, $\rho(x) = e^{-x}$, and the associated orthogonal functions $z = z_n(x)$ are given by

$$z_n(x) \; = \; L_n(x) e^{-x/2}, \qquad (0 \leq x < \infty)$$

and satisfy the Sturm–Liouville equation

$$(xz')' + \tfrac{1}{4}(2 - x) + \lambda z \; = \; 0.$$

This equation is again singular.

Exercise 13 Deduce (15) from the formula for the generating function

$$h(x,t) \equiv \exp(-t^2 + 2tx) = \sum_{n=0}^{\infty} \frac{H_n(x)}{n!} t^n.$$

Show that

$$h_x = 2th, \qquad h_t + 2(t - x)h = 0$$

and hence, that

(18) $$\qquad H_n' = 2nH_{n-1}, \qquad H_{n+1} - 2xH_n + 2nH_{n-1} = 0,$$

for $n \geq 1$. Use (18) to show that $y = H_n(x)$ satisfies (14) with $\lambda = 2n$, for every non-negative integer n.

Exercise 14 (a) Use the formula

$$p(x,t) \equiv (1 - 2xt + t^2)^{-\frac{1}{2}} = \sum_{n=0}^{\infty} P_n(x) t^n,$$

which defines the generating function for the Legendre polynomials $P_n(x)$, to derive the recurrence formula:

$$(n + 1)P_{n+1}(x) - (2n + 1)xP_n(x) + nP_{n-1}(x) = 0.$$

(b) If $w(x) = (x^2 - 1)^n$, show that

(19) $$\qquad (x^2 - 1)w' = 2nxw,$$

where $w^{(n)}(x) = 2^n n! P_n(x)$ and $P_n(x)$ is given by (12). By differentiating (19) $(n+1)$-times with respect to x, show that P_n satisfies (11) with $\lambda = n(n + 1)$.

Exercise 15 Show that the generating function l for the Laguerre polynomials, defined by

$$l(x,t) \equiv \frac{\exp(-xt/(1 - t))}{1 - t} = \sum_{n=0}^{\infty} \frac{L_n(x)}{n!} t^n,$$

gives rise to

$$(1 - t)^2 l_t = (1 - t - x)l, \qquad (1 - t)l_x = -tl,$$

and deduce the recurrence relations

$$L_{n+1} - (2n + 1 - x)L_n + n^2 L_{n-1} = 0,$$

$$L_n' - nL_{n-1}' = -nL_{n-1},$$

$$xL_n' = nL_n - n^2 L_{n-1},$$

for $n \geq 1$. Hence, show that L_n satisfies (16) with $\lambda = n$.

12.5 The Rayleigh–Ritz method

This method has been used with great success to find, successively, better and better approximations to functions sought to give stationary values to the variational integrals of the type considered in Chapter 11 and in particular in the previous section of the current chapter. The techniques involved reduce to using the differential calculus to find stationary values of functions of one or more real variables, replacing the need to solve an Euler equation.

To be more concrete, let us suppose we wish to find an appropriately differentiable function y which gives a minimum to $K = K(y)$, where K is an integral $I = I(y)$ or a quotient I/J of integrals $I = I(y)$, $J = J(y)$. The Rayleigh–Ritz method attempts to approximate y with functions of the form

$$y_n = y_0 + \sum_{i=1}^{n} c_i \varphi_i,$$

where y_0 is a first approximation to y, the c_i's are real parameters, and the functions y_0 and φ_i $(i \geq 1)$ are suitably differentiable. One then minimises the functions

$$K(c_1, \ldots, c_n) \equiv K(y_n)$$

by varying the parameters c_i. In this way, we can approach the greatest lower bound of the $K(y_n)$. However, deeper analysis than we will attempt here is usually necessary to discuss, in any particular case, whether the method will actually yield a minimum to K, even if the sequence (φ_i) is an orthonormal set consisting of all eigenfunctions of the corresponding (Euler) differential equation.

Example 1 We consider the problem of minimising the integral

$$I(y) = \int_0^1 ((y')^2 + 2xy) \, dx$$

subject to boundary conditions

$$y(0) = 0, \qquad y(1) = k.$$

This corresponds to small vibrations of a string, with linearly varying loading in the y-direction. We choose the approximations

$$y_n(x) = kx + x(1-x) \sum_{i=1}^{n} c_i x^{i-1}.$$

Here, we have taken $y_0(x) = kx$ and all the $y_n(x)$ $(n \geq 1)$ to satisfy both boundary conditions. We now seek c_1 to minimise $I(y_1)$ for the one-parameter approximations given by

$$y_1(x) = kx + c_1 x(1-x).$$

Differentiating,

$$y_1'(x) = k + c_1(1-2x),$$

so that

$$I = I(y_1) = \int_0^1 \left\{ (k + c_1(1-2x))^2 + 2x(kx + c_1 x(1-x)) \right\} dx$$

$$= \frac{c_1^2}{3} + \frac{c_1}{6} + k^2 + \frac{2k}{3}.$$

To minimise with respect to c_1, we differentiate:

$$\frac{dI}{dc_1} = 0, \quad \text{when } c_1 = -\frac{1}{4},$$

giving

$$y_1 = kx - \frac{x(1-x)}{4}.$$

The corresponding Euler equation is

$$y'' = x$$

which, once one applies the boundary conditions, yields the exact solution

$$y = kx - \frac{x(1-x^2)}{6}.$$

This agrees with the approximation y_1 at the mid-point $x = 1/2$.   □

Exercise 16 In respect of Example 1, perform the calculations to determine the minimising function

$$y_2(x) = kx + c_1 x(1-x) + c_2 x^2(1-x)$$

and compare your answer with the exact solution.

We noted at the end of section 12.3 that the Sturm–Liouville equation is the Euler equation corresponding to finding a stationary value to the quotient I_1/J_1, where

$$I_1 \equiv \int_a^b \{p(y')^2 - qy^2\} - \frac{A_1}{B_1} p(a)(y(a))^2 + \frac{A_2}{B_2} p(b)(y(b))^2$$

and

$$J_1 \equiv \int_a^b \rho y^2 = 1,$$

and the quotient satisfies the boundary conditions

$$A_1 y(a) + B_1 y'(a) = A_2 y(b) + B_2 y'(b) = 0, \qquad (B_1 \neq 0, \, B_2 \neq 0).$$

Integrating the first term of the integrand in I_1 by parts, and using the boundary conditions,

$$\frac{I_1}{J_1} = \frac{-\int_a^b y\{(py')' + qy\}}{\int_a^b \rho y^2}.$$

At a stationary value of I_1/J_1, y is an eigenfunction of

$$(py')' + qy = -\lambda \rho y$$

and hence $I_1/J_1 = \lambda$, the corresponding eigenvalue. So, finding a minimum for I_1/J_1 is finding the lowest eigenvalue of the corresponding Sturm–Liouville equation (subject to the same boundary conditions).

Exercise 17 Use the Rayleigh–Ritz method and the one-parameter approximation

$$y_1(x) = x(1 - x) + c_1 x^2 (1 - x)^2$$

to estimate the eigenvalue λ of the equation

$$y'' + \lambda y = 0$$

when subject to boundary conditions

$$y(0) = y(1) = 0,$$

by considering the quotient

$$\int_0^1 (y')^2 \left/ \int_0^1 y^2 \right. .$$

13 Series Solutions

In this chapter, we investigate a special technique which provides solutions to a wide class of differential equations. Again, we concentrate on the homogeneous linear second-order equation

$$(1) \qquad\qquad p_2 y'' + p_1 y' + p_0 y = 0,$$

where p_0, p_1, p_2 are continuous functions *which we shall suppose throughout this chapter to have no common zeros*. We do not, for the moment, specify the domain of these functions, though it will in general be taken to be an interval, finite or infinite.

Hitherto, we have often specified that the function p_2 is never zero. A point x_0 in the domain of p_2 for which $p_2(x_0) \neq 0$ is called an *ordinary point* of the differential equation (1). We shall discuss how solutions y to (1) in a neighbourhood of an ordinary point x_0 can, in rather general circumstances, be represented as a *power series*

$$y = \sum_{n=0}^{\infty} a_n (x - x_0)^n.$$

For example, the geometric series

$$(2) \qquad\qquad y = \sum_{n=0}^{\infty} x^n,$$

convergent for $|x| < 1$, is the solution of the equation

$$(1 - x^2) y'' - 4xy' - 2y = 0,$$

satisfying $y(0) = y'(0) = 1$, in the open interval $(-1, 1)$ about the ordinary point 0. The reader can easily check this by first noting that the series solution (2) can be represented in $(-1, 1)$ in the *closed* (or *finite*) form

$$y = \frac{1}{1-x}.$$

However, there are many interesting and important equations where solutions are required near points which are not ordinary. A point x_0 is a *singular point* of (1) if $p_2(x_0) = 0$. Now, in general, power series do not suffice as solutions. However, when x_0 is a *regular singular point* (defined in section 13.2), a solution in a neighbourhood of x_0 can be found by considering *extended power series* in the form

$$y = (x - x_0)^c. \sum_{n=0}^{\infty} a_n(x - x_0)^n = \sum_{n=0}^{\infty} a_n(x - x_0)^{n+c},$$

where c is a real number. (In this connection, the reader will recall the definition

$$x^c \equiv e^{c \log x}$$

which is in general only a real number if $x > 0$.) A classical example is Bessel's equation of order $\frac{1}{2}$,

(3) $$x^2 y'' + xy' + (x^2 - \tfrac{1}{4}) = 0,$$

which has a (regular) singular point at $x = 0$ and the general solution

$$y = Ax^{-\frac{1}{2}} \sum_{n=0}^{\infty} \frac{(-1)^n x^{2n+1}}{(2n+1)!} + Bx^{-\frac{1}{2}} \sum_{n=0}^{\infty} \frac{(-1)^n x^{2n}}{(2n)!}, \qquad (x \neq 0)$$

where A and B are constants. This solution can be expressed in the finite form

$$y = x^{-\frac{1}{2}}(A \sin x + B \cos x)$$

and solves (3) for every non-zero real (and complex) value of x. We return to this in Example 5 of section 13.5.

Some of the equations we shall consider will be of the important Sturm–Liouville type (see Chapter 12). We shall in particular be discussing solutions of the Legendre and Bessel equations further.

An appendix will extend the discussion to the case of complex-valued functions of a single complex variable. This turns out to be the ideal unifying context for the theory of series solutions, but requires more sophisticated techniques, including a discussion of branch points of complex functions.

Those readers whose primary interest is in technique, rather than theory, and who are familiar with the elementary theory of power series, may (after first familiarising themselves with the definition of regular singular point in section 13.2) turn directly to section 13.5, supplemented by section 13.6.

A summary of the techniques employed in this chapter can be found in section 13.7. Much of the work in this chapter was initiated by Fuchs and Frobenius.

13.1 Power series and analytic functions

In this section, we review, without the inclusion of proofs, some basic results about (real) *power series*, that is, series of the form

$$\sum_{n=0}^{\infty} a_n (x - x_0)^n,$$

where x is real and x_0 and all a_n are real constants. This is a *power series about x_0*. For simplicity of exposition, both in this section and in the rest of this chapter, we shall deal in the most part with the case $x_0 = 0$; that is, with the case of power series

$$\sum_{n=0}^{\infty} a_n x^n$$

about 0. Note that, in general, discussion of power series about x_0 can be conveniently reduced to the case $x_0 = 0$ by the substitution $x' = x - x_0$. Then, $dx'/dx = 1$ implies, for functions $y = y(x)$, that

$$\frac{dy}{dx} = \frac{dy}{dx'} \quad \text{and} \quad \frac{d^2y}{dx^2} = \frac{d^2y}{dx'^2}.$$

Thus the reduction to $x_0 = 0$ provides the slightest of technical obstacles.

The classical results on power series we record here for future use are contained in the following theorems.

Theorem 1 For the power series

$$\sum_{n=0}^{\infty} a_n x^n,$$

just one of the following three statements is true:

  (i) the series converges only when $x = 0$,

  (ii) the series converges for every real x,

  (iii) there exists a positive real number R such that the series converges for $|x| < R$ and diverges for $|x| > R$.

The number R given in Theorem 1(iii) is called the *radius of convergence*, and the set

$$\{x \in \mathbb{R} : |x| < R\},$$

the *interval of convergence*, of the power series. By convention, we say that when case (i) of Theorem 1 occurs, the power series has *radius of convergence zero*, and when case (ii) occurs, *radius of convergence infinity*. Thus, Theorem 1 states that every power series has a radius of convergence R, where $0 \le R \le \infty$. If the limit exists, the radius of convergence can be determined from the following formula

(4)
$$R = \lim_{n \to \infty} \left| \frac{a_{n-1}}{a_n} \right|.$$

Theorem 2 Suppose that R is the radius of convergence of the power series

$$f(x) = \sum_{n=0}^{\infty} a_n x^n.$$

(a) The function $f : (-R, R) \to \mathbb{R}$ is differentiable on the open interval $(-R, R)$ and its derivative

$$f'(x) = \sum_{n=1}^{\infty} n a_n x^{n-1}$$

also has radius of convergence R. Hence f is n-times differentiable on $(-R, R)$ and $f^{(n)}$ has a power series representation with radius of convergence R, for every positive integer n. As the derivative of $a_n x^n$ is $n a_n x^{n-1}$, we may describe the situation by saying that the power series is *n-times differentiable term-by-term* within its interval of convergence, for every positive integer n.

(b) The function $f : (-R, R) \to \mathbb{R}$ is integrable on $(-R, R)$ and

$$\int^x f = \sum_{n=0}^{\infty} \frac{a_n}{n+1} x^{n+1}$$

which also has radius of convergence R. Hence, f is *n-times integrable term-by-term* within its interval of convergence, for every positive integer n.

Theorem 3 Suppose that R and, respectively, S are the radii of convergence of the power series

$$f(x) = \sum_{n=0}^{\infty} a_n x^n, \text{ respectively, } g(x) = \sum_{n=0}^{\infty} b_n x^n.$$

Then,

(a) $f(x) + g(x) = \sum_{n=0}^{\infty} (a_n + b_n) x^n$, whenever $|x| < \min(R, S)$,

(b) $f(x) \cdot g(x) = \sum_{n=0}^{\infty} c_n x^n$, whenever $|x| < \min(R, S)$, where

$$c_n = \sum_{r=0}^{n} a_r b_{n-r}, \qquad (n \geq 0)$$

(c) provided $g(0) = b_0 \neq 0$, $f(x)/g(x)$ may be expressed as a power series about 0, with non-zero radius of convergence,

(d) $f(x) = g(x)$ for $|x| < \min(R, S)$ implies $a_n = b_n$, for every $n \geq 0$, and

(e) in particular, if $f(x) = 0$ for $|x| < R$, then $a_n = 0$ for every $n \geq 0$.

In order to take advantage of these theorems, it is useful if certain combinations of the coefficient functions p_0, p_1, p_2 in our differential equation

(1) $$p_2 y'' + p_1 y' + p_0 y = 0$$

have power series representations. In this connection, it is convenient to make the following definition.

Definition 1 The real-valued function f of a real variable is *analytic at x_0* in $\mathbb{R}$ if it can be expressed in the form of a power series about x_0,

$$(5) \qquad\qquad f(x) \;=\; \sum_{n=0}^{\infty} a_n (x - x_0)^n,$$

convergent in the open interval $(x_0 - R, \, x_0 + R)$ for some strictly positive R, or for all real x.

Notes

(a) The definition insists that the radius of convergence of the power series is non-zero.

(b) Theorem 2 tells us that a function f, analytic at x_0, is n-times differentiable in a neighbourhood of x_0, for every positive integer n, that is, *infinitely differentiable at x_0*. Further, by differentiating formula (5) term-by-term n-times and putting $x = x_0$, the reader will see immediately that $f^{(n)}(x_0) = n! a_n$ and hence that (5) may be re-written

$$(6) \qquad\qquad f(x) \;=\; \sum_{n=0}^{\infty} \frac{f^{(n)}(x_0)}{n!} \, (x - x_0)^n.$$

Conversely, using Taylor's Theorem the reader may quickly deduce that if f is infinitely differentiable in a neighbourhood of x_0, then formula (6) holds there and hence f is analytic at x_0.

(c) The above definition and all the theorems of this section have exact counterparts in the theory of complex-valued functions $f = f(z)$ of a single complex variable z, the interval of convergence being replaced by a disc of convergence $\{z \in \mathbb{C} : |z| < R\}$. Readers familiar with this theory will note that f is analytic at every point of an open set U if and only if it is differentiable on U. We return to complex functions in the appendix to this chapter.

13.2   Ordinary and regular singular points

The reader will recall from the introductory remarks to the chapter that x_0 is an *ordinary point* of the differential equation

$$(1) \qquad\qquad p_2 y'' + p_1 y' + p_0 y \;=\; 0$$

if $p_2(x_0) \neq 0$ and a *singular point* (or *singularity*) of (1), if it is not an ordinary point. We are now in a position to fulfill the promise made earlier, to define when such a singular point is regular. For $x \neq x_0$, let

(7) $$P_1(x) = (x - x_0) \cdot \frac{p_1(x)}{p_2(x)} \quad \text{and} \quad P_0(x) = (x - x_0)^2 \cdot \frac{p_0(x)}{p_2(x)}.$$

Definition 2 The point x_0 is a *regular singular point* of (1) if

(i) x_0 is a singular point,
(ii) $\lim_{x \to x_0} P_1(x)$ and $\lim_{x \to x_0} P_0(x)$ both exist,
(iii) the functions P_1 and P_0, defined by letting $P_1(x)$ and $P_0(x)$ be the values given by (7) when $x \neq x_0$ and

$$P_1(x_0) = \lim_{x \to x_0} P_1(x) \quad \text{and} \quad P_0(x_0) = \lim_{x \to x_0} P_0(x),$$

are analytic at x_0.

A singular point which is not regular is an *irregular singular point*.

Notes

(a) For complex-valued functions of a complex variable, (ii) and (iii) may be replaced by saying that x_0 is 'at worst' a simple pole of p_1/p_2 and a double pole of p_0/p_2.

(b) If p_2 is a polynomial, in order that a singular point x_0 is regular, it is sufficient for 'P_0, P_1 analytic at x_0' to be replaced by 'p_0, p_1 analytic at x_0'. We leave it to the interested reader to check this. Of course, p_0 and p_1 are analytic in the special case when they themselves are also polynomials.

Exercise 1 For the following equations, determine which points are ordinary, which are regular singular, and which are neither:

(a) $$x(1 + x)y'' + (\alpha + 4x)y' + 2y = 0,$$

where α is a real constant,

(b) $$x^3 y'' + xy' - y = 0.$$

[We shall return to both these equations later (see Exercises 10 and 14).]

Exercise 2 Suppose that $x_0 = 0$ is a regular singular point of (1) and that p_0, p_1, p_2 are all polynomials in x. Suppose further that p_i and p_2 have no common zeros ($i = 0, 1$). Prove that the radius of convergence of the functions P_0, P_1 of Definition 2 is the least modulus of the non-zero zeros of p_2. Deduce that, when p_0, p_1, p_2 are polynomials, a singular point is regular if (ii) alone is satisfied.

[HINT: Express p_2 in terms of its linear factors and consider each factor's binomial expansion.]

Exercise 3 With the above notation, show that 'P_0, P_1 analytic at x_0' can be replaced by 'p_0, p_1 analytic at x_0', when x_0 is a regular singular point of (1) and p_2 is a polynomial.

[HINT: Note that if, for $i = 0, 1, 2$,

$$p_i(x) = (x - x_0)^{n_i} \varphi_i(x), \qquad (\varphi_i(0) \neq 0)$$

where φ_2 is a polynomial and φ_0, φ_1 are power series, then

$$P_i(x) = (x - x_0)^{s_i} \cdot \frac{p_i(x)}{p_2(x)} = (x - x_0)^{n_i - n_2 + s_i} \cdot \frac{\varphi_i(x)}{\varphi_2(x)}, \qquad (i = 0, 1)$$

where $s_1 = 1$ and $s_0 = 2$. For $\lim_{x \to x_0} P_i(x)$ to exist, $n_i - n_2 + s_i \geq 0$. Now apply Theorem 3(c).]

13.3   Power series solutions near an ordinary point

Theorem 4 of Chapter 2 already establishes that if x_0 is an ordinary point of the differential equation

(1) $$p_2 y'' + p_1 y' + p_0 y = 0$$

and the quotient functions p_1/p_2 and p_0/p_2 are both analytic at x_0, then (1) has a unique solution in a neighbourhood of x_0, satisfying $y(x_0) = c$, $y'(x_0) = d$, where c, d are given constants. In fact, Picard iteration, which forms the basis of our proof of that theorem, provides a power series representation of the solution, convergent where both p_1/p_2 and p_0/p_2 are. Thus, the solution is also analytic at x_0.

Rather than belabour the above idea, we shall later in the chapter and in the context of 'extended' power series solutions near a regular singular point prove a more general result. In this section, we shall encourage the reader to start to sharpen technique by developing a scheme for determining a power series solution and considering some specific examples.

The scheme we now introduce, which rests on the results quoted in section 13.1, runs as follows. For simplicity, we take $x_0 = 0$ and p_0, p_1, p_2 to be power series.

(i) Use Theorem 1 to seek a power series solution

$$y = \sum_{n=0}^{\infty} a_n x^n$$

of equation (1), convergent for $|x| < R$.

(ii) Use Theorem 2 to substitute in (1) for y, for

$$y' = \sum_{n=1}^{\infty} n a_n x^{n-1}$$

and for

$$y'' = \sum_{n=2}^{\infty} n(n-1) a_n x^{n-2}$$

when $|x| < R$.

(iii) Use Theorem 3(b) to multiply p_0 and y, p_1 and y', p_2 and y'', again for $|x| < R$.

(iv) Use Theorem 3(a) to re-organise the terms, so that (1) may be re-written in the form

$$\sum_{n=0}^{\infty} d_n x^n = 0,$$

for $|x| < R$. Note that the d_n's must be linear functions of the a_n's.

(v) Use Theorem 3(e) to equate each d_n to zero.

(vi) If possible, solve the equations $d_n = 0$, to find the a_n's (in terms of, as it turns out, at most two undetermined a_n's) and hence write down solutions in the form $\sum a_n x^n$.

(vii) Use (4) to find the radius/radii of convergence of the series thus determined.

It will be apparent to the reader that our initial examples can most easily be solved using other methods. They do however provide a useful introduction to the use of series.

Example 1 Use power series about the origin to solve

(a) $\quad y'' = 0,$

(b) $\quad y'' + y = 0.$

For both equations, we try the series for y given in (i) above and substitute for y'' given in (ii).

Equation (a) becomes

$$\sum_{n=2}^{\infty} n(n-1)a_n x^{n-2} = 0$$

and hence, as both n and $n-1$ are non-zero for $n \geq 2$, we have (using (v)) that $a_n = 0$ for $n \geq 2$. We are thus led to the solution

$$y(x) = a_0 + a_1 x,$$

where the undetermined constants a_0 and a_1 may be found if boundary or initial conditions are given. The solution is valid everywhere $(R = \infty)$.

On the other hand, equation (b) becomes

$$\sum_{n=2}^{\infty} n(n-1)a_n x^{n-2} + \sum_{n=0}^{\infty} a_n x^n = 0$$

or, alternatively, following (iv), as

$$\sum_{n=2}^{\infty} \{n(n-1)a_n + a_{n-2}\}x^{n-2} = 0.$$

So, using (v),

(8) $$a_n = -\frac{1}{n(n-1)} a_{n-2}, \qquad (n \geq 2).$$

We now consider separately the two cases n even and n odd. When $n = 2m$,

$$a_{2m} = -\frac{1}{2m(2m-1)} a_{2(m-1)}$$

$$= \frac{1}{2m(2m-1)(2m-2)(2m-3)} a_{2(m-2)} = \cdots$$

$$= \frac{(-1)^m}{(2m)!} a_0, \qquad\qquad\qquad (m \geq 1)$$

whereas, when $n = 2m + 1$,

$$a_{2m+1} = -\frac{1}{(2m+1)(2m)} a_{2(m-1)+1}$$

$$= \frac{1}{(2m+1)2m(2m-1)(2m-2)} a_{2(m-2)+1} = \cdots$$

$$= \frac{(-1)^m}{(2m+1)!} a_1, \qquad\qquad\qquad (m \geq 1).$$

So,

$$y = a_0 \sum_{m=0}^{\infty} \frac{(-1)^m}{(2m)!} x^{2m} + a_1 \sum_{m=0}^{\infty} \frac{(-1)^m}{(2m+1)!} x^{2m+1},$$

a solution in terms of the undetermined constants a_0 and a_1. Finally, we use (4) to determine the radii of convergence of the two series. The radius of convergence of the series multiplying a_0 is

$$\lim_{m\to\infty} \left| \frac{(-1)^{m-1}}{(2(m-1))!} \cdot \frac{(2m)!}{(-1)^m} \right| = \lim_{m\to\infty} 2m$$

which is infinite, as is (similarly) the radius of convergence of the other series. Thus, the solution is valid everywhere. Of course it can be expressed in the closed form

$$y = a_0 \cos x + a_1 \sin x. \qquad\qquad \square$$

Example 2 (Legendre's equation - see sections 7.4 and 10.4) Solve

$$(1 - x^2)y'' - 2xy' + k(k+1)y = 0,$$

where k is a constant, in a neighbourhood of the origin.

The origin is clearly an ordinary point. Substituting for y, y', y'', in terms of their power series, supposed convergent for $|x| < R$, we get

$$(1 - x^2) \sum_{n=2}^{\infty} n(n-1)a_n x^{n-2} - 2x \sum_{n=1}^{\infty} na_n x^{n-1} + k(k+1) \sum_{n=0}^{\infty} a_n x^n = 0$$

and hence

$$\sum_{n=2}^{\infty} n(n-1)a_n x^{n-2} - \sum_{n=0}^{\infty} \{n(n+1) + 2n - k(k+1)\}a_n x^n = 0.$$

So, for $n \geq 2$,

(9) $n(n-1)a_n = \{(n-2)(n-1) - k(k+1)\}a_{n-2}$

and hence

$$a_n = -\frac{(k-n+2)(k+n-1)}{n(n-1)} a_{n-2}, \qquad (n \geq 2).$$

As in Example 1(b), we consider n even and n odd separately. For $n = 2m$,

$$a_{2m} = -\frac{(k-2m+2)(k+2m-1)}{2m(2m-1)} a_{2(m-1)} \qquad (m \geq 1)$$

$$= \ldots$$

$$= \frac{(-1)^m}{(2m)!} b_{k,m} c_{k,m} a_0,$$

where

$$b_{k,m} = (k-2m+2)(k-2m+4)\ldots(k-4)(k-2)k,$$

$$c_{k,m} = (k+2m-1)(k+2m-3)\ldots(k+3)(k+1).$$

Noting that $b_{k,n+p} = 0$ for $p \geq 1$ when $k = 2n$, we see that the corresponding solution

$$y_1(x) = a_0 \left(1 + \sum_{m=1}^{\infty} \frac{(-1)^m}{(2m)!} b_{k,m} c_{k,m} x^{2m} \right)$$

reduces to a polynomial whenever k is a non-negative even integer. The *Legendre polynomial* $P_n = P_n(x)$, corresponding to $k = n$, is this polynomial solution once one chooses a_0 to ensure that $P_n(1) = 1$. The reader will easily check that $P_0(x) = 1$ and $P_2(x) = \frac{1}{2}(3x^2 - 1)$, in agreement with (42) of Chapter 7.

We leave the similar consideration of the case n odd as an exercise.   □

Notes

(a) Once one (power) series solution of (1) has been located, a second solution can be found by the 'method of reduction of order', as given in (5) of the Appendix. This is particularly efficient if the first solution can be expressed in closed form.

(b) The reader will have noticed that in the above examples, the recurrence relations relating the coefficients (for example in (8) and (9)) fortuitously contained only two terms (in the above, only a_n and a_{n-2}). This need not by any means be always the case, and when there are three or more terms, it may not be possible to solve the recurrence relation explicitly. We give below (in Exercise 12) a second-order differential equation, which gives rise to a three-term relation, which can be explicitly solved.

Exercise 4 Find two linearly independent power series solutions to the equation

$$y'' - y = 0,$$

find their radii of convergence and express the solutions in closed form.

Exercise 5 Find two linearly independent power series solutions of *Airy's equation*

$$y'' + xy = 0.$$

Show that these two solutions can be written as constant multiples of

$$x^{\frac{1}{2}} J_{\pm\frac{1}{3}}\left(\tfrac{2}{3} x^{\frac{3}{2}}\right)$$

when the Bessel function J_ν of the first kind of order ν is given in the form

$$J_\nu(t) = \sum_{m=0}^{\infty} \frac{(-1)^m t^{2m+\nu}}{2^{2m+\nu} m! \, (\nu+m)!}.$$

[The factorial $(\nu+m)!$ can be defined to be $(\nu+m)(\nu+m-1)\ldots(\nu+1)$. The reader may care to show, by integrating by parts, that it is the quotient of gamma functions,

$$(\nu+m)! = \frac{\Gamma(\nu+m+1)}{\Gamma(\nu+1)},$$

where the *gamma function* $\Gamma = \Gamma(\alpha)$ is given by

$$\Gamma(\alpha) = \int_0^{\infty} e^{-t} t^{\alpha-1} \, dt \qquad (\alpha > 0).$$

We note that, as $\Gamma(1) = 1$, it follows that $\Gamma(m+1) = m!$.]

Exercise 6 Find a power series solution of the differential equation

$$y'' + e^{-x}y = 0$$

and show that it can be written in the form

$$y(x) = \sum_{n=0}^{\infty} \frac{(-1)^n b_n}{n!} x^n,$$

where b_n satisfies the recurrence relation

$$b_{n+2} = -\sum_{r=0}^{n} \binom{n}{r} b_{n-r}$$

and

$$\binom{n}{r} = \frac{n!}{r!(n-r)!}$$

is the binomial coefficient.

[You will need to use Theorem 3(b) of section 13.1 to multiply two power series together.]

Exercise 7 (Tchebycheff Polynomials) Find two linearly independent power series solutions about zero of the differential equation

$$(10) \qquad\qquad (1 - x^2)y'' - xy' + k^2 y = 0,$$

where k is an arbitrary real constant. Show that if k is a non-negative integer, there is a polynomial solution of degree n. Denoting the polynomial solution y of degree n satisfying $y(1) = 1$, the *Tchebycheff polynomial of degree n*, by $y = T_n(x)$, write down $T_n(x)$ for $n = 0, 1, 2, 3$. Verify that

$$T_n(x) = \cos(n \cos^{-1} x),$$

for these values of n.

Verify also that

$$y(x) = \sin(k \sin^{-1} x)$$

solves (10) and hence show that

$$\sin 10° = \frac{1}{2} \sum_{n=0}^{\infty} \frac{(3n)!}{(2n+1)!\, n!\, 3^{3n+1}}.$$

Note We could similarly have asked the reader to show that the Laguerre polynomials

$$L_n(x) = \frac{e^x}{n!} \frac{d^n}{dx^n} (x^n e^{-x}) \qquad (n = 0, 1, 2, \ldots)$$

satisfy the differential equation

$$xy'' + (1 - x)y' + ny = 0.$$

However, we shall leave this fact to our discussion of the Laplace transform in the next chapter. (See Examples 7 and 10 of Chapter 14 and also section 10.4.)

Exercise 8 (Legendre polynomials) (a) Perform the analysis, corresponding to our work for n even in Example 2, for the case n odd, $n = 2m + 1$. Find a series solution in the form

$$y_2(x) = a_1 \left(x + \sum_{m=1}^{\infty} a_m x^{2m+1} \right)$$

and show that, when k is the odd positive integer $2n + 1$, y_2 is a polynomial of degree $2n + 1$. Show further that, with the notation of Example 2,

$$P_1(x) = x \quad \text{and} \quad P_3(x) = \tfrac{1}{2}(5x^3 - 3x),$$

in agreement with (42) of Chapter 7.

(b) Using the method of reduction of order ((5) of the Appendix), find a solution to the Legendre equation with $k = 1$, independent of the function P_1 given in (a) above.

(c) (*Harder*) Prove that the Legendre polynomial of degree n can be expressed as

(11)
$$P_n(x) = \frac{1}{2^n} \sum_{m=0}^{M} \frac{(-1)^m (2n - 2m)!}{m! \, (n - m)! \, (n - 2m)!} x^{n-2m},$$

where the integer M is $\tfrac{1}{2}n$ or $\tfrac{1}{2}(n - 1)$ according as n is even or odd.

(d) Establish Rodrigues's Formula

$$P_n(x) = \frac{1}{2^n n!} \frac{d^n}{dx^n} (x^2 - 1)^n.$$

[HINT: Expand $(x^2 - 1)^n$ by using the binomial theorem, differentiate n times and compare with the identity (11) in (c) above.]

(e) Show that

(12)
$$\frac{1}{\sqrt{1 - 2xt + t^2}} = \sum_{n=0}^{\infty} P_n(x) t^n,$$

whenever $|t(2x - t)| < 1$. [The left-hand side of (12) is known as the *generating function* for the Legendre polynomials.]

13.4   Extended power series solutions near a regular singular point: theory

By an *extended power series about* x_0, we shall mean a power series multiplied by $(x-x_0)^c$ for some real number c; that is, a series in the form

$$(x - x_0)^c \sum_{n=0}^{\infty} a_n (x - x_0)^n \;=\; \sum_{n=0}^{\infty} a_n (x - x_0)^{n+c},$$

where x is a real variable and a_n $(n \geq 0)$, x_0 and c are real constants. Recalling that (by definition) $x^c = \exp(c \log x)$, in order that $(x - x_0)^c$ is always a well-defined real number, we must have that $x > x_0$. (The definition has already been used to write $(x - x_0)^c.(x - x_0)^n = (x - x_0)^{n+c}$.)

Using extended power series, rather than power series alone, much enlarges the scope of the scheme for solving the differential equations discussed in section 13.3, in particular to finding solutions in neighbourhoods of regular singular points. It thus allows us to solve many of the more important differential equations which occur in classical mathematical physics.

Notes

(a) Of course, we do not always need to use a logarithm to define $(x-x_0)^c$; for example, when c is a positive integer. But we wish to be in a position to allow c to be even an irrational number.

(b) We shall return later in the chapter to the possibility, by change of variable, of using extended power series to solve equation (1) when $x < x_0$.

For the rest of this section, we shall restrict ourselves to considering the case when $x_0 = 0$ is a regular singular point. Non-zero values of x_0 can be dealt with by making a linear change of variables, as indicated in the chapter's introductory remarks. The discussion above leads us to seek solutions to

(1) $$p_2 y'' + p_1 y' + p_0 y \;=\; 0$$

in the form

$$y(x) \;=\; x^c \sum_{n=0}^{\infty} a_n x^n \;=\; \sum_{n=0}^{\infty} a_n x^{n+c}, \qquad (a_0 \neq 0)$$

convergent in $0 < x < R$, where R is the radius of convergence of the power series $\sum_{n=0}^{\infty} a_n x^n$.

Since for any real c, $(d/dx)x^c = cx^{c-1}$,

$$y'(x) = cx^{c-1} \sum_{n=0}^{\infty} a_n x^n + x^c \sum_{n=0}^{\infty} n a_n x^{n-1}$$

$$= \sum_{n=0}^{\infty} (n+c) a_n x^{n+c-1},$$

the series being convergent for $0 < x < R$, as the power series are. Repeating the process,

$$y''(x) = \sum_{n=0}^{\infty} (n+c)(n+c-1) a_n x^{n+c-2},$$

convergent again for $0 < x < R$.

As the origin is a regular singular point, the functions P_1, P_0, defined by

$$P_1(x) = \frac{x p_1(x)}{p_2(x)}, \qquad P_0(x) = \frac{x^2 p_0(x)}{p_2(x)}$$

near x and for $x \neq 0$, and by their limits as $x \to 0$, have power series representations

$$P_1(x) = \sum_{n=0}^{\infty} b_n x^n, \qquad P_0(x) = \sum_{n=0}^{\infty} d_n x^n,$$

for $|x| < S$, some $S > 0$. Re-writing (1) as

$$y''(x) + \frac{1}{x} y'(x) \cdot \frac{x p_1(x)}{p_2(x)} + \frac{1}{x^2} y(x) \cdot \frac{x^2 p_0(x)}{p_2(x)} = 0,$$

for $0 < x < R$, and substituting for the series involved, we see that for $0 < x < T$, where $T \equiv \min(R, S)$,

$$\sum_{n=0}^{\infty} (n+c)(n+c-1) a_n x^{n+c-2} + \left(\sum_{n=0}^{\infty} (n+c) a_n x^{n+c-2} \right) \left(\sum_{n=0}^{\infty} b_n x^n \right)$$

$$+ \left(\sum_{n=0}^{\infty} a_n x^{n+c-2} \right) \left(\sum_{n=0}^{\infty} d_n x^n \right) = 0$$

and hence, using the formula for the multiplication of power series (Theorem 3(b)), that

$$x^{c-2} \sum_{n=0}^{\infty} \left\{ (n+c)(n+c-1)a_n + \sum_{r=0}^{n} (r+c)a_r b_{n-r} + \sum_{r=0}^{n} a_r d_{n-r} \right\} x^n = 0.$$

We next cancel the non-zero term x^{c-2} and note that the coefficient of x^n inside the curly brackets may be written

$$\left\{ (c+n)(c+n-1) + (c+n)b_0 + d_0 \right\} a_n + \sum_{r=0}^{n-1} (c+r)a_r b_{n-r} + \sum_{r=0}^{n-1} a_r d_{n-r}.$$

Defining $I(t) = t(t-1) + tb_0 + d_0$, we can deduce, equating the different powers of x to zero, that

$$I(c) \cdot a_0 = 0$$

and

(13) $$I(c+n)a_n + \sum_{r=0}^{n-1} \left\{ (c+r)b_{n-r} + d_{n-r} \right\} a_r = 0, \qquad (n \geq 1).$$

As we have taken $a_0 \neq 0$, c must satisfy the *indicial equation*

(14) $$I(c) \equiv c^2 + (b_0 - 1)c + d_0 = 0,$$

a quadratic, its roots c_1 and c_2 being called the *exponents* at the regular singularity $x_0 = 0$. As we require c to be real, $(b_0 - 1)^2 \geq 4d_0$ is necessary. The equations (13) allow us, successively and for $c = c_1$ and $c = c_2$, to determine uniquely the coefficients $a_1, a_2, \ldots, a_n, \ldots$ in terms of a_0 *unless* the coefficient $I(c+n)$ of a_n in (13) vanishes for some n. The two solutions will clearly be linearly independent if $c_1 \neq c_2$, as the leading terms are then x^{c_1} and x^{c_2}.

When $c_1 - c_2 = N$, a positive integer, we can hope for a series solution, convergent for some non-zero radius of convergence, when $c = c_1$. However, $I(c_2 + N) = 0$, and so, $c = c_2$ will not in general furnish us with a solution. In order for us to find a second independent solution, both in this case and when $c_1 = c_2$, further discussion is needed. We should first, however, consider the matter of convergence, though we leave the proof of the next theorem to section 13.8, in the context of complex-valued solutions. We denote the set of all integers, positive and negative, by $\mathbb{Z}$.

Theorem 4 With the above notation and when $c_1 - c_2 \notin \mathbb{Z}$, there is, corresponding to each c_i, a series solution

$$y(x) = x^{c_i} \sum_{n=0}^{\infty} a_n x^n, \qquad (a_0 \neq 0)$$

such that the power series $\sum a_n x^n$ is, like P_0, P_1, convergent for $|x| < S$.

We now seek a second solution, independent of the solution

$$y = u(x) = x^{c_1} f(x),$$

where the coefficients in the power series

$$f(x) = \sum_{n=0}^{\infty} a_n x^n$$

have been determined by using (13) at $c = c_1$, in the case when $c_1 - c_2 = N$ and N is a non-negative integer. We use the method of reduction of order (section (5) in the Appendix) and try to find a second solution $y = y(x)$ to (1) in the form $y = uv$. As $a_0 \neq 0$, we know that $u(x)$ is non-zero near the origin, where we shall be working. The derivative v' must, in these circumstances (*loc. cit.*), satisfy

$$(p_2 u)v'' + (2p_2 u' + p_1 u)v' = 0$$

and hence be given by

(15) $$v'(x) = \frac{A}{(u(x))^2} e^{-\int^x p}$$

where

$$p(x) \equiv \frac{p_1(x)}{p_2(x)} = \frac{P_1(x)}{x} = \frac{b_0}{x} + g(x) \qquad (b_0 \text{ constant})$$

and $g = g(x)$ is the function given, near the origin, by the convergent power series

$$g(x) = \sum_{n=1}^{\infty} b_n x^{n-1}.$$

From

$$\int^x p \;=\; b_0 \log x + \int^x g,$$

we deduce

$$e^{-\int^x p} \;=\; x^{-b_0} \cdot e^{-\int^x g}$$

and, from (15),

$$v'(x) \;=\; \frac{A}{x^{N+1}} \cdot \frac{\exp(-\int^x g)}{(f(x))^2}, \qquad (A \text{ constant}, A \neq 0)$$

since the sum $c_1 + c_2$ of the roots of the indicial equation is $-(b_0 - 1)$ and $c_1 - c_2 = N$. As $\exp(-\int^x g)/(f(x))^2$ is clearly analytic in a neighbourhood of zero $(a_0 \neq 0)$, we can integrate v term-by-term (Theorem 2) to give a series with leading term a non-zero multiple of $x^{-N} = x^{-c_1+c_2}$. Hence, the second solution $y = uv$ will have leading term a non-zero multiple of $x^{c_1} \cdot x^{-N} = x^{c_2}$. The $(1/x)$-term in v' will give a $\log x$ term in v, and hence $u(x) \log x$ in the solution. A second solution can therefore be found in the form

$$(16) \qquad y(x) \;=\; Bu(x) \log x + x^{c_2} \sum_{n=0}^{\infty} e_n x^n.$$

The constant B may possibly be zero, but cannot take that value when $c_1 = c_2$ and it is included in the leading term. Otherwise, $e_0 \neq 0$.

The above method does not give a prescription for determining the constants e_n explicitly. To determine the e_n, we could substitute y in the form (16) into equation (1). A more efficient method due to Frobenius will be discussed later in the chapter. But first, we consider the practical matter of determining solutions in particular cases.

13.5   Extended power series solutions near a regular singular point: practice

In this section, we shall concentrate on sharpening the technique necessary to use extended power series to solve the differential equation

$$(1) \qquad p_2 y'' + p_1 y' + p_0 y \;=\; 0$$

which has a regular singular point at the origin, in the cases when

(A) the roots of the indicial equation do not differ by an integer and, in particular, are not equal,

(B) the roots of the indicial equation are equal or differ by a non-zero integer and the method of reduction of order provides a second solution efficiently,

(C) the roots of the indicial equation differ by a non-zero integer and the indeterminacy of a coefficient a_n, for some $n \geq 1$, leads to a complete solution being generated by the 'lower' root (in a manner to be described).

Notes

(a) If $y = u(x)$ is the solution corresponding, as in section 13.4, to the 'upper' root, the second (independent) solution is the only one that may contain the term $u(x) \log x$ in case (B).

(b) We are in general, as we have commented earlier, considering examples which produce only real exponents. This permits us to use the words 'upper' and 'lower' above. Complex roots, which can result (see Exercise 13) in such terms as $\sin(\log x)$ and $\cos(\log x)$, do not correspond, by and large, to observations in the natural world.

(c) By 'efficiently' in (B) above, we mean: because the first solution, corresponding to the 'upper' root of the indicial equation, is available in closed form.

(d) We shall, in general, consider examples which produce two-term recurrence relations between the coefficients; that is, each identity (13) will contain no more than two of the a_n's (but see Exercise 12).

The first example here is rather straightforward technically, but should help to 'fix ideas'. We again recall the *method of reduction of order* (see (5) of the Appendix) that, when one solution $y = u(x)$ of the equation

(1)
$$p_2 y'' + p_1 y' + p_0 y = 0$$

is known and a second solution is sought in the form $y = uv$, the derivative v' of the function $v = v(x)$ is given by

(17)
$$v'(x) = \frac{A}{(u(x))^2} e^{-\int^x p(t)\, dt}$$

where $p(t) = p_1(t)/p_2(t)$. (Note that we can only be certain of solutions in this form when both p_2 and u do not vanish.)

Example 3 Solve the differential equation

$$x^2 y'' - (\alpha + \beta - 1)xy' + \alpha\beta y = 0, \qquad (x > 0)$$

where α and β are real constants satisfying $\alpha \geq \beta$.

The equation is of course in Euler's form and may be solved directly by seeking solutions in the form $y = x^\lambda$, or indirectly by first making the substitution $x = e^t$ to transform the equation into one with constant coefficients (see (7) in the Appendix). In either case, the auxiliary equation is

$$\lambda^2 - (\alpha + \beta)\lambda + \alpha\beta = 0$$

with roots $\lambda = \alpha, \beta$, and this gives rise to the solutions

$$y = \begin{cases} Ax^\alpha + Bx^\beta, & (\alpha \neq \beta) \\ \\ (A \log x + B)x^\alpha, & (\alpha = \beta) \end{cases}$$

where A and B are constants. Here, however, our concern will be to show how the extended power series method works (and gives the same result!).

We therefore seek a solution in the form

$$y = \sum_{n=0}^\infty a_n x^{n+c}, \qquad (a_0 \neq 0)$$

convergent for $0 < x < R$. With this substitution and range of x, the differential equation becomes

$$\sum (n+c)(n+c-1)a_n x^{n+c} + \sum (-(\alpha+\beta-1)(n+c))a_n x^{n+c} + \sum \alpha\beta a_n x^{n+c} = 0.$$

Equating coefficients of powers of x to zero:

$$x^c \qquad\qquad \{c(c-1) - (\alpha+\beta-1)c + \alpha\beta\}a_0 = 0$$

giving rise, as $a_0 \neq 0$, to the indicial equation

$$c^2 - (\alpha+\beta)c + \alpha\beta = (c-\alpha)(c-\beta) = 0,$$

identical to the auxiliary equation above and giving exponents $c = \alpha, \beta$, and

$$x^{n+c} \qquad\qquad (n+c-\alpha)(n+c-\beta)a_n = 0, \qquad (n \geq 1).$$

(A) When $\alpha - \beta$ is not an integer, neither $n + c - \alpha$ nor $n + c - \beta$ can be zero for $c = \alpha$ or β and $n \geq 1$. So, $a_n = 0$ for $n \geq 1$ in these cases, and we have, corresponding to $c = \alpha$ and $c = \beta$, the linearly independent solutions

$$y = a_0 x^\alpha, \qquad y = a_0 x^\beta$$

or, equivalently, the complete solution

$$y = A x^\alpha + B x^\beta,$$

where A, B are constants.

(B) When $\alpha = \beta + N$ and N is a non-negative integer, $n + c - \alpha$ and $n + c - \beta$ remain non-zero for $c = \alpha$ and $n \geq 1$, giving one solution in the form

$$y = u(x) = a_0 x^\alpha.$$

We use the method of reduction of order to find a second solution in the form $y = uv$. Then, as

$$u = x^\alpha, \quad p_2 = x^2, \quad p_1 = -(\alpha + \beta - 1)x,$$

$v = v(x)$ is found, using (17), from

$$v'(x) = \frac{A}{x^{2\alpha}} e^{-\int^x (-(\alpha + \beta - 1)/t)\, dt},$$

where A is a constant. This may be simplified quickly to give

$$v'(x) = A x^{\beta - \alpha - 1} = A x^{-(N+1)}, \qquad (A \text{ constant}).$$

When $\alpha = \beta$ and therefore $N = 0$, one further integration gives us quickly the complete solution

$$y(x) = u(x) \cdot v(x) = (A \log x + B) x^\alpha.$$

When $N = \alpha - \beta$ is non-zero, integrating v' this time gives

$$y(x) = u(x) \cdot v(x) = \left(A \frac{x^{\beta - \alpha}}{\beta - \alpha} + B \right) x^\alpha = \frac{A}{\beta - \alpha} x^\beta + B x^\alpha,$$

the same form of solution as we found for $\alpha - \beta$ non-integral.

(C) An alternative route, when $N = \alpha - \beta$ is a strictly positive integer, so that

$$n + c - \alpha = 0, \quad \text{when } c = \beta \text{ and } n = N,$$

runs as follows. The coefficient a_N must, as a result, be undetermined. So, the complete solution is given (by consideration solely of the lower root $c = \beta$ of the indicial equation) as

$$y \;=\; x^{\beta}(a_0 + a_N x^N) \;=\; x^{\beta}(a_0 + a_{\alpha-\beta}x^{\alpha-\beta}) \;=\; a_0 x^{\beta} + a_{\alpha-\beta}x^{\alpha}$$

in this case.

Of course, the solutions are valid for all positive x. $\qquad\qquad\qquad\qquad\qquad$ □

The labelling (A), (B), (C) in the above example corresponds to the scheme outlined at the start of the section. Similar considerations apply in the next example which develops technique a little further.

Example 4 Show that $x = 0$ is a regular singular point of the differential equation

$$(18) \qquad\qquad x(1-x)y'' + (\beta - 3x)y' - y \;=\; 0,$$

where $\beta \geq 0$ is a constant.

(i) Assuming that β is *not* an integer, obtain two linearly independent series solutions.

(ii) Find solutions in closed form for the cases $\beta = 0, 1, 2$.

As the coefficients in the equation are polynomials, it is sufficient, in showing that the singular point $x = 0$ is regular, to check the limits in Definition 2(ii) (see Exercise 2 above):

$$\lim_{x \to 0} x \frac{(\beta - 3x)}{x(1-x)} \;=\; \beta, \qquad \lim_{x \to 0} x^2 \frac{(-1)}{x(1-x)} \;=\; 0.$$

We now seek a solution in the form

$$y \;=\; \sum_{n=0}^{\infty} a_n x^{n+c}, \qquad (a_0 \neq 0)$$

convergent for $0 < x < R$, and substitute in the differential equation (18) to give

$$(19) \quad \sum (n+c)(n+c-1)a_n x^{n+c-1} + \sum(-(n+c)(n+c-1))a_n x^{n+c}$$

$$+ \sum \beta(n+c)a_n x^{n+c-1} + \sum(-3(n+c))a_n x^{n+c} + \sum(-a_n)x^{n+c} \;=\; 0.$$

Equating coefficients of powers of x to zero:

$$x^{c-1} \qquad\qquad \{c(c-1) + \beta c\}a_0 \;=\; 0.$$

As $a_0 \neq 0$, this gives the indicial equation

$$c(c - 1 + \beta) \;=\; 0$$

and, therefore, exponents $c = 0, \, 1 - \beta$. For $n \geq 1$,

$$x^{n+c-1} \;\; (n+c)(n+c-1+\beta)a_n - \{(n+c-1)(n+c-2) + 3(n+c-1) + 1\}\, a_{n-1} \;=\; 0.$$

Hence,

$$(20) \qquad\qquad (n+c)(n+c-1+\beta)a_n \;=\; (n+c)^2 a_{n-1}, \qquad (n \geq 1).$$

It follows that, for $n \geq 1$,

$$(21) \qquad a_n \;=\; \frac{n+c}{n+c-1+\beta}\,a_{n-1} \;=\; \ldots \;=\; \frac{(n+c)(n+c-1)\ldots(1+c)}{(n+c-1+\beta)(n+c-2+\beta)\ldots\beta}\,a_0$$

when neither $n + c$ nor any terms in the denominators are zero. Since

$$\lim_{n\to\infty} \left| \frac{a_{n-1}}{a_n} \right| \;=\; \lim_{n\to\infty} \left| \frac{n+c-1+\beta}{n+c} \right| \;=\; 1$$

in all cases, the radius of convergence of solutions is unity.

(A) The equations (21) hold when β is not an integer and $c = 0$ or $1 - \beta$. Noting that, in these circumstances, the roots of the indicial equation are neither zero nor differ by an integer, we can determine two linearly independent solutions y_1 and y_2 to (18) by substituting $c = 0$ and $c = 1 - \beta$ in (21). This gives

$$y_1(x) \;=\; a_0 \sum_{n=0}^{\infty} \frac{n!}{(n-1+\beta)\ldots\beta}\,x^n$$

and

$$y_2(x) \;=\; a_0 x^{1-\beta} \sum_{n=0}^{\infty} \frac{(n+1-\beta)\ldots(2-\beta)}{n!}\,x^n.$$

(B) When $\beta = 1$, the indicial equation has equal roots $c = 0$. As $n + c \neq 0$ for $n \geq 1$, (20) gives

$$a_n = a_{n-1} = \ldots = a_0 \qquad (n \geq 1)$$

and the solution

$$y = u(x) = a_0 \sum_{n=0}^{\infty} x^n = \frac{a_0}{1-x}.$$

To find the complete solution, we use the method of reduction of order, seek a solution in the form $y = uv$, and use (17) with

$$u = \frac{1}{1-x}, \quad p_2 = x(1-x), \quad p_1 = 1-3x,$$

to give, after a straightforward integration,

$$v'(x) = A(1-x)^2 e^{\int^x \{(1-3t)/t(1-t)\}\, dt} = \frac{A}{x}$$

and hence, the complete solution

$$y(x) = u(x).v(x) = \frac{1}{1-x}(A \log x + B),$$

where A and B are constants.

When $\beta = 0$, the exponents are $c = 0, 1$. We note that at $c = 0$, equation (20) does not give a finite value for a_1 (put $n = 1$) and hence no series solution in this form for that value of c. However, for $c = 1$, equation (21) gives

$$a_n = (n+1)a_0$$

and hence, using Theorem 2(a),

$$y = u(x) = a_0 x \sum_{n=0}^{\infty} (n+1)x^n = a_0 x \sum_{n=0}^{\infty} \frac{d}{dx} x^{n+1} = a_0 x \frac{d}{dx} \sum_{n=0}^{\infty} x^{n+1}$$

$$= a_0 x \frac{d}{dx}\left(\frac{x}{1-x}\right) = \frac{a_0 x}{(1-x)^2}.$$

Try $y = uv$ with

$$u = \frac{x}{(1-x)^2}, \quad p_2 = x(1-x), \quad p_1 = -3x.$$

Then (17) gives

$$v'(x) = \frac{A(1-x)^4}{x^2} e^{-\int^x (-3/(1-t))\, dt} = \frac{A(1-x)}{x^2}, \quad (A \text{ constant}).$$

So,

$$v(x) = A'\left(\log x + \frac{1}{x}\right) + B \qquad (A' = -A \text{ and } B \text{ constant})$$

and we obtain the complete solution

$$y(x) = u(x).v(x) = \frac{x}{(1-x)^2}\left\{A'\left(\log x + \frac{1}{x}\right) + B\right\},$$

where A' and B are constants.

(C) For $\beta = 2$, the roots of the indicial equation are $c = -1, 0$. Equation (20) with $\beta = 2$ and $c = -1$, namely

$$n(n-1)a_n = (n-1)^2 a_{n-1},$$

does not allow us to determine a_1, but only subsequent a_n's in terms of a_1:

$$a_n = \frac{n-1}{n}a_{n-1} = \frac{(n-1)(n-2)}{n(n-1)}a_{n-2} = \frac{n-2}{n}a_{n-2} = \ldots = \frac{1}{n}a_1 \qquad (n \geq 2).$$

Thus, we have the complete solution

$$y = \frac{1}{x}\left\{a_0 + a_1\sum_{n=1}^{\infty}\frac{1}{n}x^n\right\} = \frac{1}{x}\left\{a_0 - a_1\log(1-x)\right\}.$$

The reader is invited to check that $c = 0$ gives one part of this solution (only), and therefore no additional solution (as was to be expected). A complete solution can then be found by the method of reduction of order – a particularly rewarding exercise. □

Notes

(a) Notice that it is technically simpler to substitute a series directly into (1) in its original form than to determine P_0, P_1 first, as was convenient in our theoretical considerations in section 13.4.

(b) The reader will note our 'book-keeping', in writing the powers of x in the left margin, beside its coefficient when equated to zero.

(c) In determining the coefficient of x^{n+c-1}, it was necessary to replace 'n' by '$n-1$' in terms occurring in three of the summations in (19), from which the coefficient was derived.

(d) We have chosen the coefficient of x^{n+c-1} as the general term, because it gives a_n in terms of a_k's with lower indices k. (If we had chosen x^{n+c}, we would have an expression for a_{n+1}, which is less useful.) The lowest power x^{c-1} gives us a multiple of a_0; so, adding n to its index gives an expression for a_n.

(e) In Example 4, we only had to consider one power of x, namely x^{c-1}, different from the 'general power' x^{n+c-1}, in order to determine all the a_n. This was fortuitous. Consider again Airy's equation (Exercise 5)

$$y'' + xy = 0$$

and substitute the extended power series

$$y = \sum_{n=0}^{\infty} a_n x^{n+c} \qquad (a_0 \neq 0)$$

and its second derivative in it, to give

$$\sum (n+c)(n+c-1)a_n x^{n+c-2} + \sum a_n x^{n+c+1} = 0.$$

Then, we clearly need

$$x^{c-2} \qquad\qquad\qquad\qquad\qquad c(c-1)a_0 = 0,$$

$$x^{c-1} \qquad\qquad\qquad\qquad\qquad (c+1)ca_1 = 0,$$

$$x^c \qquad\qquad\qquad\qquad\qquad (c+2)(c+1)a_2 = 0,$$

$$x^{n+c-2} \qquad\qquad (n+c)(n+c-1)a_n + a_{n-3} = 0, \qquad (n \geq 3)$$

and hence

$$a_n = -\frac{a_{n-3}}{(n+c)(n+c-1)}, \qquad (n \geq 3)$$

to determine the coefficients. As a_1 cannot be determined for the lower exponent $c = 0$, we find ourselves with case (C) of our general scheme, where the complete solution is given by substituting only for that exponent. This entails

$$a_n = -\frac{a_{n-3}}{n(n-1)}, \qquad (n \geq 3)$$

and hence

$$a_{3m} = \frac{(-1)^m}{3m(3m-1)\ldots 3.2}\, a_0, \qquad (m \geq 1),$$

$$a_{3m+1} = \frac{(-1)^m}{(3m+1)3m\ldots 4.3}\, a_1, \qquad (m \geq 1),$$

$$a_{3m+2} = 0, \qquad\qquad\qquad\qquad (m \geq 0).$$

We needed to consider three 'initial powers', as well as the 'general power', of x, in order to determine all the coefficients a_n. The general rule is that we need to consider 'initial powers' until terms from all the different summations (occurring when we have substituted for y, y', y'' in the differential equation) come into play.

Example 5 (Bessel's Equation) Bessel's equation of order ν is the differential equation

(22) $$x^2 y'' + xy' + (x^2 - \nu^2)y = 0,$$

where ν is a real constant.

(i) Show that, when ν is *not* an integer, the function $J_\nu = J_\nu(x)$, defined by

$$J_\nu(x) = \left(\tfrac{1}{2}x\right)^\nu \sum_{m=0}^{\infty} \frac{(-1)^m \left(\tfrac{1}{2}x\right)^{2m}}{m!\,\Gamma(\nu+m+1)},$$

is a solution of (22), convergent for every non-zero x. This is the *Bessel function of the first kind of order ν*. [Recall from Exercise 5 that the gamma function $\Gamma = \Gamma(\alpha)$ satisfies, for $\alpha > 0$,

$$\Gamma(\nu+m+1) = (\nu+m)(\nu+m-1)\ldots(\nu+1)\Gamma(\nu+1).]$$

(ii) Find the complete solution of Bessel's equation of order $\tfrac{1}{2}$ in closed form.

We seek a solution in the form

$$y = \sum_{n=0}^{\infty} a_n x^{n+c}, \qquad (a_0 \neq 0)$$

convergent for $0 < x < R$, and substitute in the differential equation:

$$\sum (n+c)(n+c-1)a_n x^{n+c} + \sum (n+c)a_n x^{n+c} + \sum a_n x^{n+c+2} + \sum (-\nu^2)a_n x^{n+c} = 0.$$

Then

$$x^c \qquad\qquad\qquad \{(c(c-1)+c-\nu^2\}a_0 = 0.$$

As $a_0 \neq 0$, this gives the indicial equation

$$c^2 - \nu^2 = 0$$

and, therefore, exponents $c = \pm\nu$.

$$x^{c+1} \qquad\qquad\qquad \{(c+1)^2 - \nu^2\}a_1 = 0.$$

As $(c+1)^2 - \nu^2$ is non-zero for $c = \pm\nu$, we have $a_1 = 0$ in all cases.

$$x^{n+c} \qquad\qquad \{(n+c)^2 - \nu^2\}a_n + a_{n-2} = 0, \qquad (n \geq 2).$$

(i) When ν is not an integer and $c = \nu > 0$,

(23)
$$a_n = -\frac{1}{n(n+2\nu)}a_{n-2}, \qquad (n \geq 2)$$

and hence, when $n = 2m$,

$$a_{2m} = -\frac{1}{2^2 m(m+\nu)}a_{2(m-1)} = \cdots = \frac{(-1)^m}{2^{2m}m!\,(m+\nu)!}a_0,$$

where we have defined

$$(m+\nu)! = \frac{\Gamma(\nu+m+1)}{\Gamma(\nu+1)},$$

the quotient of two gamma functions, at the end of Exercise 5. *Bessel's function of the first kind of order* ν, $J_\nu = J_\nu(x)$, is the solution determined for $c = \nu > 0$ with

$$a_0 = \frac{1}{2^\nu \Gamma(\nu+1)}.$$

Hence,

(24)
$$J_\nu(x) = \sum_{m=0}^{\infty} \frac{(-1)^m x^{2m+\nu}}{2^{2m+\nu}m!\,\Gamma(\nu+m+1)}.$$

The coefficients, and hence the solution, remain valid when $c = \nu$ is a non-negative integer.

Bessel's equation (22) remains the same when ν is replace by $-\nu$. Then (24) becomes

$$J_{-\nu}(x) = \sum_{m=0}^{\infty} \frac{(-1)^m x^{2m-\nu}}{2^{2m-\nu} m! \, \Gamma(-\nu + m + 1)},$$

which is another solution, when ν is positive and *not* an integer, which ensures that $\Gamma(-\nu + m + 1)$ remains finite. In this case, J_ν and $J_{-\nu}$ are linearly independent – their first terms are non-zero multiples, respectively, of x^ν and $x^{-\nu}$. Hence, for non-integral ν, the complete solution of Bessel's equation is $y = y(x)$, where

$$y(x) = A J_\nu(x) + B J_{-\nu}(x)$$

and A, B are constants. As

$$\lim_{m \to \infty} \left| \frac{a_{2(m-1)}}{a_{2m}} \right| = \lim_{m \to \infty} |2^m m(m + \nu)|$$

is infinite, y solves (22) for $x \neq 0$. This covers most of the possibilities under (A) and (B) in our scheme. The other cases under (A) and (B), including the case of equal exponents when $\nu = 0$, are most easily discussed by using the method of Frobenius (in the next section).

(ii) When $c = -\nu = -\frac{1}{2}$, the roots of the indicial equation differ by unity, the coefficient a_1 is indeterminate and the complete solution can be found by using $c = -\frac{1}{2}$ alone. Then, equation (23) reduces to

(8)
$$a_n = -\frac{1}{n(n-1)} a_{n-2}, \qquad (n \geq 2).$$

So, using our work in Example 1(b), the complete solution of (22) for $\nu = \frac{1}{2}$ is

$$y = x^{-\frac{1}{2}} (a_0 \cos x + a_1 \sin x), \qquad (x \neq 0). \qquad \square$$

Exercise 9 Find two linearly independent series solutions to each of the following equations and express your answers in closed form:

(a) $\quad 2xy'' + y' - y = 0$, $\qquad\qquad (x > 0)$,

(b) $\quad 4x^2 y'' - 4xy' + (3 + 16x^2)y = 0$, $\quad (x > 0)$,

(c) $\quad x(1-x)y'' - 3y' - y = 0$, $\qquad\quad (0 < x < 1)$.

In the case of equation (b),

(b1) determine the most general real-valued function $y = y(x)$ defined on the whole real line, which satisfies the equation for all non-zero x and is continuous at $x = 0$;

(b2) prove that, if y satisfies the equation for all non-zero x and is differentiable at $x = 0$, then $y'(0) = 0$.

Can you find solutions of (c) which are valid for $x > 1$?

[In (a), you might find it helpful to write $b_n = 2^{-n}a_n$.]

Exercise 10 Show that, when $0 < \alpha < 1$, there are series solutions of the equation

$$x(1 + x)y'' + (\alpha + 4x)y' + 2y = 0$$

and find their radii of convergence. In the case $\alpha = 1$, show that the equation has a series solution which may be written in the form $(1 + x)^k$ for suitable k. Hence, or otherwise, find a second solution to the equation in that case.

Exercise 11 (Alternative method for Examples 4 and 5) By first finding a solution corresponding to the highest exponent and then using the method of reduction of order, find the complete solution of the following equations:

(a) $\quad x^2y'' + xy' + (x^2 - \frac{1}{4})y = 0$,

(b) $\quad x(1 - x)y'' + (2 - 3x)y' - y = 0$.

Exercise 12 (3-term recurrence relation) Show that the differential equation

(25) $$xy'' + (1 - 2x)y' + (x - 1)y = 0 \qquad (x > 0)$$

has exactly one solution $y = y(x)$ of the form

$$y(x) = x^c \sum_{n=0}^{\infty} a_n x^n$$

with c a constant and $a_0 = 1$. Determine the constant c and calculate the coefficients a_1, a_2, a_3 explicitly. Establish a general formula for a_n and hence show that $y(x) = e^x$. Find a second solution of (25) which is linearly independent of y.

Exercise 13 (Complex exponents) By seeking an extended power series solution, find two linearly independent real-valued solutions to the Euler equation

$$x^2y'' + xy + y = 0.$$

Exercise 14 (Irregular singular point) By first seeking a solution of the differential equation

$$x^3y'' + xy' - y = 0, \qquad (x > 0)$$

in the form of an extended power series, find two linearly independent solutions of the equation, one of which *cannot* be expanded as a power series in ascending powers of x.

Exercise 15 (Bessel functions of the second kind) (a) Assuming that n is a positive integer and that $\Gamma(\nu + m + 1) \to \infty$ as $\nu \to -n$ when $m < n$, show that

$$J_{-n}(x) = (-1)^n J_n(x) \quad \text{for } n = 1, 2, \ldots$$

[Hence, when n is an integer, a complete solution of Bessel's equation of order n *cannot* be represented as a linear combination of J_n and J_{-n}.]

(b) Consider Bessel's equation of order zero, which may be written

(26) $$xy'' + y' + xy = 0$$

and has equal exponents $c = 0$ and hence a solution containing the function $\log x$ (see section 13.4). Show that (26) has the solution

$$y_B(x) = J_0(x) \log x + \sum_{m=1}^{\infty} \frac{(-1)^{m-1} \gamma_m}{2^{2m} (m!)^2} x^{2m}$$

where

$$\gamma_m = \sum_{k=1}^{m} \frac{1}{k}.$$

[The function $Y_0 = Y_0(x)$, defined by

$$Y_0(x) = \frac{2}{\pi} \{ (\gamma - \log 2) J_0(x) + y_B(x) \}$$

$$= \frac{2}{\pi} \left\{ J_0(x)(\log \tfrac{x}{2} + \gamma) + \sum_{m=1}^{\infty} \frac{(-1)^m \gamma_m}{2^{2m} (m!)^2} x^{2m} \right\},$$

where γ is Euler's constant

$$\gamma = \lim_{m \to \infty} (\gamma_m - \log m),$$

is called *Bessel's function of the second kind of order zero*. Clearly, J_0 and Y_0 are linearly independent. *Bessel's function of the second kind of order ν*, for arbitrary real ν, can be defined similarly.]

13.6   The method of Frobenius

The method provides a straightforward and direct way of finding extended power series solutions. It is particularly useful in providing a clear line of attack which covers all possible cases and in finding second solutions easily when the first cannot be expressed in closed form. Its only 'disadvantage' is the need to differentiate what can often be elaborate coefficients; here, 'logarithmic differentiation' can lighten the burden.

We start, as in section 13.4 (which the reader may care to review), by substituting an extended power series $\sum a_n x^{n+c}$ into the differential equation

$$(1) \qquad\qquad p_2 y'' + p_1 y' + p_0 y \;=\; 0$$

and equating the coefficients of powers of x to zero, allowing us to determine the coefficients $a_n = a_n(c)$ for $n \geq 1$ as functions of c (and the non-zero constant a_0). We next consider the expansion

$$(27) \qquad\qquad y(x,c) \;=\; x^c \left\{ a_0 + \sum_{n=1}^{\infty} a_n(c) x^n \right\},$$

where we avoid, for the time being, substituting for the exponents $c = c_1, c_2$ derived from the indicial equation. As the $a_n(c)$ have been specifically chosen to make the coefficients of all but the lowest power of x zero when substituting $y(x,c)$ into (1), we have immediately that

$$(28) \qquad\qquad \left(p_2 \frac{\partial^2}{\partial x^2} + p_1 \frac{\partial}{\partial x} + p_0 \right) y(x,c) \;=\; a_0 x^{c-2} (c - c_1)(c - c_2).$$

The right-hand side of this equation is the coefficient $I(c).a_0$ of the lowest power of x (see section 13.4).

We now differentiate (28) partially with respect to c. The interchange of partial differentiation with respect to c and x, and differentiation of (27) 'term-by-term' with respect to c are both justified: the latter best proved in the complex analysis context. So,

$$(29) \qquad \left(p_2 \frac{\partial^2}{\partial x^2} + p_1 \frac{\partial}{\partial x} + p_0 \right) \frac{\partial y}{\partial c}(x,c) \;=\; a_0 \frac{\partial}{\partial c} \{ x^{c-2}(c - c_1)(c - c_2) \},$$

where

$$(30) \qquad \frac{\partial y}{\partial c}(x,c) \;=\; x^c \log x \left\{ a_0 + \sum_{n=1}^{\infty} a_n(c) x^n \right\} + x^c \sum_{n=1}^{\infty} \frac{\partial a_n}{\partial c}(c) x^n.$$

We reinforce our earlier remark that the differential coefficient $\partial a_n / \partial c$ can be efficiently calculated by using logarithmic differentiation. (The reader may revise this method by consulting Example 7 below.)

We have already dealt in section 13.4 with the straightforward case when the exponents are neither equal nor differ by an integer.

(i) When the exponents are equal, $c_1 = c_2$

In this case, the right-hand sides of equation (28) and (29) are both zero at $c = c_1$ and two linearly independent solutions of (1), $y = y_1(x)$ and $y = y_2(x)$, are given by

$$y_1(x) = y(x, c_1) \quad \text{and} \quad y_2(x) = \frac{\partial y}{\partial c}(x, c_1).$$

Notice that, as $a_0 \neq 0$, we have established that the coefficient of $\log x$ is always non-zero in this case.

(ii) When the exponents differ by an integer, $c_1 = c_2 + N$ (N a positive integer)

With the notation of section 13.4, since $I(c_1) = 0$ and $I(c) = (c - c_1)(c - c_2)$, we have

$$I(c_2 + N) = N(c_2 + N - c_1) = 0.$$

Therefore, the expressions for a_n ($n \geq N$) derived from (13) would, in general, have zeros in the denominators. (The exception is when there are matching factors in the numerators.) To avoid this difficulty, we define a new function

$$Y(x, c) = (c - c_2) . y(x, c)$$

which is just a constant multiple (with respect to x) of $y(x, c)$. Hence,

$$\left(p_2 \frac{\partial^2}{\partial x^2} + p_1 \frac{\partial}{\partial x} + p_0 \right) Y(x, c) = x^{c-2}(c - c_1)(c - c_2)^2$$

and so, our differential equation (1) is solved by

$$y = Y(x, c_1), \quad y = Y(x, c_2) \quad \text{and} \quad y = \frac{\partial Y}{\partial c}(x, c_2).$$

The reader will quickly see that $Y(x, c_2)$ is just a constant multiple of $Y(x, c_1)$. (Notice that the additional multiplicative factor $c - c_2$ ensures that all the coefficients before $(c - c_2)a_N(c)$ are zero at $c = c_2$. Further, the resulting leading term is a multiple of $x^{c_2} . x^N = x^{c_1}$ and the factors $c_2 + r$ appearing in (13) are just $c_1 + r - N$.)

So, the equation (1), when $c_1 = c_2 + N$ and N is a positive integer, has linearly independent solutions $y = y_1(x)$ and $y = y_2(x)$, where

$$y_1(x) = \lim_{c \to c_1} Y(x, c) \quad \text{and} \quad y_2(x) = \lim_{c \to c_2} \frac{\partial Y}{\partial c}(x, c).$$

Example 6 For the Euler equation

$$x^2 y'' - (\alpha + \beta - 1)xy' + \alpha\beta y = 0, \qquad (x > 0)$$

with exponents α and β $(\alpha \geq \beta)$, the function $y = y(x, c)$ is given by

$$y(x, c) = a_0 x^c$$

(see Example 3). Hence,

$$\frac{\partial y}{\partial c} = a_0 x^c \log x$$

and, when

$$Y(x, c) = a_0(c - \beta)x^c,$$

we see that

$$\frac{\partial Y}{\partial c}(x, c) = a_0\{(c - \beta)x^c \log x + x^c\}.$$

So, linearly independent solutions when $\alpha = \beta$ are

$$y_1(x) = a_0 x^\alpha \quad \text{and} \quad y_2(x) = a_0 x^\alpha \log x,$$

and when $\alpha = \beta + N$, N a positive integer,

$$y_1(x) = a_0(\alpha - \beta)x^\alpha \quad \text{and} \quad y_2(x) = a_0 x^\beta,$$

in agreement with our results in Example 3.   □

Example 7 (Bessel's equation of order 0) From Example 5, Bessel's equation of order zero is

(26) $$xy'' + y' + xy = 0, \qquad (x > 0)$$

it has equal exponents $c = 0$, and

$$y(x, c) = a_0 x^c \left(1 + \sum_{m=1}^{\infty} a_m(c)x^{2m}\right)$$

where

$$a_m(c) = \frac{(-1)^m}{2^{2m} m! \, (m + c) \dots (1 + c)} a_0, \qquad (m \geq 1).$$

We employ logarithmic differentiation:

$$\log a_m(c) \;=\; \log\left(\frac{(-1)^m}{2^m m!}\,a_0\right) - \sum_{k=1}^{m}\log(k+c)$$

and hence

$$\frac{1}{a_m(c)}\cdot\frac{\partial a_m}{\partial c}(c) \;=\; -\sum_{k=1}^{m}\frac{1}{k+c}.$$

So,

$$\frac{\partial a_m}{\partial c}(0) \;=\; \frac{(-1)^{m+1}\gamma_m}{2^{2m}}\,a_0, \quad \text{where } \gamma_m \;=\; \sum_{k=1}^{m}\frac{1}{k}.$$

Thus, linearly independent solutions to (26) are

$$y(x,0) \;=\; a_0\sum_{m=0}^{\infty}\frac{(-1)^m}{2^{2m}(m!)^2}\,x^{2m}$$

and

$$\frac{\partial y}{\partial c}(x,0) \;=\; y(x,0)\log x + a_0\sum_{m=1}^{\infty}\frac{(-1)^{m+1}\gamma_m}{2^{2m}(m!)^2}\,x^{2m},$$

in agreement with Example 5 and Exercise 15. □

We deal finally in this section with the exception under (ii), when in the coefficient of x^n ($n \geq N$), there is a factor $c - c_2$ in the numerator of the expression for a_n as derived from (13). In this case, a_N cannot be determined, and two linearly independent solutions are given at $c = c_2$ with first terms $a_0 x^{c_2}$ and $a_N x^{c_2+N} = a_N x^{c_1}$. For essentially the same reasons as $Y(x, c_2)$ is a multiple of $Y(x, c_1)$ given above, the series headed by $a_N x^{c_1}$ is a multiple of the series given by using the exponent c_1 directly. Examples and exercises concerning this case were given in section 13.5 and will not be found below.

Exercise 16 Demonstrate the advantages of Frobenius' method by using it to find two linearly independent solutions of the equation

(18) $$x(1-x)y'' + (\beta - 3x)y' - y \;=\; 0,$$

discussed in Example 4, for the cases $\beta = 0$ and $\beta = 1$.

Exercise 17 (Bessel's function of the second kind of order 1) Show that

$$y(x) = a_0 x \sum_{m=0}^{\infty} \frac{(-1)^m}{2^{2m} m! \, (m+1)!} x^m$$

is a solution of Bessel's equation of order 1,

$$x^2 y'' + xy' + (x^2 - 1)y = 0, \qquad (x > 0).$$

Using Frobenius' method, show that the first four terms of a second and linearly independent solution are

$$\frac{1}{x} + \frac{x}{4} - \frac{5x^3}{64} + \frac{5x^5}{1152}.$$

13.7 Summary

We now summarise the methods of this chapter for obtaining series solutions in powers of a real variable x of the homogeneous differential equation

(1) $$p_2 y'' + p_1 y' + p_0 y = 0,$$

where p_0, p_1, p_2 are continuous real-valued functions of x, with no common zeros.

Radius of Convergence

The range of values for which a series solution is valid can normally be found by using the formula

$$R = \lim_{n \to \infty} \left| \frac{a_{n-1}}{a_n} \right|$$

for the radius of convergence of a power series $\sum a_n x^n$.

Method of reduction of order

One method for finding a second solution once one solution $y = u(x)$ is known, especially useful when $u(x)$ is expressed in closed form and when p_2 and u are non-zero, is to seek a solution in the form $y = uv$, which leads to $v = v(x)$ satisfying

$$v'(x) = \frac{A}{(u(x))^2} e^{-\int^x (p_1/p_2)}$$

with A constant.

(α) The origin is an ordinary point ($p_2(0) \neq 0$) and both p_1/p_2 and p_0/p_2 are analytic functions

Substituting a power series

$$y(x) = \sum_{n=0}^{\infty} a_n x^n$$

into (1) and equating the coefficients of powers of x to zero determine solutions in terms of the undetermined coefficients and hence two linearly independent series solutions, with non-zero radius of convergence.

(β) The origin is a regular singular point (see Definition 2)

Substitution of an extended power series

$$y(x) = \sum_{n=0}^{\infty} a_n x^{n+c} \qquad (a_0 \neq 0)$$

into (1) leads to c having to satisfy the *indicial equation* $I(c) = 0$, a quadratic equation, and to the a_n's satisfying recurrence relations. Denoting the roots of the indicial equation, or *exponents*, by $c = c_1$ and $c = c_2$, we can distinguish four cases. We refer to

$$y(x, c) = x^c \left\{ a_0 + \sum_{n=1}^{\infty} a_n(c) x^n \right\},$$

where the $a_n(c)$ have been determined from the recurrence relations for the coefficients, *before* any substitution is made for c. The linearly independent solutions are given by $y = y_1(x)$ and $y = y_2(x)$ as follows. They are valid for $0 < x < R$, some $R > 0$.

(β1) $c_1 - c_2 \notin \mathbb{Z}$

$$y_1(x) = y(x, c_1) \quad \text{and} \quad y_2(x) = y(x, c_2)$$

(β2) $c_1 = c_2$

$$y_1(x) = y(x, c_1) \quad \text{and} \quad y_2(x) = \frac{\partial y}{\partial c}(x, c_1).$$

The function $y = y_2(x)$ contains a multiple of $y_1(x) \log x$.

(β3) $c_1 = c_2 + N$, where N is a positive integer and the recurrence relation for the coefficients makes a_N infinite

In this case, define

$$Y(x, c) = (c - c_2) \cdot y(x, c).$$

Then

$$y_1(x) = \lim_{c \to c_1} Y(x,c) \quad \text{and} \quad y_2(x) = \lim_{c \to c_2} \frac{\partial Y}{\partial c}(x,c).$$

(β4) $c_1 = c_2 + N$, where N is a positive integer and the recurrence relation for the coefficients makes a_N indeterminate

The complete solution is given by $y = y(x)$, where

$$y(x) = y(x, c_2).$$

Then $y_1(x)$ has first term $a_0 x^{c_2}$ and $y_2(x)$ has first term $a_N x^{c_2+N}$.

Alternative method to (β2), (β3), (β4), where $c_1 = c_2 + N$ and N is a non-negative integer

Substitute $y = y(x)$, given by

$$y(x) = B y(x, c_1) \log x + x^{c_2} \sum_{n=0}^{\infty} e_n x^n,$$

where e_n $(n \geq 0)$ and B are constants, in equation (1), and equate the coefficients of $\log x$ and powers of x to determine B and the e_n. The constant B is, as above, non-zero when $c_1 = c_2$.

Solutions for $-S < x < 0$

Making the change of variable $x = -t$ and using relations (13) quickly establishes that, if we replace x^c by $|x|^c$ in $y(x,c)$ and $\log x$ by $\log |x|$, we have real solutions valid in $-R < x < 0$, as well as in $0 < x < R$. This can also be derived directly from the complex variable solutions discussed in the Appendix.

13.8   Appendix: the use of complex variables

Because of the relative ease in working with power series with complex terms and because of the equivalence (in open sets) of 'analytic' and 'differentiable' for complex-valued functions of a single complex variable (see below), there are considerable advantages in broadening our discussion of power series solutions to incorporate complex variables. There are further unifying gains: for example, in the treatment of exponential, trigonometric and hyperbolic functions, through such identities as

$$e^x = \cos x + i \sin x \quad \text{and} \quad \sin ix = i \sinh x.$$

Technical difficulties (and further advantages!) centre around the need to consider branch points of multifunctions, in particular of $z^c = \exp(c \log z)$, because of the many values of the complex logarithm.

We shall use the word *holomorphic* to describe a complex-valued function of a complex variable which is differentiable on an open set and the fact (Theorem 1 below) that such functions are analytic on the set. We note that all the results on power series in section 13.1 carry over unaltered, once one replaces the real open interval $\{x \in \mathbb{R} : |x| < R\}$ by the complex 'disc' $\{z \in \mathbb{C} : |z| < R\}$, or *disc of convergence*, and allows the coefficients a_n and the exponent c to be complex constants.

Theorem 5 (Taylor's Theorem for holomorphic functions) Suppose that f is holomorphic when $|z| < R$. Then f may be expressed, for $|z| < R$, as the convergent power series expansion, the *Taylor expansion*,

$$(31) \qquad f(z) = \sum_{n=0}^{\infty} a_n z^n,$$

where

$$a_n = \frac{1}{2\pi i} \int_C \frac{f(z)}{z^{n+1}} \, dz, \qquad (n \geq 0)$$

the contour C is $\{z \in \mathbb{C} : z = se^{i\theta}, 0 \leq \theta \leq 2\pi\}$ and s is any real number satisfying $0 < s < R$.

Let $M = M(s)$ denote the maximum value of the differentiable, and therefore continuous, function f in Theorem 5 on the closed and bounded set C. Then, by putting $z = se^{i\theta}$ ($0 \leq \theta \leq 2\pi$), the value of z on C, into the integral for each a_n, one quickly deduces that

$$(32) \qquad |a_n| \leq \frac{M}{s^n}, \qquad (n \geq 0).$$

We are now in a position to prove the complex variable, and stronger, version of Theorem 4 of section 13.4. We use the notation of that section, but replacing the real variable x by the complex variable z and intervals by discs of convergence, as above. Since Taylor's Theorem is now available, we can just presume that $P_1 = P_1(z)$ and $P_0 = P_0(z)$ are holomorphic.

Theorem 6 Suppose that P_1 and P_0 are holomorphic for $|z| < S$ and that $c = c_1, c_2$ are the exponents corresponding to substituting the extended power series

$$(33) \qquad y(z) \;=\; z^c \sum_{n=0}^{\infty} a_n z^n \qquad (a_0 \neq 0)$$

in the differential equation

$$(34) \qquad p_2(z)y'' + p_1(z)y' + p_0(z) \;=\; 0.$$

Then for $c = c_1, c_2$ and $p \equiv |c_1 - c_2| \notin \mathbb{Z}$, the series (33) is a solution of (34), convergent for $0 < |z| < S$.

Proof From our discussion above, there is a constant $M = M(r)$ such that, whenever $0 < s < S$,

$$(35) \qquad |b_n| \;\leq\; \frac{M}{s^n}, \quad |d_n| \;\leq\; \frac{M}{s^n}, \qquad (n \geq 0).$$

As $I(c) = (c - c_1)(c - c_2)$, we see immediately that $I(c_1 + n) = n(n + c_1 - c_2)$. We concentrate our attention on the series corresponding to the exponent $c = c_1$. Whenever $n > p$, from (13) we can deduce that

$$(36) \qquad |a_n| \;\leq\; \frac{M}{n(n - \lambda)} \sum_{r=0}^{n-1} \frac{1}{s^{n-r}} (|c_1| + r + 1)|a_r|.$$

Define positive constants A_n $(n \geq 0)$ inductively by

$$(37) \qquad A_n \;=\; \begin{cases} |a_n|, & (0 \leq n \leq p) \\[2ex] \dfrac{M}{n(n - \lambda)} \displaystyle\sum_{r=0}^{n-1} \dfrac{1}{s^{n-r}} (|c_1| + r + 1)A_r, & (p < n) \end{cases}$$

so that, using (36), $|a_n| \leq A_n$ for every $n \geq 0$. When $n - 1 > p$, the definition (37) gives

$$A_n - \frac{(n-1)(n-1-p)}{n(n-p)} \cdot \frac{A_{n-1}}{s} \;=\; \frac{m(|c_1| + n)}{n(n-p)} \cdot \frac{A_{n-1}}{s}$$

and hence

$$\lim_{n \to \infty} \frac{A_{n-1}}{A_n} = s.$$

So, the power series $\sum A_n z^n$ has radius of convergence s. Thus, $\sum a_n z^n$ has radius of convergence at least s. However, as s is any real number satisfying $0 < s < S$, $\sum a_n z^n$ must converge for $|z| < S$ and the extended power series $z^c \sum a_n z^n$ for $0 < |z| < S$. $\square$

We conclude the chapter with a few elementary observations about branch points.

When the exponents are neither equal nor differ by an integer, a solution of the form $y(z) = z^c \sum a_n z^n$ has, when c is not an integer, a branch point at the origin: $y\left(z e^{2\pi i}\right) = \exp(2\pi i c) y(z)$. Using the notation of Theorem 6 and in any simply connected region D contained in

$$U = \{z : 0 < |z| < S\},$$

the two solutions given by the theorem continue to be linearly independent and solve (34). Therefore, as in Proposition 6 of Chapter 3 (which continues to hold for complex variables), any solution of the differential equation can be expressed as a linear combination of these two solutions in D.

When the origin is an ordinary point, analytic continuation can be used to extend any pair of linearly independent solutions near 0 to any other point where p_1/p_2 and p_0/p_2 are holomorphic.

In the case of the origin being a regular singular point, a circuit around the origin within U will change two linearly independent solutions to new linearly independent solutions: each solution will be a linear combination of functions with branches like those of z^c (or possibly those of z^c and $\log z$ when $c_1 - c_2 \in \mathbb{Z}$).

Exercise 18 (Ordinary and regular singular points at infinity) Consider the equation

$$(38) \qquad\qquad y'' + Q_1(z) y' + Q_0(z) y = 0.$$

Show that $z = \infty$ is

(i) an ordinary point of (38) if the functions

$$f_1(z) = 2z - z^2 Q_1(z), \quad f_2(z) = z^4 Q_0(z)$$

are holomorphic or have removable singularities at ∞,

(ii) a regular singular point of (38) if the functions

$$f_3(z) = z Q_1(z), \quad f_4(z) = z^2 Q_0(z)$$

are holomorphic or have removable singularities at ∞.

[HINT: Put $w = 1/z$ and transform (38) to an equation in $Y = Y(w)$ by means of $y(z) = Y(1/z)$.]

Exercise 19 Use the notation of Exercise 18. Show that

(a) when $Q_0 \equiv 0$ and Q_1 has a simple pole at $z = 0$, but no other singularities, then the general solution of (38) is $y = Az^{-1} + B$, where A and B are constants, and

(b) when $z = 0$ and $z = \infty$ are regular singular points, but there are no others, equation (38) can be re-written in the form

$$z^2 y'' - (\alpha + \beta - 1)zy' + \alpha\beta y = 0$$

(which has already been solved for real variables in Example 3).

14 Transform Methods

This chapter introduces a powerful technique for solving differential equations, both ordinary and partial, and integral equations, involving methods quite different in nature from those discussed earlier. The technique involves applying an integral operator $\mathcal{T} = \mathcal{T}_{x \to v}$ to 'transform' a function $y = y(x)$ of the independent real variable x to a new function $\mathcal{T}(y) = \mathcal{T}(y)(v)$ of the real or complex variable v. If y solves, say, a differential equation (DE) – *and satisfies conditions allowing the integral operator to be applied* – the function $\mathcal{T}(y)$ will be found to solve another equation ($\mathcal{T}$DE), perhaps also (but not necessarily) a differential equation. This gives us a new method for solving a differential equation (DE):

(a) supposing y satisfies conditions for the integral operator $\mathcal{T}$ to be applied, transform (DE) to ($\mathcal{T}$DE),

(b) solve ($\mathcal{T}$DE) to find $\mathcal{T}(y)$,

(c) find a function y the transform of which is $\mathcal{T}(y)$, which will then solve (DE).

We will find that, in solving differential equations, there can be at most one function y corresponding to a transformed function $\mathcal{T}(y)$, and hence that we will have a one-to-one correspondence between transformable y's solving (DE) and transformed functions $\mathcal{T}(y)$'s solving ($\mathcal{T}$DE).

We should note that, as with the Green's function method discussed in section 4.2, the method can allow us to build in initial or boundary conditions from the start and hence produce (following our discussion in the last paragraph) the unique solution of a problem consisting of a differential equation and side conditions.

We shall limit our discussions to considerations involving the two most important and widely used transform operators, associated with the names of Fourier and Laplace. In each case, we shall first present (unproved) some basic facts about the transform before going on to discuss in detail the application to solve differential and integral equations. We thus avoid going deeply into integration theory, a discussion which would take us too far from the central topic of this book. The reader may in any case be familiar with this part of the work from other sources, and references to relevant books are given in the Bibliography.

The final section of this chapter concerns itself with the use of complex analysis, and in particular the residue calculus, to find the function y corresponding to $\mathcal{T}(y)$. This section presumes knowledge of the theory of complex-valued functions of a single complex variable.

An appendix on similarity solutions gives an efficient alternative method for tackling some partial differential equations when subject to convenient boundary conditions.

Throughout the chapter we shall need to integrate over an infinite interval I. When we say that a function f is 'integrable' on I, we allow the alternative interpretations:

(i) f is Lebesgue integrable over I,

(ii) f is piecewise continuous on every closed and bounded interval in I and the improper Riemann integrals of both f and $|f|$ exist over I.

We recall that f is *piecewise continuous*, respectively *piecewise smooth*, on I if it is continuous, respectively smooth (that is, continuously differentiable), at all but a finite number of points of I.

14.1   The Fourier transform

Suppose that $f : \mathbb{R} \to \mathbb{R}$ is integrable over the whole real line $\mathbb{R}$. Then the *Fourier transform* $\hat{f} : \mathbb{R} \to \mathbb{R}$ of f is defined by

$$(1) \qquad \hat{f}(s) \;=\; \int_{-\infty}^{\infty} e^{-isx} f(x)\, dx \qquad (s \in \mathbb{R}).$$

The number i is $\sqrt{-1}$. For convenience, we may use the alternative notations

$$\hat{f} \;=\; \mathcal{F}(f) \quad \text{or} \quad \hat{f}(s) \;=\; \mathcal{F}(f(x))(s),$$

or even, in the case of a function $u = u(x, y)$ of two variables,

$$\hat{u}(s, y) = \mathcal{F}_{x \to s}(u(x, y))(s, y)$$

to indicate that it is the independent variable x, not the independent variable y, that is being transformed.

Direct integration will allow the reader to check the entries in the following table, save the last which is best deduced using contour integration (see section 14.5) and the well-known integral

$$(2) \qquad \int_{-\infty}^{\infty} e^{-x^2} \, dx = \sqrt{\pi}.$$

Recall that, given a subset A of $\mathbb{R}$, the *characteristic function* $\chi_A : \mathbb{R} \to \mathbb{R}$ is defined by

$$(3) \qquad \chi_A(x) = \begin{cases} 1 & \text{if } x \in A, \\ 0 & \text{otherwise.} \end{cases}$$

Let k denote a positive real constant.

$f(x)$	$\hat{f}(s)$
$e^{-kx} \cdot \chi_{(0,\infty)}(x)$	$\dfrac{1}{k + is}$
$e^{+kx} \cdot \chi_{(-\infty,0)}(x)$	$\dfrac{1}{k - is}$
$e^{-k\lvert x \rvert}$	$\dfrac{2k}{k^2 + s^2}$
$\chi_{(-k,k)}(x)$	$\dfrac{2\sin(ks)}{s}$
e^{-kx^2}	$\sqrt{\dfrac{\pi}{k}} \, e^{-s^2/4k}$

Table 1

The following compound theorem contains a number of useful and famous results about the Fourier transform. Some parts of it are essentially trivial, others represent deep results.

Theorem 1 Suppose that $f : \mathbb{R} \to \mathbb{R}$ and $g : \mathbb{R} \to \mathbb{R}$ are integrable over $\mathbb{R}$, and that a, b, k are real constants. Then, for each s in $\mathbb{R}$,

(i) (linearity)
$$\mathcal{F}(af + bg) = a\hat{f} + b\hat{g}$$

(ii) (translation)
$$\mathcal{F}(f(x + k))(s) = e^{iks}\hat{f}(s)$$

(iii) (scaling)
$$\mathcal{F}(f(kx))(s) = \frac{1}{|k|}\hat{f}\left(\frac{s}{k}\right) \qquad (k \neq 0)$$

(iv) if f' is integrable, then
$$\mathcal{F}(f'(x))(s) = is\hat{f}(s)$$

(v) (Convolution Theorem) if $f * g = f *_{\mathcal{F}} g$ denotes the Fourier *convolution* integral defined as
$$(f * g)(x) = \int_{-\infty}^{\infty} f(u)g(x - u)\, du \qquad (x \in \mathbb{R})$$
then
$$\mathcal{F}((f * g)(x))(s) = \hat{f}(s) . \hat{g}(s) \qquad (s \in \mathbb{R})$$

(vi) (Inversion Theorem) if f is continuous and piecewise smooth, then

(4)
$$f(x) = \frac{1}{2\pi} \int_{-\infty}^{\infty} e^{isx}\hat{f}(s)\, ds$$

for each x in $\mathbb{R}$.

Corollary 1 (i) If $f : \mathbb{R} \to \mathbb{R}$ is integrable, then for all s in $\mathbb{R}$,

(5)
$$\mathcal{F}(\hat{f}(x))(s) = 2\pi f(-s).$$

(ii) If the r-th derivative function $f^{(r)} : \mathbb{R} \to \mathbb{R}$ is integrable for $r = 0, 1, \ldots, n$, then for all s in $\mathbb{R}$,

(6)
$$\mathcal{F}(f^{(n)}(x))(s) = (is)^n \hat{f}(s).$$

Notes

(a) Some authors define the Fourier transform $\mathcal{F}_D$ of integrable f as

$$\hat{f}_D(s) = k \int_{-\infty}^{\infty} e^{isx} f(x)\, dx, \qquad (k \text{ constant}, k \neq 0).$$

This makes essentially no change to the theory, but one must note that then the Inversion Theorem becomes

$$f(x) = \frac{1}{2\pi k} \int_{-\infty}^{\infty} e^{-isx} \hat{f}_D(s)\, ds$$

and, for integrable f, f', we have, for example,

$$\mathcal{F}_D(f'(x))(s) = (-is)\hat{f}_D(s).$$

(b) Note that, for integrable f, g,

$$\int_{-\infty}^{\infty} f(u)g(x-u)\, du = \int_{-\infty}^{\infty} f(x-u)g(u)\, du$$

(just make the change of variable $v = x - u$). So, $f * g = g * f$ and the situation described by the Convolution Thorem (Theorem 1(v)) is a symmetric one.

(c) If in Theorem 1(vi) f is not continuous at x, then $f(x)$ on the left-hand side of the inversion formula is replaced by $\frac{1}{2}\left(f(x+) + f(x-)\right)$.

(d) One by-product of the Inversion Theorem (Theorem 1(vi)) is that, for continuous, piecewise smooth, integrable functions, there can be only one $f = f(x)$ corresponding to $\hat{f} = \hat{f}(s)$; namely, the one given by the inversion formula (4). This tells us that, in these circumstances, if one can find a function $f = f(x)$, the transform of which is $\hat{f} = \hat{f}(s)$, then one need look no further.

Example 1 For any positive constant k,

$$\mathcal{F}\left(\frac{2k}{k^2 + x^2}\right) = 2\pi e^{-k|s|}.$$

The result is deduced immediately from the third line of the table and Corollary (i) above. $\qquad\qquad\qquad\qquad\qquad\qquad\qquad\qquad\qquad\qquad\qquad\qquad\qquad\qquad\square$

The 'symmetry' represented by Example 1 and by the fifth line of Table 1 are worthy of note.

14.2   Applications of the Fourier transform

We work by example: our applications cover ordinary and partial differential equations, and integral equations.

Example 2 Use the Fourier transform to find the continuous integrable function $y = y(x)$ which satisfies the differential equation

$$y' + 2y \;=\; e^{-|x|}.$$

at every non-zero x in $\mathbb{R}$.

Using Theorem 1(iv) and the third line of Table 1 in section 14.1,

$$(is + 2)\hat{y}(s) \;=\; \frac{2}{1 + s^2}\,.$$

Hence, taking partial fractions,

$$\hat{y}(s) \;=\; \frac{1}{(1 + is)} + \frac{1}{3(1 - is)} - \frac{2}{3(2 + is)}$$

and so, using Note (d) and Table 1 again,

$$y(x) \;=\; e^{-x}\chi_{(0,\infty)}(x) + \frac{1}{3} e^{x}\chi_{(-\infty,0)}(x) - \frac{2}{3} e^{-2x}\chi_{(0,\infty)}(x)$$

or, alternatively,

$$y(x) \;=\; \begin{cases} e^{-x} - \dfrac{2}{3} e^{-2x}, & (x > 0), \\[2mm] \dfrac{1}{3} e^{x}, & (x \le 0). \end{cases} \qquad \square$$

Example 3 Assuming that $f : \mathbb{R} \to \mathbb{R}$ is integrable over $\mathbb{R}$, use the Fourier transform to show that a solution of the differential equation

$$y'' - y \;=\; f(x) \qquad (x \in \mathbb{R})$$

can be expressed in the form

$$y(x) \;=\; \int_{-\infty}^{\infty} k(x - u) f(u)\, du, \qquad (x \in \mathbb{R})$$

where the function k should be determined.

Applying the Fourier transform to the differential equation and using (6),

$$((is)^2 - 1)\hat{y}(s) \ = \ \hat{f}(s).$$

Hence,

$$\hat{y}(s) \ = \ -\left(\frac{1}{1+s^2}\right)\hat{f}(s)$$

and, by the Convolution Theorem (Theorem 1(v)) and line 3 of Table 1 in the last section,

$$y(x) \ = \ \int_{-\infty}^{\infty} k(x-u)f(u)\,du \qquad (x \in \mathbb{R})$$

is a solution, where

$$k(x) \ = \ -e^{-|x|} \qquad (x \in \mathbb{R}). \qquad \square$$

The next two examples give applications to classical partial differential equations from mathematical physics.

Example 4 (The heat equation) Use the Fourier transform to find a solution $u = u(x,t)$ of the partial differential equation

$$(7) \qquad\qquad u_t \ = \ k u_{xx}, \qquad (x \in \mathbb{R}, \, t > 0)$$

when subject to the initial condition

$$(8) \qquad\qquad u(x,0) \ = \ e^{-x^2}, \qquad (x \in \mathbb{R})$$

where k is a positive constant.

Assume that, for each t, the function u satisfies, as a function of x, the conditions for the application of the Fourier transform and that

$$\mathcal{F}_{x \to s}(u_t(x,t))(s,t) \ = \ \frac{\partial}{\partial t}\,\hat{u}(s,t);$$

that is, we can interchange the operations of applying the Fourier transform and partial differentiation with respect to t. Then, applying the transform to (7) gives

$$\frac{\partial}{\partial t}\,\hat{u}(s,t) \ = \ k(is)^2\hat{u}(s,t)$$

and hence $\hat{u} = \hat{u}(s,t)$ satisfies the first-order partial differential equation

$$\frac{\partial \hat{u}}{\partial t} + ks^2 \hat{u} = 0.$$

Hence,

$$\frac{\partial}{\partial t}\left(e^{ks^2 t}\hat{u}\right) = 0$$

and so,

$$\hat{u}(s,t) = e^{-ks^2 t}g(s),$$

where g is an arbitrary function of s. However, we may also apply the transform to the initial condition (8) to give (see Table 1)

$$\hat{u}(s,0) = \sqrt{\pi}e^{-s^2/4};$$

so,

$$\hat{u}(s,t) = \sqrt{\pi}e^{-(1+4kt)s^2/4}.$$

Again referring to Table 1, our solution is

$$u(x,t) = \frac{1}{\sqrt{1+4kt}}e^{-x^2/(1+4kt)}. \qquad \square$$

Note The method 'hides' conditions on the function $u(x,t)$ in order that the transform can be applied to the differential equation. For example, in order that u, u_x, u_{xx} be integrable (conditions for us to be able to apply Corollary (ii) to Theorem 1), we must in particular have that $u(x,t)$ and $u_x(x,t)$ tend to zero as $x \to \pm\infty$, for all $t > 0$.

Example 5 (Poisson's formula as a solution of Laplace's equation)
Let

$$(9) \qquad u(x,y) = \frac{y}{\pi}\int_{-\infty}^{\infty}\frac{e^{-|t|}}{y^2 + (x-t)^2}\,dt.$$

Find the Fourier transform $\hat{u}(s,y)$ of $u(x,y)$ as a function of x for fixed $y > 0$. Hence show that $\hat{u} = \hat{u}(s,y)$ solves the problem consisting of the equation

$$(10) \qquad -s^2\hat{u} + \frac{\partial^2 \hat{u}}{\partial y^2} = 0$$

subject to

(11)
$$\hat{u}(s, y) \to \frac{2}{1 + s^2} \quad \text{as } y \to 0.$$

Of which problem (that is, of which equation with which boundary condition) is this the Fourier transform?

The right-hand side of (9) is a convolution integral and therefore, bearing in mind line 3 of Table 1 and Example 1, the Fourier transform of (9) is

$$\hat{u}(s, y) = \frac{2}{1 + s^2} e^{-y|s|}.$$

It is immediate that $\hat{u}$ satisfies both (10) and (11). As the order of application of the Fourier and $\partial^2/\partial y^2$ operators to u is immaterial and

$$\mathcal{F}_{x \to s} \left(\frac{\partial^2}{\partial x^2} u(x, y) \right)(s, y) = (is)^2 \hat{u}(s, y),$$

we see that (10) is the Fourier transform of Laplace's equation

$$\frac{\partial^2 u}{\partial x^2} + \frac{\partial^2 u}{\partial y^2} = 0$$

for the function $u = u(x, y)$. The boundary condition (11) (see line 3 of Table 1) is clearly the transform of

$$u(x, y) \to e^{-|x|} \quad \text{as } y \to 0. \qquad \square$$

Note The reader should notice how the boundary condition transforms.

We conclude this section by using the Fourier transform to solve an integral equation.

Example 6 Suppose that the function $f : \mathbb{R} \to \mathbb{R}$ is defined by

$$f(x) = e^{-x} \chi_{(0,\infty)}(x), \qquad (x \in \mathbb{R}).$$

Use the Fourier transform to find a function $y : \mathbb{R} \to \mathbb{R}$, integrable over $\mathbb{R}$ which satisfies the integral equation

$$\int_{-\infty}^{\infty} f(u) y(x - u) \, du = e^{-\lambda|x|}, \qquad (x \in \mathbb{R})$$

where λ is a positive real constant.

Using the Convolution Theorem and lines 1 and 3 of Table 1,

$$\frac{1}{1+is}\,\hat{y}(s) \;=\; \frac{2\lambda}{\lambda^2+s^2}\,, \qquad (s \in \mathbb{R}).$$

Using partial fractions,

$$\hat{y}(s) \;=\; \frac{1-\lambda}{\lambda+is} + \frac{1+\lambda}{\lambda-is}\,, \qquad (s \in \mathbb{R}).$$

So, using lines 1 and 2 of Table 1,

$$y(x) \;=\; (1-\lambda)e^{-\lambda x}\chi_{(0,\infty)}(x) + (1+\lambda)e^{\lambda x}\chi_{(-\infty,0)}(x)$$

or

$$y(x) \;=\; \begin{cases} (1-\lambda)e^{-\lambda x}, & (x>0), \\[2mm] (1+\lambda)e^{\lambda x}, & (x<0). \end{cases}$$

This solution, though not continuous at the origin, still satisfies the integral equation for all x in $\mathbb{R}$. $\qquad\qquad\qquad\qquad\qquad\qquad\qquad\qquad\qquad\qquad\qquad\qquad\square$

Exercise 1 Use the Fourier transform to find a solution of each of the following equations, leaving your answer in the form of a convolution:

(a) $\quad y'' - 2y \;=\; e^{-|x|}, \qquad (x \in \mathbb{R})$

(b) $\quad y'' + 2y' + y \;=\; g(x), \qquad (x \in \mathbb{R})$

where $g : \mathbb{R} \to \mathbb{R}$ is integrable over $\mathbb{R}$.

Exercise 2 (a) Let f^{*n} denote the convolution $f * f * \ldots * f$, with n occurrences of the integrable function $f : \mathbb{R} \to \mathbb{R}$. Assuming that the series $\sum_{n=1}^{\infty} f^{*n}$ converges to an integrable function $F = F(x)$, show that the Fourier transform $\hat{F} = \hat{F}(s)$ of F equals $\hat{f}(s)/(1 - \hat{f}(s))$.

(b) Suppose that $g : \mathbb{R} \to \mathbb{R}$ is integrable and let α be a real number with $\alpha > 2$. Use the Fourier transform to solve

$$\alpha y'(x) + y(x-1) - y(x+1) \;=\; g(x+1) - g(x-1), \qquad (x \in \mathbb{R})$$

for y in terms of g. Obtain a solution in the form $y = g * F$, where $F = \sum_{n=1}^{\infty} f^{*n}$ and f is function which you should specify.

[You may assume convergence of the series involved.]

Exercise 3 The function $u = u(x, t)$ satisfies the heat equation

$$u_t = ku_{xx}, \qquad (x \in \mathbb{R}, \, t \geq 0)$$

where k is a positive constant, and is subject to the initial condition

$$u(x, 0) = e^{-|x|}.$$

Show, using the Fourier transform, that a solution of the problem may be expressed in the form

$$u(x, t) = \frac{2}{\pi} \int_0^\infty \frac{e^{-ku^2 t} \cos(ux)}{1 + u^2} \, du.$$

Exercise 4 Prove that if the function $f : \mathbb{R} \to \mathbb{R}$ is defined by

$$f(x) = e^{-x^2/2}, \qquad (x \in \mathbb{R})$$

then the equation

$$y * y * y = y * f * f$$

has exactly three solutions integrable over $\mathbb{R}$ (when subject to suitable regularity conditions).

14.3 The Laplace transform

The Laplace transform is perhaps the most widely used transform in mathematical physics and other applications of continuous, as opposed to discrete, mathematics. It embodies a wide range of techniques, even at the elementary level we shall discuss in this chapter, and this variety should be of especial interest to the reader. It can build in initial values of a problem and thus produce the problem's unique solution. As we shall also see, the main restriction is to functions for which the transform integral exists. However, this still allows a wide range of application.

In our discussion, we shall follow the pattern of the last two sections on the Fourier transform. In this section, we shall state the basic theory (without proofs). In the next, we shall consider application to differential and integral equations, leaving to section 14.5 those problems most appropriately tackled with the help of complex variables.

Let $\mathbb{R}^+$ denote the set $[0, \infty)$ of non-negative reals. Suppose that $f : \mathbb{R}^+ \to \mathbb{R}$ satisfies the condition that $e^{-ct} f(t)$ is integrable on $\mathbb{R}^+$ for some $c \geq 0$: we shall say that f is *c-Laplaceable*. Then the *Laplace transform* $\bar{f} = \bar{f}(p)$ is defined by

$$\bar{f}(p) \;=\; \int_0^\infty e^{-pt} f(t) \, dt,$$

for all complex numbers p satisfying $\operatorname{re} p > c$. That f is c-Laplaceable ensures that $\bar{f}$ exists for $\operatorname{re} p > c$ (a domination argument), as then $|e^{-pt} f(t)| < e^{-ct}|f(t)|$. For convenience, we may use the alternative notations

$$\bar{f} \;=\; \mathcal{L}(f) \quad \text{or} \quad \bar{f}(p) \;=\; \mathcal{L}(f(t))(p)$$

or even, in the case of a function $u = u(x,t)$ of two variables,

$$\bar{u}(x,p) \;=\; \mathcal{L}_{t \to p}(u(x,t))(x,p)$$

to indicate that it is the independent variable t, not the independent variable x, that is being transformed.

Direct integration will allow the reader to check the entries in the following table. Let λ be a real number, w a complex number and n a positive integer.

$f(t)$	$\bar{f}(p)$
t^n	$\dfrac{n!}{p^{n+1}}$
e^{-wt}	$\dfrac{1}{p+w}$
$\cos \lambda t$	$\dfrac{p}{p^2 + \lambda^2}$
$\sin \lambda t$	$\dfrac{\lambda}{p^2 + \lambda^2}$
$\cosh \lambda t$	$\dfrac{p}{p^2 - \lambda^2}$
$\sinh \lambda t$	$\dfrac{\lambda}{p^2 - \lambda^2}$

Table 2

As with the Fourier transform, we now bring together a number of useful results, some of considerable significance. We use the *Heaviside function* $H : \mathbb{R} \to \mathbb{R}$, defined by

$$H(x) \;=\; \chi_{[0,\infty)}(x) \;=\; \begin{cases} 1, & \text{for } x \geq 0, \\[2mm] 0, & \text{for } x < 0. \end{cases}$$

Theorem 2 Suppose that $f : \mathbb{R}^+ \to \mathbb{R}$ and $g : \mathbb{R}^+ \to \mathbb{R}$ are c-Laplaceable for some $c \geq 0$, that a, b, λ are real constants, that w is a complex number, and n is a positive integer. Then, for $\mathrm{re}\, p > c$,

(i) (linearity) $\quad \mathcal{L}(af + bg) \;=\; a\bar{f} + b\bar{g}$

(ii) (Shift Theorem)

$$\mathcal{L}(e^{-wt} f(t))(p) \;=\; \bar{f}(p + w)$$

(iii) $\qquad\qquad \mathcal{L}(f(t - \lambda)H(t - \lambda))(p) \;=\; e^{-\lambda p}\bar{f}(p)$

(iv) if $t^r f(t)$ is c-Laplaceable for $r = 0, 1, \ldots, n$, then

$$\mathcal{L}(t^n f(t))(p) \;=\; (-1)^n \frac{d^n}{dp^n}\,\bar{f}(p)$$

(v) if $f^{(r)}(t)$ is c-Laplaceable for $r = 0, 1, \ldots, n$, then

$$\mathcal{L}(f^{(n)}(t))(p) \;=\; -f^{(n-1)}(0) - pf^{(n-2)}(0) - \ldots - p^{n-1}f(0) + p^n\bar{f}(p)$$

(vi) (Convolution Theorem) if $f * g = f *_{\mathcal{L}} g$ denotes the Laplace *convolution* integral defined by

$$(f * g)(t) \;=\; \int_0^t f(u)g(t - u)\, du \qquad (t \in \mathbb{R}^+)$$

then

$$\mathcal{L}((f * g)(t))(p) \;=\; \bar{f}(p) \cdot \bar{g}(p)$$

(vii) (Lerch's Uniqueness Theorem) if f and g are both *continuous* on $\mathbb{R}^+$ and $\bar{f}(p) = \bar{g}(p)$ whenever $\operatorname{re} p > c$, then $f = g$ on $\mathbb{R}^+$

(viii) if f is continuous and piecewise smooth, and

$$\bar{f}(p) = \sum_{n=0}^{\infty} \frac{n!\, a_n}{p^{n+1}}$$

is convergent for $|p| > T$, where T is a non-negative constant, then

$$f(t) = \sum_{n=0}^{\infty} a_n t^n,$$

for $t > 0$.

Notes

(a) The reader will have noticed our omission of an Inversion Theorem. We have delayed a statement of this result until section 14.5, as it involves a contour integral best used in conjunction with the Residue Theorem of complex analysis.

(b) Provided we are only looking for continuous $f = f(t)$ corresponding to a given $\bar{f} = \bar{f}(p)$ (as when seeking solutions to differential equations), Lerch's Theorem ((vii) above) assures us that once an $f = f(t)$ has been found, there can be no other.

(c) As with the Fourier transform, it is easy to show that

$$f *_{\mathcal{L}} g = g *_{\mathcal{L}} f.$$

(d) Theorem 1(viii) legitimises term-by-term inversion, extending Table 2, line 1.

Example 7 Determine the Laplace transform $\overline{L_n} = \overline{L_n}(p)$ of the Laguerre polynomial $L_n = L_n(t)$, defined by

$$L_n(t) = e^t \frac{d^n}{dt^n} (t^n e^{-t}), \qquad (t \in \mathbb{R}^+)$$

where n is a positive integer (see Chapters 12 and 13).

Using line 2 of Table 2 and Theorem 2(iv),

$$\mathcal{L}(t^n e^{-t})(p) \;=\; (-1)^n \frac{d^n}{dp^n}\left(\frac{1}{p+1}\right) \;=\; \frac{1}{(p+1)^{n+1}}.$$

Hence, by Theorem 2(v),

$$\mathcal{L}\left(\frac{d^n}{dt^n}\left(t^n e^{-t}\right)\right)(p) \;=\; \frac{p^n}{(p+1)^{n+1}}.$$

Therefore, using Theorem 2(ii),

$$\overline{L_n}(p) \;=\; \frac{(p-1)^n}{p^{n+1}}. \qquad\qquad \square$$

Example 8 Find the continuous function $f = f(t)$ which has the Laplace transform $\bar{f} = \bar{f}(p)$ given by

$$\bar{f}(p) \;=\; \frac{p^4 + 3p^2 - 2p + 2}{(p^2+1)^2(p+1)^2}.$$

By partial fractions,

$$\bar{f}(p) \;=\; \frac{1}{(p^2+1)^2} + \frac{1}{(p+1)^2}.$$

Using line 3 of Table 2,

$$\mathcal{L}(t\cos t) \;=\; -\frac{d}{dp}\left(\frac{p}{p^2+1}\right) \;=\; \frac{1}{p^2+1} - \frac{2}{(p^2+1)^2},$$

we can deduce that

$$\frac{1}{(p^2+1)^2} \;=\; \tfrac{1}{2}\{\mathcal{L}(\sin t) - \mathcal{L}(t\cos t)\}$$

$$\;=\; \mathcal{L}(\tfrac{1}{2}\sin t - \tfrac{1}{2} t\cos t)$$

by linearity (Theorem 2(i)) and line 4 of Table 2. Hence, using line 1 of Table 2, Theorem 2(ii) and linearity again,

$$f(t) \;=\; \tfrac{1}{2}\sin t - \tfrac{1}{2} t\cos t + te^{-t}. \qquad\qquad \square$$

14.4   Applications of the Laplace transform

Again, we work by example.

Example 9 Use the Laplace transform to solve the equation

$$ty'' + (1+t)y' + y \; = \; t^2, \qquad (t \geq 0).$$

Why does your solution contain only one arbitrary constant?

Using Theorem 2(iv) and (v), and line 1 of Table 2,

$$-\frac{d}{dp}\left(-y'(0) - py(0) + p^2\bar{y}(p)\right) + \left(-y(0) + p\bar{y}(p)\right) - \frac{d}{dp}\left(-y(0) + p\bar{y}(p)\right) + \bar{y}(p) \; = \; \frac{2!}{p^3}.$$

Hence,

$$\frac{d}{dp}\{(p+1)\bar{y}(p)\} \; = \; -\frac{2}{p^4}$$

and, after one integration,

$$\bar{y}(p) \; = \; \frac{2}{3p^3(p+1)} + \frac{A}{p+1},$$

where A is arbitrary constant. Partial fractions give

$$\bar{y}(p) \; = \; \frac{2}{3}\left(\frac{1}{p^3} - \frac{1}{p^2} + \frac{1}{p}\right) + \frac{A'}{p+1},$$

where A' is constant. So, as any solution of the differential equation must be continuous, Lerch's Theorem (Theorem 2(vii)) allows us to deduce that

$$y(t) \; = \; \tfrac{1}{3}(t^2 - 2t + 2) + A'e^{-t}.$$

The reason why this solution contains only one arbitrary constant is that the second solution does not have a Laplace transform; that is, there is no $c \geq 0$ for which it is c-Laplaceable. The method of solutions in series (see Chapter 13) in fact produces equal exponents and hence one solution containing a logarithm. □

Note The limitations of the use of the Laplace transform are well-illustrated in the paragraph before this note.

Example 10 Find a non-trivial solution of the differential equation

$$(12) \qquad\qquad ty'' + (1 - t)y' + ny = 0,$$

where n is a positive integer.

Using the same results as we did in Example 9,

$$-\frac{d}{dp}\left(-y'(0) - py(0) + p^2\bar{y}(p)\right) + \left(-y(0) + p\bar{y}(p)\right)$$

$$-\left\{-\frac{d}{dp}\left(-y(0) + p\bar{y}(p)\right)\right\} + n\bar{y}(p) = 0$$

and hence

$$\frac{d}{dp}\left(\frac{p^{n+1}}{(p-1)^n}\bar{y}(p)\right) = 0,$$

so that,

$$\bar{y}(p) = \frac{A(p-1)^n}{p^{n+1}},$$

where A is an arbitrary constant. However, in Example 7, we showed that

$$\overline{L_n}(p) = \frac{(p-1)^n}{p^{n+1}}$$

when $L_n = L_n(t)$ is the Laguerre polynomial of degree n. Hence, by Lerch's Theorem, a solution of (12) is

$$y(t) = AL_n(t),$$

where A is an arbitrary constant. $\qquad\qquad\qquad\qquad\qquad\qquad\qquad\square$

Example 11 (Bessel's equation of order zero – see Chapter 13) Suppose that $y = y(t)$ satisfies Bessel's equation of order zero

$$(13) \qquad\qquad ty'' + y' + ty = 0 \qquad (t \geq 0)$$

and $y(0) = 1$. Show that the Laplace transform $\bar{y} = \bar{y}(p)$ of y satisfies

$$(14) \qquad\qquad (p^2 + 1)\frac{d\bar{y}}{dp} + p\bar{y} = 0$$

and deduce that

(15)
$$y(t) = \sum_{n=0}^{\infty} \frac{(-1)^n t^{2n}}{2^{2n}(n!)^2}, \qquad (t \geq 0).$$

Show also that

$$\int_0^t y(u)y(t-u)\,du = \sin t, \qquad (t > 0).$$

[You may assume that the Laplace transforms of y, y', y'', ty, ty'' exist for re $p > 1$ and that the convolution is piecewise smooth.]

Applying the Laplace transform to (13), with the help of Theorem 2, gives

$$-\frac{d}{dp}(-y'(0) - py(0) + p^2\bar{y}(p)) + (-y(0) + p\bar{y}(p)) - \frac{d}{dp}\bar{y}(p) = 0$$

and hence equation (14). Therefore,

$$\frac{d}{dp}\{(p^2+1)^{\frac{1}{2}}\bar{y}(p)\} = 0$$

from which we derive, where A is an arbitrary constant,

(16)
$$\bar{y}(p) = \frac{A}{(p^2+1)^{\frac{1}{2}}}$$

$$= \frac{A}{p}\left(1 + \frac{1}{p^2}\right)^{-\frac{1}{2}}$$

$$= A\sum_{n=0}^{\infty} \frac{(-1)^n (2n)!}{2^{2n}(n!)^2} \cdot \frac{1}{p^{2n+1}}, \qquad (|p| > 1).$$

As $y = y(t)$ is continuous, we can use Theorem 2(viii):

$$y(t) = A\sum_{n=0}^{\infty} \frac{(-1)^n}{2^{2n}(n!)^2} t^{2n}, \qquad (t > 0).$$

The condition $y(0) = 1$ gives $A = 1$ and hence equation (15).

From equation (16), we have

$$(\bar{y}(p))^2 \;=\; \frac{1}{p^2 + 1} \;=\; \mathcal{L}(\sin t)(p).$$

So, by the Convolution Theorem,

$$\int_0^t y(u)y(t - u)\, du \;=\; \sin t, \qquad (t > 0). \qquad \square$$

We now solve an 'integro-differential' equation, containing an integral and a derivative.

Example 12 Find a solution of the equation

$$(17) \qquad\qquad y'(t) + 2\int_0^t y(u)e^{2(t-u)}\, du \;=\; e^{2t}, \qquad (t \ge 0)$$

subject to the condition $y(0) = 0$.

Using Theorem 2(v) and (vi),

$$-y(0) + p\bar{y}(p) + \frac{2\bar{y}(p)}{p - 2} \;=\; \frac{1}{p - 2}$$

and hence,

$$\bar{y}(p) \;=\; \frac{1}{(p - 1)^2 + 1}.$$

As any solution of (17) must be differentiable and therefore continuous, Lerch's Theorem, together with Theorem 2(ii), gives

$$y(t) \;=\; e^t \sin t. \qquad \square$$

Our last worked example in the section combines a partial differential equation and a Heaviside function.

Example 13 The function $u = u(x, t)$ is defined for $x \ge 1$, $t \ge 0$ and solves the following boundary value problem:

$$(18) \qquad\qquad x\frac{\partial u}{\partial x} + \frac{\partial u}{\partial t} \;=\; 1, \qquad u(x, 0) \;=\; u(1, t) \;=\; 1.$$

Show that the Laplace transform $\bar{u} = \bar{u}(x,p)$ of u, transforming the t variable, satisfies

(19)
$$x\frac{\partial \bar{u}}{\partial x} + p\bar{u} = \frac{1}{p} + 1.$$

Show further that

(20)
$$\bar{u}(x,p) = \frac{1}{p^2} + \frac{1}{p} - \frac{x^{-p}}{p^2}$$

and deduce that

$$u(x,t) = \begin{cases} 1 + \log x, & \text{if } e^t \geq x, \\ \\ 1 + t, & \text{if } e^t < x. \end{cases}$$

Applying the operator $\mathcal{L}_{t \to p}$ to the partial differential equation gives, for each $x \geq 1$,

$$x\frac{\partial \bar{u}}{\partial x}(x,p) + (-u(x,0) + p\bar{u}(x,p)) = \frac{1}{p},$$

which is (19), once we apply the first boundary condition. (We have assumed that our solution will allow the interchange of the $\mathcal{L}_{t \to p}$ and $\partial/\partial x$ operators.) Hence,

$$\frac{\partial}{\partial x}(x^p \bar{u}(x,p)) = x^{p-1}\left(\frac{1}{p} + 1\right).$$

Therefore,

(21)
$$\bar{u}(x,p) = \frac{1}{p^2} + \frac{1}{p} + A(p)x^{-p},$$

where $A = A(p)$ is an arbitrary function of p. Now apply $\mathcal{L}_{t \to p}$ to the second boundary condition, to give

(22)
$$\bar{u}(1,p) = \frac{1}{p}.$$

Equations (21) and (22) give $A(p) = -p^{-2}$ from which (20) follows immediately.

Recalling that $x^{-p} = e^{-p \log x}$, we deduce, using Theorem 2(iii) with $\lambda = \log x$, that

$$u(x,t) = t + 1 - (t - \log x)H(t - \log x),$$

where $H = H(x)$ is the Heaviside function. Hence,

$$u(x,t) = \begin{cases} 1 + \log x, & \text{for } t \geq \log x, \\ 1 + t, & \text{for } t < \log x, \end{cases}$$

which is the required solution. □

Exercise 5 Using the Laplace transform, find a solution to the differential equation

$$xy'' + 2y' + xy = 0 \qquad (x > 0)$$

which satisfies $y(0+) = 1$. Explain briefly why this method yields only one solution.

Exercise 6 Suppose that the function $F : (0, \infty) \to \mathbb{R}$ is defined by

$$F(y) = \int_0^\infty \frac{e^{-xy} \sin x}{x} \, dx, \qquad (y > 0).$$

Derive a differential equation satisfied by F and hence determine this function.

[You may assume that

$$\int_0^\infty \frac{\sin x}{x} \, dx = \frac{\pi}{2}.]$$

Exercise 7 Using the Laplace transform, solve the equation

$$\left(\frac{d^2}{dt^2} + a^2 \right)^2 y = \sin bt, \qquad (t \geq 0)$$

where $y^{(n)}(0) = 0$ $(n = 0, 1, 2, 3)$ and a, b are positive constants.

Exercise 8 Find a solution $y = y(x)$ to the integral equation

$$y(t) = e^{-t} + \int_0^t y(t - u) \sin u \, du, \qquad (t > 0).$$

Exercise 9 The functions $u_n : [0, \infty) \to \mathbb{R}$ $(n = 0, 1, 2, \ldots)$ are related by a system of differential equations

$$u_n' = u_{n-1} - u_n, \qquad (n = 1, 2, 3, \ldots).$$

Use the Laplace transform to express u_n in the form

$$u_n(t) = \int_0^t \varphi_{n-1}(t - \tau) u_0(\tau) \, d\tau + \sum_{r=1}^n \varphi_{n-r}(t) u_r(0), \qquad (n = 1, 2, 3, \ldots; t > 0)$$

where the functions $\varphi_0, \varphi_1, \varphi_2, \ldots$ should be specified.

Exercise 10 Let a and b be the roots of the quadratic equation $x^2 + \alpha x + \beta = 0$, where α and β are real constants satisfying $\alpha^2 > 4\beta$. Show that the Laplace transform of

$$y(t) \;=\; \frac{\cos at - \cos bt}{a - b} \qquad (t \geq 0)$$

is $\bar{y} = \bar{y}(p)$ given by

$$\bar{y}(p) \;=\; \frac{\alpha p}{(p^2 - \beta)^2 + \alpha^2 p^2} \,.$$

Use the Laplace transform to find the solution, in the case $\alpha^2 > 4\beta$, of the system of simultaneous differential equations

$$\frac{d^2 x}{dt^2} - \alpha \frac{dy}{dt} - \beta x \;=\; 0,$$

$$\frac{d^2 y}{dt^2} + \alpha \frac{dx}{dt} - \beta y \;=\; 0,$$

for $t \geq 0$, subject to the initial conditions

$$x(0) \;=\; \frac{dx}{dt}(0) \;=\; y(0) \;=\; 0, \qquad \frac{dy}{dt}(0) \;=\; 1.$$

What is $x(t)$ when $\alpha^2 = 4\beta$?

Exercise 11 (The wave equation) Use the Laplace transform to find a function $u = u(x,t)$ which is continuous on $\{(x,t) : x \geq 0,\, t \geq 0\}$ and which satisfies the partial differential equation

$$u_{xx} \;=\; \frac{1}{c^2} u_{tt}, \qquad (x > 0,\, t > 0,\, c \text{ a positive constant})$$

together with the conditions

$$u(x,0) \;=\; 0, \quad u_t(x,0) \;=\; 0, \qquad (x > 0)$$

$$\lim_{x \to \infty} u(x,t) \;=\; 0, \quad \frac{d^2}{dt^2}(u(0,t)) + 2u_x(0,t) \;=\; 0, \qquad (t > 0)$$

where

$$u(0,0) \;=\; 0, \qquad \frac{d}{dt}(u(0,t))\Big|_{t=0} \;=\; 1.$$

Exercise 12 (The heat equation) Suppose that the Laplace transform $\bar{f} = \bar{f}(p)$ of the continuous function $f : \mathbb{R}^+ \to \mathbb{R}$ is defined for $\operatorname{re} p > c > 0$, where c is a real constant, and that $g : \mathbb{R}^+ \to \mathbb{R}$ is given by

$$g(t) = \int_0^t f(s)\, ds, \qquad (t \geq 0).$$

Show that the Laplace transform $\bar{g} = \bar{g}(p)$ of g satisfies

$$\bar{g}(p) = \frac{\bar{f}(p)}{p}, \qquad (\operatorname{re} p > c).$$

Further, show that the Laplace transform of $t^{-\frac{1}{2}}$ is $(\pi/p)^{\frac{1}{2}}$.

Consider the problem consisting of the heat equation

$$u_t = u_{xx}, \qquad (x \geq 0,\, t \geq 0)$$

together with the boundary conditions

$$u(x,0) = 0 \quad (x \geq 0), \quad k u_x(0,t) = -Q \quad (t \geq 0)$$

where Q is a real constant. Assuming the Laplace transform with respect to t may be applied to the differential equation, find $u(0,t)$ for $t \geq 0$.

Note The problem of Exercise 12 corresponds to modelling a conductor in the half-space $x \geq 0$, initially having zero temperature, but to which a constant flux Q is supplied. The temperature on the face $x = 0$, for subsequent times t, is sought.

14.5 Applications involving complex analysis

In this section, we shall assume some familiarity with the theory of complex-valued functions of a single complex variable, in particular, Cauchy's Theorem and the Residue Theorem. We shall therefore find ourselves needing to calculate residues, to use Jordan's Lemma, to consider limiting arguments as contours 'expand' to infinity and to have due regard to branch-points.

As earlier in the chapter, we begin by looking at the Fourier transform before moving on to consideration of the Laplace transform. In both cases, the discussion will centre on the application of the relevant inversion theorem.

First of all, we fulfil a promise made in section 14.1.

Lemma Suppose that the constant $k > 0$. Then

$$\mathcal{F}(e^{-kx^2})(s) = \sqrt{\frac{\pi}{k}}\, e^{-s^2/4k}, \qquad (s \in \mathbb{R}).$$

Proof We need to evaluate the integral

$$\mathcal{F}(e^{-kx^2})(s) = \int_{-\infty}^{\infty} e^{-isx} e^{-kx^2}\, dx, \qquad (s \in \mathbb{R})$$

$$= e^{-s^2/4k} \int_{-\infty}^{\infty} e^{-k\left(x+\frac{is}{2k}\right)^2}\, dx.$$

Integrate $f(z) = e^{-kz^2}$ around the following contour, where $s > 0$ and R, S are positive real numbers.

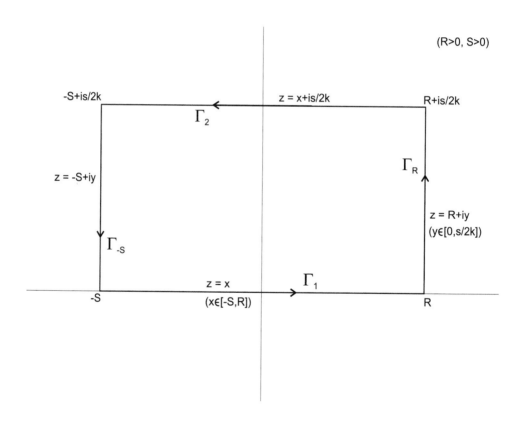

Consider the integral of f along the contour Γ_T, where $z = T + iy$ and $0 \le y \le s/2k$.

$$E \equiv \left| \int_{\Gamma_T} f \right| = \left| \int_0^{\frac{s}{2k}} e^{-k(T+iy)^2} i\, dy \right| \le e^{-kT^2} \int_0^{\frac{s}{2k}} e^{ky^2}\, dy.$$

As the integral of e^{ky^2} is bounded on $[0, s/2k]$, $E \to 0$ as $T \to \pm\infty$. So, the integrals along the 'vertical' sides of the contour tend to zero as $R, S \to \infty$. As f has no singularities, its integral around the whole contour must be zero. Hence, letting $R, S \to \infty$,

$$\int_{-\infty}^{\infty} e^{-kx^2}\, dx + \int_{\infty}^{-\infty} e^{-k\left(x + \frac{is}{2k}\right)^2}\, dx \;=\; 0$$

and therefore

$$\mathcal{F}(e^{-kx^2})(s) \;=\; e^{-s^2/4k} \int_{-\infty}^{\infty} e^{-kx^2}\, dx,$$

for $s > 0$. We now assume knowledge of the result

$$\int_{-\infty}^{\infty} e^{-y^2}\, dy \;=\; \sqrt{\pi}$$

which, after the substitution $y = \sqrt{k}x$ $(k > 0)$, gives

$$\int_{-\infty}^{\infty} e^{-kx^2}\, dx \;=\; \sqrt{\frac{\pi}{k}},$$

itself the result for $s = 0$. As minor modifications to the argument give the result for $s < 0$, the Lemma follows at once. $\qquad\square$

We next apply the Fourier Inversion Theorem (Theorem 1(vi)) to give an alternative solution to an example in section 14.2.

Example 14 Use the Fourier Inversion Theorem to find the continuous integrable function $y = y(x)$ which satisfies the differential equation

$$y' + 2y \;=\; e^{-|x|}$$

at every non-zero x in $\mathbb{R}$.

From our working in Example 2,

$$\hat{y}(s) \;=\; \frac{2}{(is + 2)(1 + s^2)}.$$

The Inversion Theorem gives

$$y(x) \;=\; \frac{1}{i\pi} \int_{-\infty}^{\infty} \frac{e^{isx}}{(s - 2i)(s^2 + 1)}\, ds.$$

The integrand is $O(s^{-3})$ and we may therefore use the principal value integral $\lim_{R\to\infty}\int_{-R}^{R}$ to evaluate it. We integrate the function

$$f(z) \;=\; \frac{e^{ixz}}{(z-2i)(z^2+1)},$$

considering separately the two cases $x > 0$ and $x < 0$.

For $x > 0$, we integrate $f(z)$ around the contour

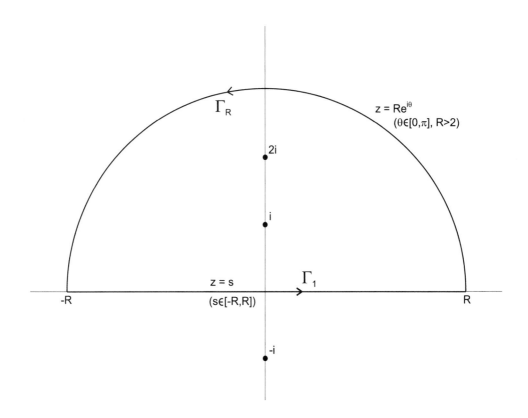

There are two simple poles, at i and $2i$, with respective residues

$$\left(\frac{e^{ixz}}{(z-2i).2i}\right)\Bigg|_{z=i} \;=\; \frac{e^{-x}}{2}, \qquad \left(\frac{e^{ixz}}{z^2+1}\right)\Bigg|_{z=2i} \;=\; -\frac{e^{-2x}}{3}.$$

Further, on the semi-circle Γ_R, as $R > 2$,

$$F \equiv \left| \int_{\Gamma_R} f \right| = \left| \int_0^\pi \frac{e^{ixR(\cos\theta + i\sin\theta)}}{(Re^{i\theta} - 2i)(R^2 e^{2i\theta} + 1)} iRe^{i\theta}\, d\theta \right|$$

$$\leq \frac{R}{(R-2)(R^2-1)} \int_0^\pi e^{-xR\sin\theta}\, d\theta.$$

Since $x > 0$, $xR\sin\theta \geq 0$ when $\theta \in [0, \pi]$, and $F \to 0$ as $R \to \infty$. Therefore, using the Residue Theorem and letting $R \to \infty$, we have that the principal value integral

$$\mathrm{P}\!\int_{-\infty}^{\infty} \frac{e^{ixs}}{(s-2i)(s^2+1)}\, ds = 2\pi i \left(\frac{e^{-x}}{2} - \frac{e^{-2x}}{3} \right).$$

Hence,

(23) $$y(x) = e^{-x} - \tfrac{2}{3} e^{-2x}, \qquad (x > 0).$$

For the case $x < 0$, we integrate $f(z)$ around the following contour.

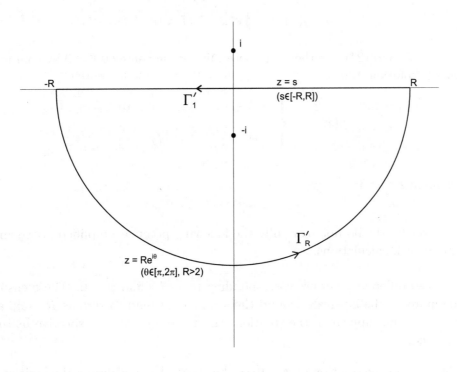

This time there is just one (simple) pole within the contour at $z = -i$ of residue

$$\left(\frac{e^{ixz}}{(z-2i).2z}\right)\bigg|_{z=-i} = -\frac{e^x}{6}.$$

On Γ'_R ($R > 2$), we have (as above)

$$F' \equiv \left|\int_{\Gamma'_R} f\right| \leq \frac{R}{(R-2)(R^2-1)}\int_0^\pi e^{-xR\sin\theta}\,d\theta.$$

Since both $x < 0$ and $\theta \in [\pi, 2\pi]$, we retain $xR\sin\theta \geq 0$, and therefore $F' \to 0$ as $R \to \infty$. Noting that integration along the line segment $[-R, R]$ is from right to left, use of the Residue Theorem gives, as $R \to \infty$,

$$\mathrm{P}\int_\infty^{-\infty} \frac{e^{ixs}}{(s-2i)(s^2+1)}\,ds = 2\pi i\left(-\frac{e^x}{6}\right)$$

and hence,

(24) $y(x) = \tfrac{1}{3}e^x, \qquad (x < 0).$

Putting (23) and (24) together, and determining the value at $x = 0$ by continuity, we arrive at the solution

$$y(x) = \begin{cases} e^{-x} - \tfrac{2}{3}e^{-2x}, & (x > 0) \\[2mm] \tfrac{1}{3}e^x, & (x \leq 0) \end{cases}$$

as in Example 2.   □

Note The reader should note carefully the following points exemplified above and commonplace in such calculations.

(a) We need different contours corresponding to $x > 0$ and $x < 0$. These ensure that the limits of the integrals around the semi-circles tend to zero as $R \to \infty$ and the functions that appear in the solution tend to zero on their respective half-lines as $|x| \to \infty$.

(b) The value of the solution at $x = 0$ can be sought by considering the limit as $x \to 0$.

(c) We have used the following formula for calculating the residue of a holomorphic function of the form

$$f(z) = \frac{P(z)}{Q(z)}$$

at $z = a$, where P is holomorphic in a neighbourhood of a with $P(a) \neq 0$, and Q has a simple zero at a:

$$\operatorname{res}(f, a) = \frac{P(a)}{Q'(a)}.$$

The rest of this section and its exercises concern the Laplace transform.

Theorem 3 (Laplace Inversion Theorem) Suppose that the continuous and piecewise smooth function $f : \mathbb{R}^+ \to \mathbb{R}$ is c-Laplaceable for some $c \geq 0$. Then, for $t > 0$, f may be expressed as the principal value contour integral,

$$f(t) = \frac{1}{2\pi i} \mathrm{P} \int_{\sigma-i\infty}^{\sigma+i\infty} e^{pt} \bar{f}(p) \, dp \equiv \frac{1}{2\pi i} \lim_{R \to \infty} \int_{\sigma-iR}^{\sigma+iR} e^{pt} \bar{f}(p) \, dp,$$

for any real $\sigma > c$.

Note As with the Fourier Inversion Theorem, it f is not continuous at t, then $f(t)$ on the left-hand side of the above formula is replaced by $\frac{1}{2}(f(t+) + f(t-))$.

Corollary Suppose in addition, $\bar{f}$ is holomorphic save for at most a finite sequence of poles $(p_n)_{n=1}^N$, where $\operatorname{re} p_n < c$, and that there are real positive constants M and d such that

$$|\bar{f}(p)| \leq \frac{M}{|p|^d}$$

as $p \to \infty$, then

$$f(t) = \sum_{n=1}^N \operatorname{res}(e^{pt} \bar{f}, p_n),$$

the sum of the residues at the poles p_n.

Proof We integrate the function

$$g(p) = e^{pt} \bar{f}(p)$$

around the contour

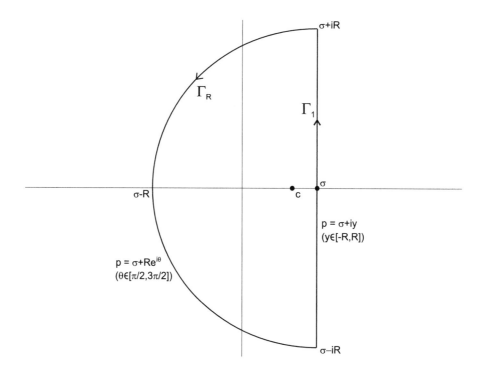

where R is so large that all the p_n lie inside the contour. Then, considering only integration along the semi-circular arc,

$$E \equiv \left| \int_{\Gamma_R} g \right| = \left| \int_{\frac{\pi}{2}}^{\frac{3\pi}{2}} e^{t(\sigma + R(\cos\theta + i\sin\theta))} \bar{f}(\sigma + Re^{i\theta}) iRe^{i\theta}\, d\theta \right|$$

$$\leq \frac{Me^{t\sigma}}{R^{d-1}} \int_{\frac{\pi}{2}}^{\frac{3\pi}{2}} e^{tR\cos\theta}\, d\theta$$

$$= \frac{2Me^{t\sigma}}{R^{d-1}} \int_0^{\frac{\pi}{2}} e^{-tR\sin\varphi}\, d\varphi$$

when one puts $\theta = \varphi + \frac{\pi}{2}$ and notes that $\sin\varphi = \sin(\pi - \varphi)$. Using Jordan's Lemma (which implies $-\sin\theta \leq -2\theta/\pi$ on $[0, \frac{\pi}{2}]$) and as we have taken $t > 0$, we can deduce that

$$\int_0^{\frac{\pi}{2}} e^{-tR\sin\varphi}\, d\varphi \leq \int_0^{\frac{\pi}{2}} e^{-2tR\varphi/\pi}\, d\varphi = \frac{\pi}{2tR}\left(1 - e^{-tR}\right).$$

Therefore, $E \to 0$ as $R \to \infty$ (as $d > 0$), and using the Residue Theorem

$$\lim_{R \to \infty} \int_{\sigma - iR}^{\sigma + iR} g(p)\, dp \;=\; 2\pi i \sum_{n=1}^{N} \operatorname{res}(g, p_n).$$

The corollary now follows from the Inversion Theorem. □

In the context of the Laplace Inversion Theorem, we now re-visit Example 9.

Example 15 Use the Laplace Inversion Theorem to solve the equation

$$ty'' + (1 + t)y' + y \;=\; t^2, \qquad (t \geq 0).$$

Our working in Example 9 gave

$$\bar{y}(p) \;=\; \frac{3Ap^3 + 2}{3p^3(p+1)},$$

where A is an arbitrary constant. This function satisfies the conditions of the Corollary to Theorem 3 (with $d = 1$), and has a triple pole at $p = 0$ and a simple pole at $p = -1$. Considering now the function

$$g(p) \;=\; e^{pt}\bar{y}(p) \;=\; \left(\frac{2 + 3Ap^3}{3p^3(p+1)} \right) e^{pt},$$

for $t > 0$,

$$\operatorname{res}(g, -1) \;=\; \left. \left(\frac{2 + 3Ap^3}{3p^3(p+1)} \right) e^{pt} \right|_{p=-1} \;=\; A'e^{-t},$$

where A' is an arbitrary constant. Expanding near $p = 0$, we find

$$g(p) \;=\; \frac{1}{3p^3}\,(2 + 3Ap^3)\left(1 + pt + \frac{p^2 t^2}{2} + \ldots \right)(1 - p + p^2 + \ldots).$$

Hence, the residue at $p = 0$, which is the coefficient of p^{-1} in this unique Laurent expansion, is

$$\tfrac{1}{3}(t^2 - 2t + 2).$$

Therefore, by the Corollary to Theorem 3,

$$y(t) \;=\; \tfrac{1}{3}(t^2 - 2t + 2) + A'e^{-t},$$

for $t > 0$, which extends to $t = 0$ by continuity. This agrees with our result in Example 2. □

The next example is important and involves a branch-point and a 'keyhole' contour.

Example 16 (The heat equation) Find the real-valued function $u = u(x,t)$ which is defined and continuous on $\{(x,t) : x \geq 0,\, t \geq 0\}$, and satisfies the equation

$$\kappa u_{xx} = u_t,$$

when subject to

$$u(x,0) = 0 \quad (x \geq 0), \quad u(0,t) = u_0 \quad (t \geq 0),$$

and such that

$$\lim_{x \to \infty} u(x,t) \text{ exists},$$

as a (finite) real number, for each $t \geq 0$.

Applying the operator $\mathcal{L}_{t \to p}$ to the differential equation (and assuming it can be interchanged with $\partial^2/\partial x^2$),

$$\kappa \frac{\partial^2}{\partial x^2}\, \bar{u}(x,p) = -u(x,0) + p\bar{u}(x,p).$$

So,

$$\frac{\partial^2 \bar{u}}{\partial x^2} + \frac{p}{\kappa}\, \bar{u} = 0.$$

Hence,

(25)
$$\bar{u}(x,p) = A(p) e^{(p/k)^{\frac{1}{2}} x} + B(p) e^{-(p/k)^{\frac{1}{2}} x},$$

where we have cut the plane, so that $-\pi < \arg p \leq \pi$, to cope with the branch-point at the origin. Then $\mathrm{re}\,(p^{\frac{1}{2}}) \geq 0$. Applying the same operator to the second condition,

(26)
$$\bar{u}(0,p) = \frac{u_0}{p}, \qquad (\mathrm{re}\, p > 0).$$

For the last condition to hold, we must have $\bar{u}(x,p) \to 0$ as $x \to \infty$, and therefore $A(p) \equiv 0$; so, combining (25) and (26), we have

$$\bar{u}(x,p) = \frac{u_0}{p}\, e^{-(p/k)^{\frac{1}{2}} x}, \qquad (\mathrm{re}\, p > 0).$$

The Laplace Inversion now implies, for a positive constant σ, that

$$u(x,t) = \frac{u_0}{2\pi i}\, \mathrm{P}\!\int_{\sigma-i\infty}^{\sigma+i\infty} \frac{1}{p}\, e^{pt - (p/k)^{\frac{1}{2}} x}\, dp, \qquad (t > 0).$$

In order to express this solution in terms of real variables alone, we integrate the function

$$g(p) \;=\; \frac{1}{p}\, e^{pt-(p/k)^{\frac{1}{2}}x}$$

around the *keyhole contour*

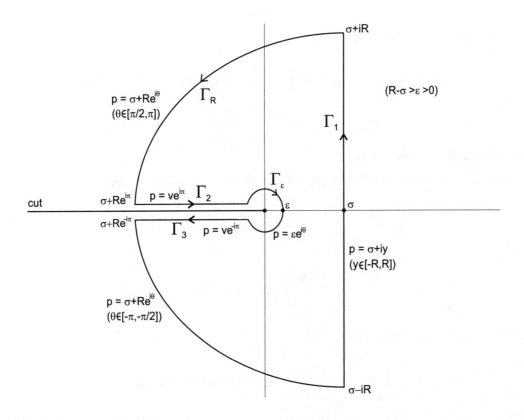

Then,

$$\int_{\Gamma_R} g \;=\; \int_{\frac{\pi}{2}}^{\pi} \frac{1}{\sigma + Re^{i\theta}} \exp(P(x,t,R,\theta,\sigma))iRe^{i\theta}\, d\theta,$$

where

$$P(x,t,R,\theta,\sigma) \;=\; (\sigma + Re^{i\theta})t - ((\sigma + Re^{i\theta})/k)^{\frac{1}{2}}x.$$

There is a real constant M such that, as $R \to \infty$,

$$|\exp(-((\sigma + Re^{i\theta})/k)^{\frac{1}{2}}x)| \;\leq\; M, \qquad (\theta \in [\pi/2, \pi]).$$

So,

$$\left| \int_{\Gamma_R} g \right| \leq \frac{MRe^{\sigma t}}{R-\sigma} \int_{\pi/2}^{\pi} e^{tR\cos\theta}\, d\theta,$$

which tends to zero as $R \to \infty$, for $t > 0$, as deduced above in our proof of the Corollary to Theorem 3.

Further, considering the small circle Γ_ϵ, the reader will easily find that

$$\exp(P(x,t,\epsilon,\theta,0)) - 1 = N\epsilon^{\frac{1}{2}},$$

for a real positive constant N. Thus,

$$\left| 2\pi i + \int_{\Gamma_\epsilon} g \right| = \left| 2\pi i + \int_{\pi}^{-\pi} \frac{1}{\epsilon e^{i\theta}} \exp(P(x,t,\epsilon,\theta,0)) i\epsilon e^{i\theta}\, d\theta \right|$$

$$\leq 2\pi N\epsilon^{\frac{1}{2}},$$

which tends to zero as $\epsilon \to 0$.

Along Γ_2, $p = ve^{i\pi}$ and hence $p^{\frac{1}{2}} = iv^{\frac{1}{2}}$, where $\epsilon \leq v \leq R - \sigma$. So,

$$\int_{\Gamma_2} g = \int_{R-\sigma}^{\epsilon} \frac{1}{-v} \exp(-vt - i(v/k)^{\frac{1}{2}}x)\,(-dv)$$

$$= -\int_{\epsilon}^{R-\sigma} \frac{1}{v} \exp(-vt - i(v/k)^{\frac{1}{2}}x)\, dv.$$

Similarly, along Γ_3, where $p^{\frac{1}{2}} = -iv^{\frac{1}{2}}$,

$$\int_{\Gamma_3} g = \int_{\epsilon}^{R-\sigma} \frac{1}{v} \exp(-vt + i(v/k)^{\frac{1}{2}}x)\, dv.$$

So, integrating g around the whole keyhole contour and using Cauchy's Theorem gives, on letting $R \to \infty$,

$$\lim_{R\to\infty} \left(\int_{\Gamma_1+\Gamma_2+\Gamma_3} g \right) - 2\pi i = 0$$

and hence

$$u(x,t) = \frac{u_0}{2\pi i} \lim_{R\to\infty} \int_{\sigma-iR}^{\sigma+iR} g = u_0 \left\{ 1 - \frac{1}{\pi} \int_0^{\infty} \frac{e^{-vt}}{v} \sin\left(\left(\frac{v}{k}\right)^{\frac{1}{2}} x \right) dv \right\},$$

where we have used the elementary fact that $e^{i\lambda} - e^{-i\lambda} = 2i\sin\lambda$.

We have achieved our objective of expressing our solution of the heat equation in terms of real variables alone. In fact, this solution can be expressed very neatly as

$$(27) \qquad u(x,t) = u_0 \left\{ 1 - \mathrm{erf}\left(\frac{x}{2\sqrt{kt}} \right) \right\}$$

in terms of the *error function*, defined by

$$\mathrm{erf}(w) = \frac{2}{\sqrt{\pi}} \int_0^w e^{-v^2} \, dv,$$

as the reader will be asked to demonstrate in the following exercise. □

Note (a) The motto here is that the reader should have the inversion theorems at hand as a last resort, but before using them would be wise first to seek another method!

(b) A much speedier solution will be found in the next section.

Exercise 13 Let $F : \mathbb{R} \to \mathbb{R}$ be the function defined, for $\lambda > 0$, by

$$F(x) = 2 \int_0^\infty \frac{e^{-\lambda y^2}}{y} \sin(xy) \, dy.$$

By first showing that

$$F'(x) = \mathrm{re}\left(\mathcal{F}_{y \to x}\left(e^{-\lambda y^2} \right)(x) \right),$$

Deduce from the Lemma of this section that

$$F'(x) = \sqrt{\frac{\pi}{\lambda}} e^{-x^2/4\lambda}$$

and hence that

$$F(x) = 2\sqrt{\pi} \int_0^{x/2(kt)^{\frac{1}{2}}} e^{-w^2} \, dw.$$

Deduce the form (27) for the solution to the heat equation problem of Example 16.

Exercise 14 Use the Laplace transform to show that the solution of

$$ku_{xx} \;=\; u_t, \qquad (x \geq 0,\, t \geq 0,\, k \text{ a positive constant})$$

with $u = 0$ for $t = 0$ and $u = f(t)$ for $x = 0$, is given by

$$u(x,t) \;=\; \int_0^t f'(t - u)h(x,u)\,du + f(0)h(x,t),$$

where $h(x,t)$ satisfies

$$\int_0^\infty e^{-pt}h(x,t)\,dt \;=\; \frac{1}{p}\,e^{-(p/k)^{\frac{1}{2}}x}.$$

By considering the case $f(t) \equiv 1$, show that

$$h(x,t) \;=\; 1 - \operatorname{erf}\left(\frac{x}{2\sqrt{kt}}\right).$$

Exercise 15 Show that the Laplace transform $\bar{y} = \bar{y}(p)$ of a smooth bounded solution $y = y(x)$ of the differential equation

$$(xy')' - (x - \alpha)y \;=\; 0 \qquad (\alpha \text{ constant})$$

is given by

$$\bar{y}(p) \;=\; K\frac{(p - 1)^{\frac{\alpha - 1}{2}}}{(p + 1)^{\frac{\alpha + 1}{2}}},$$

for some constant K.

Use the Laplace Inversion Theorem to show that when $(\alpha - 1)/2$ is a positive integer n, the solution $y = y(x)$ is a multiple of

$$\left(\frac{d}{dp}\right)^n \{(p - 1)^n e^{px}\}\bigg|_{p=-1}.$$

[HINT: You may wish to use one of Cauchy's integral formulae.]

Exercise 16 (Heat in a finite bar, with no heat flow across one end and the other end kept at a constant temperature)

Consider the problem consisting of the equation

$$ku_{xx} \;=\; u_t \qquad (0 \leq x \leq l,\, t \geq 0,\, k \text{ a positive constant})$$

subject to

$$u_x(0,t) \;=\; 0 \quad \text{and} \quad u(l,t) \;=\; u_1 \quad (t \geq 0), \qquad u(x,0) \;=\; u_0 \quad (0 \leq x \leq l).$$

Show that the Laplace transform $\bar{u} = \bar{u}(x, p)$ is given by

$$\bar{u}(x,p) \;=\; \frac{u_0}{p} + \left(\frac{u_1 - u_0}{p} \right) \left(\frac{\cosh((p/k)^{\frac{1}{2}} x)}{\cosh((p/k)^{\frac{1}{2}} l)} \right).$$

By using the Inversion Theorem and a contour of the type considered in the Corollary to Theorem 3 but not passing through any pole of $\bar{u}(x, p)$, show that

$$u(x,t) \;=\; u_0 + \frac{4(u_1 - u_0)}{\pi} \sum_{n=1}^{\infty} \frac{(-1)^n}{2n - 1} e^{-k(n - \frac{1}{2})^2 \pi^2 t / l^2} \cos \frac{(2n - 1)\pi x}{2l}.$$

Compare this result to those of section 7.2 and its exercises.

[You may assume that the integral of $\bar{u}(x, p)e^{px}$ around the circular arc tends to zero as its radius increases to infinity.]

14.6   Appendix: similarity solutions

In this short appendix, we introduce a quite different technique which can provide, remarkably efficiently, solutions to such equations as the heat equation, *provided* the initial boundary conditions are convenient. We give one example to describe the method, and two exercises.

Example 17 Find a solution to the problem consisting of the heat equation

(28) $$\qquad\qquad ku_{xx} \;=\; u_t \qquad (x \geq 0,\, t > 0,\, k \text{ a positive constant})$$

when subject to

(29)
$$u(0, t) \;=\; u_0 \quad (t > 0),$$
$$u(x, t) \to 0 \text{ as } t \to 0 \quad (x \geq 0), \qquad u(x, t) \to 0 \text{ as } x \to \infty \quad (t > 0).$$

The substitution $\eta = x/(kt)^{\frac{1}{2}}$, $F(\eta) = u(x, t)$ reduces the problem to

(30) $$\qquad\qquad F'' + \tfrac{1}{2}\eta F' \;=\; 0$$

subject to

(31) $$\qquad\qquad F(0) \;=\; u_0, \qquad F(\eta) \to 0 \text{ as } \eta \to \infty.$$

The equation (30) may be re-written

$$\frac{d}{d\eta}\left(e^{\eta^2/4}F'(\eta)\right) = 0,$$

has first integral

$$F'(\eta) = Ae^{-\eta^2/4}$$

and hence general solution

$$F(\eta) = A\int^{\eta} e^{-u^2/4}\,du + B$$

$$= A'\int^{\eta/2} e^{-v^2}\,dv + B,$$

where A, A', B are constants. Applying the boundary conditions (31) and using

$$\int_0^{\infty} e^{-v^2}\,dv = \frac{\sqrt{\pi}}{2},$$

the reader may quickly derive the problem's solution

$$F(\eta) = \frac{2u_0}{\sqrt{\pi}}\int_{\eta/2}^{\infty} e^{-v^2}\,dv$$

or

$$u(x,t) = \frac{2u_0}{\sqrt{\pi}}\int_{x/2\sqrt{kt}}^{\infty} e^{-v^2}\,dv = u_0(1 - \mathrm{erf}\,(x/2\sqrt{kt})),$$

where the *error function* erf $= \mathrm{erf}\,(w)$, introduced in section 14.5, is given by

$$\mathrm{erf}\,(w) = \frac{2}{\sqrt{\pi}}\int_0^{w} e^{-v^2}\,dv, \qquad (w \geq 0). \qquad \square$$

Note The applicability of the method turns on the fact that, when one puts $\eta = x/(kt)^{\frac{1}{2}}$, $F(\eta) = u(x,t)$, not only does the heat equation reduce to an ordinary differential equation for F, but the conditions

$$u(x,t) \to 0 \text{ as } t \to 0 \quad (x \geq 0), \quad u(x,t) \to 0 \text{ as } x \to \infty \quad (t > 0)$$

'collapse' to the single condition

$$F(\eta) \to 0 \text{ as } \eta \to \infty.$$

Exercise 17 Consider the problem consisting of the partial differential equation

$$(32) \qquad\qquad t^2 u_{xx} \ = \ x u_t, \qquad (x \geq 0,\ t > 0)$$

involving the function $u = u(x, t)$ and subject to

$$u(x, t) \to 0 \text{ as } t \to 0 \quad (x \geq 0), \quad u(t, t) \ = \ 1 \quad (t > 0), \quad u(x, t) \to 0 \text{ as } x \to \infty \quad (t > 0).$$

By considering the substitution $\eta = x/t^\lambda$, $F(\eta) = u(x, t)$, find a value for the constant λ for which (32) reduces to an ordinary differential equation in F. Hence, solve the problem.

Exercise 18 Consider the problem consisting of the partial differential equation

$$(33) \qquad\qquad t^\alpha u_{xx} \ = \ u_t, \qquad (0 < t^2 < x)$$

involving the function $u = u(x, t)$ and the real constant α, when the equation its subject to the conditions

$$u(x, t) \to 0 \text{ as } t \to 0 \quad (x > 0), \quad u(t^2, t) \ = \ 1 \quad (t > 0), \quad u(x, t) \to 0 \text{ as } x \to \infty \quad (t > 0).$$

By making the substitution $\eta = x/t^\beta$, $F(\eta) = u(x, t)$, find values of α and β for which (33) reduces to an ordinary differential equation in F and, for these values of α and β, solve the problem.

15 Phase-Plane Analysis

The systems with which this chapter is concerned are those of the form

$$\text{(A)} \qquad \dot{x} = X(x, y), \qquad \dot{y} = Y(x, y)$$

satisfied by the functions $x = x(t)$, $y = y(t)$, where X and Y are continuous real-valued functions and dot denotes differentiation with respect to the real variable t. They are often referred to as 'plane autonomous systems of ordinary differential equations'. Here, 'plane' indicates that just two functions x, y, dependent on the variable t, are involved and 'autonomous' that neither of the functions X, Y contains this variable explicitly.

It is not in general possible to find an analytic solution to an arbitrary differential equation and, even when it is, the solution is often expressed in a form (such as a series or integral, or implicitly) which does not allow easy discussion of its more important properties. This chapter provides an introduction to a qualitative study in which one can locate properties of a differential equation without having to find a solution. Every student of differential equations should be aware of phase-plane methods.

15.1 The phase-plane and stability

When the real plane $\mathbb{R}^2$, with axes x and y, is used to plot solutions $x = x(t)$, $y = y(t)$ of (A), it will be referred to as the *phase-plane*. The curve traced out by $(x(t), y(t))$ as t varies (usually over some finite or infinite interval) is called a *phase-path* or *trajectory*. A diagram depicting the phase-paths is called a *phase-diagram*.

Whenever $X \neq 0$, (A) gives rise to

(B)
$$\frac{dy}{dx} = \frac{Y(x,y)}{X(x,y)}$$

which the phase-paths must satisfy. Theorem 1 of Chapter 2 tells us that, when Y/X is sufficiently 'well-behaved' and subject to an appropriate boundary condition, (B) has a unique solution. So, save for 'singular points', we should expect one and only one phase-path through each point in the phase-plane.

Constant solutions $x(t) = x_0$, $y(t) = y_0$ of (A) correspond to (singular) points (x_0, y_0) satisfying

(C)
$$X(x_0, y_0) = Y(x_0, y_0) = 0.$$

Such points are called *critical points*, or *equilibrium points*.

The nature of the phase-paths near a critical point can often give important information about the solutions of the system (A). Intuitively, if near a critical point (a, b) the phase-paths $(x, y) = (x(t), y(t))$ have all their points close to (a, b), for t greater than some t_0 in the domain of (x, y), then the point (a, b) is called *stable*. We may formalise this idea as follows. For $(a, b) \in \mathbb{R}^2$ and a solution $(x, y) = (x(t), y(t))$ of (A), let

$$F(x, y, a, b, t) \equiv \sqrt{(x(t) - a)^2 + (y(t) - b)^2}.$$

Definition A critical point (a, b) is *stable* if and only if given $\epsilon > 0$, there exist $\delta > 0$ and a real t_0 such that for any solution $(x, y) = (x(t), y(t))$ of (A) for which $F(x, y, a, b, t_0) < \delta$,

$$F(x, y, a, b, t) < \epsilon$$

for every t in the domain of (x, y) which is greater than t_0. A critical point is *unstable* if and only if it is not stable.

Before exemplifying the terms introduced in this section, we prove two results concerning closed phase-paths. The first provides a reason for their being of especial interest. The second gives a sufficient condition for there to be *no* closed phase-paths in a domain of the phase-plane.

We assume conditions have been placed on $X = X(x, y)$ and $Y = Y(x, y)$, sufficient to necessitate the uniqueness of solution of the system (A), via Theorem 2 of Chapter 2. Note that a closed phase-path is a closed and bounded subset of the real plane.

Theorem 1 A phase-path traced out by $(x(t), y(t))$, as t varies, is closed if and only if the functions $x = x(t)$, $y = y(t)$ are both periodic functions of t.

Proof If the phase-path is closed, there exist constants t_0, t_1 $(t_1 > t_0)$ such that $x(t_0) = x(t_1)$ and $y(t_0) = y(t_1)$. Letting $T = t_1 - t_0$, define new functions $x_1 = x_1(t)$, $y_1 = y_1(t)$ by

$$x_1(t) = x(t + T), \qquad y_1(t) = y(t + T).$$

Then, using (A),

$$\frac{d}{dt} x_1(t) = \frac{d}{dt} x(t + T) = X(x(t + T), y(t + T)) = X(x_1(t), y_1(t))$$

and, similarly,

$$\frac{dy_1}{dt} = Y(x_1, y_1).$$

Further,

$$x_1(t_0) = x(t_0 + T) = x(t_1) = x(t_0)$$

and, similarly,

$$y_1(t_0) = y(t_0).$$

The uniqueness given by Theorem 2 of Chapter 2 now assures us that $x_1 = x$ and $y_1 = y$ identically; that is,

$$x(t + T) = x(t) \quad \text{and} \quad y(t + T) = y(t)$$

as t varies. Thus, $x = x(t)$ and $y = y(t)$ are periodic functions of t.

The converse being trivial, Theorem 1 is proved. $\qquad\square$

Theorem 2 (Bendixson–Dulac Negative Criterion) Suppose that X, Y and $\rho = \rho(x, y)$ are continuously differentiable in a simply connected domain D in the phase-plane. Then, a sufficient condition for there to be no simple closed[1] (non-constant) phase-paths in D is that

$$\frac{\partial}{\partial x} (\rho X) + \frac{\partial}{\partial y} (\rho Y)$$

is of one sign in D.

---

[1] By 'simple closed' is meant 'homeomorphic to the unit circle'; that is, which is closed and does not cross itself.

Proof Suppose that C is a simple closed phase-path in D, given by $x = x(t)$, $y = y(t)$, and let I denote the interior of C. The hypotheses assure us that we may apply Green's Theorem in the Plane, so that

$$E \equiv \iint_I \left\{ \frac{\partial}{\partial x} (\rho X) + \frac{\partial}{\partial y} (\rho Y) \right\} dx dy$$

$$= \int_C \left(-\rho Y \frac{dx}{dt} + \rho X \frac{dy}{dt} \right) dt$$

$$= \int_C \rho(-YX + XY) \, dt$$

$$= 0$$

since $(x(t), y(t))$ satisfies (A). This contradicts the fact that E cannot be zero, as it is the integral of a *continuous* function of one sign.   □

Example 1 Draw phase-diagrams for the system

(1) $\dot{x} = y, \qquad \dot{y} = cx,$

where c is a non-zero real constant.

The system has $(0,0)$ as its only critical point and its phase paths satisfy

$$\frac{dy}{dx} = \frac{cx}{y},$$

whenever y is non-zero. The phase-paths are therefore of the form

$$y^2 - cx^2 = k,$$

where k is another real constant.

When $c < 0$, the phase-paths form a family of ellipses, each with centre $(0,0)$; whereas, when $c > 0$, they form a family of hyperbolae with asymptotes $y = dx$, where $d^2 = c$. The corresponding phase-diagrams are

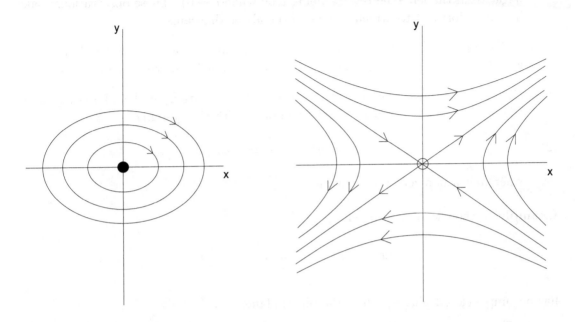

For $c < 0$, the origin is thus a stable critical point, whereas it is unstable in the case $c > 0$.   □

The system arises from the second-order differential equation

$$\ddot{x} - cx = 0,$$

once one defines a new function y by $y = \dot{x}$. For $c < 0$, this is the equation of motion of the one-dimensional harmonic oscillator.

Notes

(i) The drawing of a phase-diagram involves the determination of critical points as well as the various types of phase-path that may occur. On such a diagram, it is customary to represent stable critical points with 'solid dots' and unstable ones with 'hollow dots' (as above). An arrow should also be attached (as above) to each phase-path to indicate the direction $(\dot{x}(t), \dot{y}(t))$ in which the point $(x(t), y(t))$ traverses the path as t increases. From (A), it is clear that this direction can be determined by consideration of the vector $(X(x, y), Y(x, y))$. In Example 1, $X(x, y) = y$ is always positive in the upper half-plane and always negative in the lower; whereas $Y(x, y) = cx$ is, when $c > 0$, positive in the right half-plane and negative in the left (the reverse being true when $c < 0$). These observations make it straightforward to attach arrows to the above diagrams.

(ii) In Example 1, the critical point for $c < 0$, surrounded as it is by closed curves, is called a *centre*. For $c > 0$, the critical point is what is known as a *saddle-point*.

(iii) Different systems of the form (A) can give rise to rather similar phase-diagrams. The reader should compare (1) of Example 1 with the system

(2) $$\dot{x} = y^2, \qquad \dot{y} = cxy,$$

where c is a non-zero real constant.

Example 2 Show that the system

$$\dot{x} = y, \qquad \dot{y} = x - y + \frac{1}{2} y^2$$

has no simple closed phase-path in the phase-plane.

By Theorem 2, it is sufficient to show that there is a function $\rho = \rho(x, y)$ such that

$$F \equiv \frac{\partial}{\partial x} (\rho y) + \frac{\partial}{\partial y} (\rho(x - y + \tfrac{1}{2} y^2))$$

is of one sign in $\mathbb{R}^2$. Now, when ρ_x, ρ_y exist,

$$F = \rho_x y + \rho_y (x - y + \tfrac{1}{2} y^2) + \rho(-1 + y).$$

Choosing $\rho = e^{-x}$, we see that F reduces to

$$F = -e^{-x},$$

which is negative throughout $\mathbb{R}^2$. The Bendixson–Dulac criterion applies.   □

Exercise 1 Draw phase-diagrams for the system (2) above. (Be careful to identify all the critical points, to note which are stable, and to be precise about the directions of the phase-paths in each quadrant of $\mathbb{R}^2$.) Compare the diagrams you have drawn with those obtaining to Example 1.

Exercise 2 Draw phase-diagrams for the system

(3) $$\dot{x} = ax, \qquad \dot{y} = ay,$$

where a is a non-zero real constant. When is the origin a stable critical point? Show that there are no simple closed phase-paths. What happens when $a = 0$?

Note The critical point of Exercise 2, when $a \neq 0$, is called a *star-point*.

Exercise 3 Draw phase-diagrams for the system

(4) $$\dot{x} = by, \qquad \dot{y} = dy,$$

where b and d are non-zero real constants. When is the origin a stable critical point? What happens when $d = 0$?

Note The system (4) is known as the 'degenerate case' amongst the linear systems

$$\dot{x} = ax + by, \qquad \dot{y} = cx + dy,$$

where a, b, c, d are real constants. These systems will be discussed in greater detail in the next section. The systems (1) and (3) above are other special cases.

The system (4) arises, when $b = 1$, from the second-order differential equation

$$\ddot{x} - d\dot{x} = 0,$$

when $y = \dot{x}$, and represents, for $d < 0$, the motion of a particle in a medium providing resistance proportional to velocity (as in fluid motion). It is very unusual in that critical points are not isolated.

Exercise 4 Show that each of the following systems has no periodic solutions:

(5) $$\dot{x} = y, \qquad \dot{y} = -a \sin x - by(1 + cy^2),$$

where a, b and c are positive real constants,

(6) $$\dot{x} = y + x + x^3, \qquad \dot{y} = x + y^5,$$

(7) $$\dot{x} = x + y + x^2 y, \qquad \dot{y} = x.$$

15.2   Linear theory

In this section, we analyse in some detail the phase-diagrams representing the *linear system*

(L) $$\dot{x} \; = \; ax + by, \qquad \dot{y} \; = \; cx + dy,$$

where a, b, c, d are real constants. We first search for solutions of (L) in the form

(8) $$x \; = \; Ae^{\lambda t}, \qquad y \; = \; Be^{\lambda t},$$

where A, B are (possibly complex) constants. Substituting (8) in (L), we find that

(9) $$\begin{cases} (a - \lambda)A + bB \; = \; 0, \\[2mm] cA + (d - \lambda)B \; = \; 0. \end{cases}$$

For not both of A, B to be zero, it is necessary that

$$\begin{vmatrix} a - \lambda & b \\[2mm] c & d - \lambda \end{vmatrix} = 0,$$

that is, that

(10) $$\lambda^2 - (a + d)\lambda + (ad - bc) \; = \; 0.$$

As with the discussion of the second-order linear equation with constant coefficients (see the Appendix, section (6)), the remaining analysis turns on the nature of the roots λ_1, λ_2 of this *characteristic equation*.

(i) λ_1, λ_2 real, unequal and of the same sign

Using equations (9), we see that the root $\lambda = \lambda_1$ of the characteristic equation (10) gives rise to a (fixed) ratio $A : B$. Leaving the exceptional cases for consideration by the reader, we suppose this ratio is $1 : C_1$, with C_1 non-zero, giving the (real) solution

$$x(t) \; = \; K_1 e^{\lambda_1 t},$$

$$y(t) \; = \; K_1 C_1 e^{\lambda_1 t},$$

where K_1 is an arbitrary real constant. If the ratio $A:B$ is $1:C_2$ when $\lambda = \lambda_2$, there is similarly the (real) solution

$$x(t) = K_2 e^{\lambda_2 t},$$

$$y(t) = K_2 C_2 e^{\lambda_2 t},$$

where K_2 is an arbitrary real constant. The general solution is the sum of these two solutions, namely,

$$(11) \qquad \begin{cases} x(t) = K_1 e^{\lambda_1 t} + K_2 e^{\lambda_2 t}, \\[2mm] y(t) = K_1 C_1 e^{\lambda_1 t} + K_2 C_2 e^{\lambda_2 t}, \end{cases}$$

where K_1, K_2 are arbitrary real constants. (Note that the existence and uniqueness Theorem 2 of Chapter 2 would allow us to determine, on any interval, the constants K_1, K_2 by using initial conditions for x and y.) When $\dot{x} \neq 0$, the phase-paths satisfy

$$(12) \qquad \frac{dy}{dx} = \frac{\dot{y}}{\dot{x}} = \frac{\lambda_1 K_1 C_1 e^{\lambda_1 t} + \lambda_2 K_2 C_2 e^{\lambda_2 t}}{\lambda_1 K_1 e^{\lambda_1 t} + \lambda_2 K_2 e^{\lambda_2 t}}.$$

We now suppose that λ_1, λ_2 are both negative and, without loss of generality, that $\lambda_1 > \lambda_2$. When $K_1 \neq 0$, $K_2 = 0$, we have the phase-path $y = C_1 x$; and when $K_1 = 0$, $K_2 \neq 0$, we have the phase-path $y = C_2 x$. Further, for each phase-path,

$$\frac{dy}{dx} \to C_1 \text{ as } t \to \infty, \qquad \frac{dy}{dx} \to C_2 \text{ as } t \to -\infty.$$

The directions of the phase-paths and the behaviour of solutions at zero and infinity follow directly from consideration of $x, y, \dot{x}, \dot{y}$. We are now in a position to draw the phase-diagram. The origin is a stable critical point. When λ_1, λ_2 are both positive, the phase-diagram is similar, but the arrows point away from the origin and the origin is an unstable critical point. In both cases, this type of critical point is called a *nodal point*, or *node*.

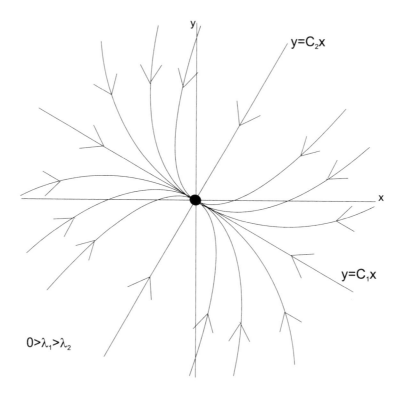

(ii) $\lambda_1 = \lambda_2 \neq 0$

When b, c are not both zero, the real general solution of the system in parametric form is

$$x(t) \;=\; (K_1 + K_2 t)e^{\lambda_1 t},$$

$$y(t) \;=\; (K_1 C_1 + K_2 C_2 t)e^{\lambda_1 t},$$

where C_1, C_2, K_1, K_2 are real constants (with the same provisos concerning C_1, C_2 as in case (i)). When λ_1 is negative, the phase-diagram can be drawn as follows. It can be considered as the limiting case of the diagram in (i) as $\lambda_2 \to \lambda_1$. The origin is a stable critical point. When λ_1 is positive, the origin is unstable and the arrows in the phase-diagram point away from the origin. This type of critical point is called an *inflected nodal point*, or *inflected node*.

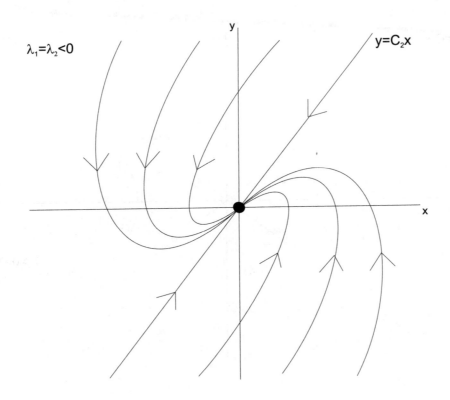

The reader will already have discussed the case $\lambda_1 = \lambda_2 \neq 0$, $b = c = 0$, which necessitates $a = d$, in carrying out Exercise 2 above. The critical point is a *star-point*.

(iii) λ_1, λ_2 real and of opposite sign

Without loss of generality, assume $\lambda_1 > 0 > \lambda_2$. The equations (11), (12) of (i) apply. In this case (and as the reader should check), $y = C_1 x$ and $y = C_2 x$ are the only phase-paths through the origin, the first being described from the origin to infinity, the second from infinity to the origin, as t increases. The phase-diagram may be drawn as below. The origin is an unstable critical point, and is a *saddle-point* as in the special case considered in Example 1, when $c > 0$ there.

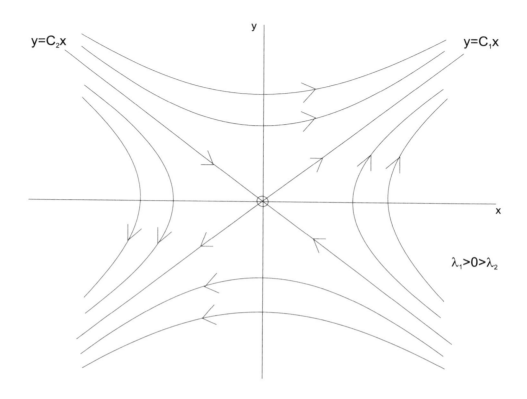

(iv) $\lambda_1 = \overline{\lambda_2} = \xi + i\eta$, where ξ, η are both non-zero real numbers

The real general solution may be written

$$x = Ke^{\xi t}\sin(\eta t + \delta),$$

$$y = KCe^{\xi t}\sin(\eta t + \epsilon),$$

where K, C, δ, ϵ are real constants. The periodicity of the sine function indicates that the phase-paths are spirals, approaching the origin as t increases if ξ is negative, but approaching infinity if ξ is positive. The phase-diagram is therefore as below. The origin is thus stable if and only if ξ is negative. This type of critical point is called a *spiral-point*, or *focus*.

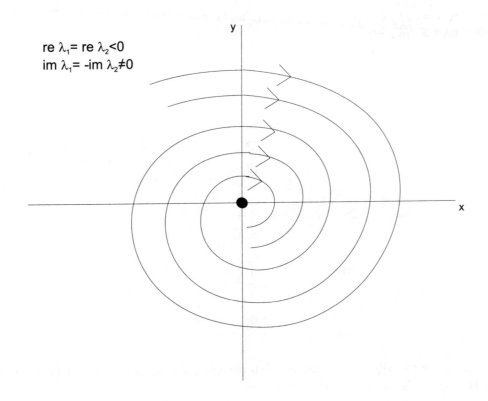

re $\lambda_1 =$ re $\lambda_2 < 0$
im $\lambda_1 = $ -im $\lambda_2 \neq 0$

(v) $\lambda_1 = \overline{\lambda_2} = i\eta$, where η is a non-zero real number

Now the system has the real general solution

$$x = K\sin(\eta t + \delta),$$

$$y = KC\sin(\eta t + \epsilon),$$

where K, C, δ, ϵ are real constants. The phase-paths are ellipses and the phase-diagram is as below.

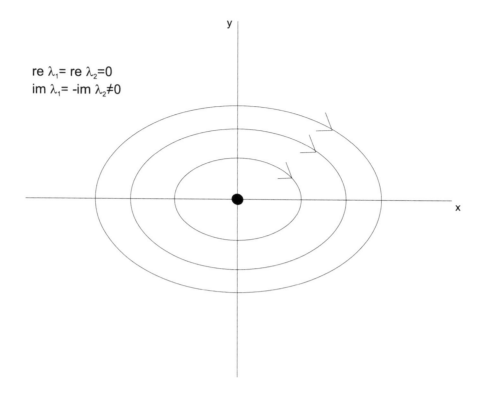

re λ_1= re λ_2=0
im λ_1= -im $\lambda_2 \neq$0

The origin is necessarily stable and, as with the special case in Example 1 (when $c < 0$), the critical point is called a *centre*, or *vortex*.

(vi) $\lambda_1 \neq 0$, $\lambda_2 = 0$

The phase-paths form a set of parallel lines. The reader will have discussed a special case in carrying out Exercise 3. Stable and unstable cases can arise depending on how the line through the origin is described.

(vii) $\lambda_1 = \lambda_2 = 0$

In this case, there is either a non-singular linear transformation of $x = x(t)$, $y = y(t)$ to functions $u = u(t)$, $v = v(t)$ such that $\dot{u} = \dot{v} = 0$, or one for which $\ddot{u} = 0$.

Summary

We now summarise our results on the linear theory of plane autonomous systems

$$\dot{x} = ax + by,$$
$$\dot{y} = cx + dy,$$

where a, b, c, d are real constants. Suppose that λ_1, λ_2 are the roots of the characteristic equation

$$\lambda^2 - (a + d)\lambda + (ad - bc) = 0.$$

<table>
<tr><td>λ_1, λ_2 real unequal, of same sign</td><td>node, stable iff $\lambda_1 < 0$</td></tr>
<tr><td>$\lambda_1 = \lambda_2 \neq 0 \begin{cases} (b, c) \neq (0, 0) \\ \\ b = c = 0 \end{cases}$</td><td>inflected node, stable iff $\lambda_1 < 0$

star-point, stable iff $\lambda_1 < 0$</td></tr>
<tr><td>λ_1, λ_2 real, of opposite sign</td><td>saddle-point, always unstable</td></tr>
<tr><td>$\lambda_1 = \overline{\lambda_2} = \xi + i\eta$, ξ, η real and non-zero</td><td>spiral point, stable iff $\xi < 0$</td></tr>
<tr><td>$\lambda_1 = \overline{\lambda_2} = i\eta$, η real and non-zero</td><td>centre, always stable</td></tr>
<tr><td>$\lambda_1 \neq 0, \lambda_2 = 0$</td><td>parallel lines, stable iff $\lambda_1 < 0$</td></tr>
<tr><td>$\lambda_1 = \lambda_2 = 0$</td><td>corresponds to either $\ddot{u} = 0$ or $\dot{u} = \dot{v} = 0$ for appropriate u, v.</td></tr>
</table>

The reader can easily check that the above table covers all possible cases.

Notes

(i) An elegant presentation is provided by the use of the theory of non-singular linear transformations[2]. Special cases still require careful treatment.

(ii) As with much classical applied mathematics, so here too in the theory of plane autonomous systems, a linear theory can be used to approximate a situation modelled by a non-linear equation. Such linearisation is considered in section 15.4.

Exercise 5 Determine the nature of the critical point of each of the following systems.

(a) $\dot{x} = x + 2y, \qquad \dot{y} = 4x - y.$

(b) $\dot{x} = 5x - 3y, \qquad \dot{y} = 3x - y.$

(c) $\dot{x} = x - y, \qquad \dot{y} = 5x - y.$

(d) $\dot{x} = x, \qquad\qquad \dot{y} = 2x + 3y.$

(e) $\dot{x} = x - 2y, \qquad \dot{y} = x + 3y.$

Exercise 6 Consider the system

$$\dot{x} = by + ax,$$

$$\dot{y} = ay - bx,$$

where a, b are constants and $b > 0$. Show that, if $a = 0$, the phase-paths are circles, and that, if $a \neq 0$, the phase-paths are spirals. What is the nature of the critical point in the cases (i) $a < 0$, (ii) $a = 0$, (iii) $a > 0$?

Exercise 7 Suppose that $p \equiv -(a + d)$ and $q \equiv ad - bc$ are the coefficients in the characteristic equation (10) of the system (L). Show that stable points of the system correspond to points in the positive quadrant $p \geq 0$, $q \geq 0$ of the pq-plane, which are not at the origin.

Exercise 8 Prove that, if $(x, y) = (x(t), y(t))$ is a solution of the system (L), then $x = x(t)$ and $y = y(t)$ are both solutions of the second-order equation

(13) $\ddot{w} - (a + d)\dot{w} + (ad - bc)w = 0.$

Conversely, show that, if $w = w(t)$ solves (13), then $(x, y) = (x(t), y(t))$, defined either by

$$x = bw, \quad y = \dot{w} - aw, \quad \text{if } b \neq 0,$$

---

[2]See, for example, G. Birkhoff and G.C. Rota, *Ordinary Differential Equations*, 3rd Edition, pages 114–126.

or by

$$x = \dot{w} - dw, \quad y = cw, \quad \text{if } c \neq 0,$$

solves (L).

Note The differential equation (13) is known as the *secular equation* of the system (L). The auxiliary equation of this secular equation (in the sense of (6) of the Appendix) is just the characteristic equation (10) of the system (L).

Exercise 9 (a) With the notation of Exercise 7, show that each of the linear transformations

$$X = x, \qquad Y = ax + by, \qquad \text{if } b \neq 0,$$

$$X = y, \qquad Y = cx + dy, \qquad \text{if } c \neq 0,$$

$$X = x + y, \qquad Y = ax + dy, \qquad \text{if } b = c = 0 \text{ and } a \neq d,$$

converts (L) to the system

$$\dot{X} = Y, \qquad \dot{Y} = -qX - pY.$$

(b) Show that the secular equation (in the sense of the above Note) of the system

$$\dot{x} = ax, \qquad \dot{y} = ay$$

(where $a = d$ and $b = c = 0$) is the same as that of the system

$$\dot{x} = ax, \qquad \dot{y} = x + ay,$$

but exhibit a solution of the latter system which does not solve the former.

15.3   Some non-linear systems

Phase-plane analysis can be used to discuss second-order ordinary differential equations. Classically, where $x = x(t)$ denotes position of a function at time t, the *Poincaré substitution* $y = \dot{x}$ takes velocity as the second function of t. For example, the equation

$$x\ddot{x} + \dot{x}^2 - x = 0$$

may be converted to the system

$$\dot{x} = y, \qquad \dot{y} = 1 - y^2/x, \qquad (x \neq 0).$$

We should at the outset point out that there can be other, perhaps even simpler, 'equivalent' systems. In the above instance, another system representing the second order equation is

$$\dot{x} \;=\; y/x, \qquad \dot{y} \;=\; x, \qquad (x \neq 0).$$

Notice that, with the Poincaré substitution, $\dot{x}$ is always positive in the upper half-plane and negative in the lower. So, the arrows on the corresponding phase-diagram must point to the right in the upper half-plane and to the left in the lower. Further, as $\dot{x}$ is zero on the x-axis and only zero there, all phase-paths which meet the x-axis must cut it at right angles (provided the point of cutting is not a critical point) and all critical points must lie on that axis.

Conservative systems The equation of motion of a one-dimensional conservative mechanical system can be written

$$\ddot{x} \;=\; F(x),$$

where the continuous function F can be regarded as the force per unit mass. When $y = \dot{x}$ and as $\ddot{x}$ then equals $y \, (dy/dx)$, one integration with respect to x gives

(14) $$\tfrac{1}{2}(y(x))^2 + V(x) \;=\; C,$$

where $V'(x) = -F(x)$ and C is a real constant. In our interpretation, V is the potential energy function and (14) the energy equation which may be re-written as

(15) $$y(x) \;=\; \pm\sqrt{2(C - V(x))}.$$

The critical points of the equivalent plane autonomous system

$$\dot{x} \;=\; y, \qquad \dot{y} \;=\; F(x)$$

occur at turning points of V. A local minimum of V gives rise to a centre, a local maximum to a saddle-point, and a point of inflexion to a *cusp* (a type of critical point we have not up to now encountered). These situations are illustrated by the following diagrams.

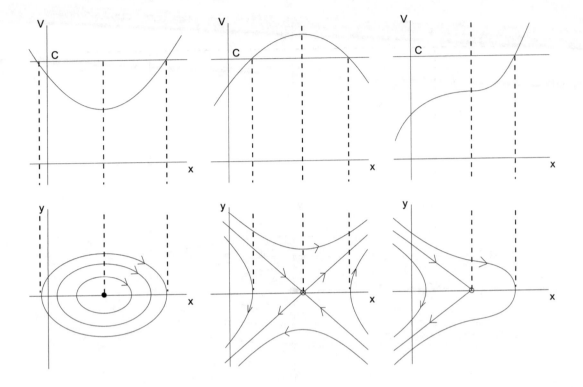

The diagrams also indicate how the (x, V)-graphs may be used to draw the phase-paths: for a given C, the difference $C - V(x)$ may be read off, and the corresponding values of $y(x)$, calculated from (15), inserted in the phase-diagram.

Example 3 Use the Poincaré substitution to draw a phase-diagram for the equation

$$\ddot{x} = -x + \tfrac{1}{6}x^3.$$

Proceeding as above, with $V'(x) = -F(x) = x - x^3/6$, we find that

$$V(x) = \frac{x^2}{2} - \frac{x^4}{24},$$

up to an additive constant which can, without loss of generality, be taken to be zero. The (x, V)- and phase-diagrams are drawn below. There is a centre at the origin and there are saddle-points at $(\pm\sqrt{6}, 0)$. $\qquad\square$

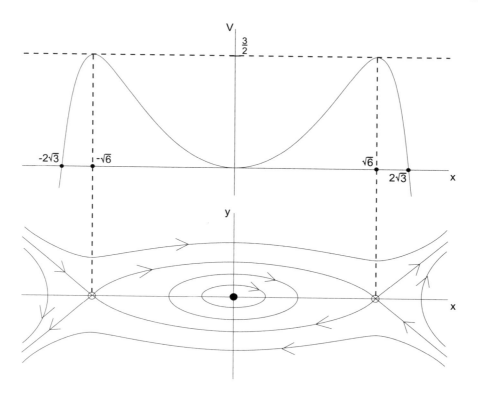

Note The function F in Example 3 is a good approximation to $-\sin x$ for small x. The reader is asked to draw the phase-diagram corresponding to $-\sin x$ in Exercise 14: it corresponds to the motion of a simple pendulum.

The use of polar co-ordinates Polar co-ordinates (r, θ) can be usefully employed to draw phase-diagrams. Often the form of the plane autonomous system will indicate their applicability. The co-ordinates satisfy

$$(16) \qquad\qquad r^2 \;=\; x^2 + y^2, \qquad \tan\theta \;=\; y/x$$

and so,

$$(17) \qquad\qquad \dot{r} \;=\; \frac{x\dot{x} + y\dot{y}}{r}, \qquad \dot{\theta} \;=\; \frac{x\dot{y} - y\dot{x}}{r^2}$$

whenever r is non-zero.

Example 4 Draw a phase-diagram for the system

$$\dot{x} \;=\; -y + x(1 - x^2 - y^2), \qquad \dot{y} \;=\; x + y(1 - x^2 - y^2).$$

Using (16) and (17), we find, for $r \neq 0$,

$$\dot{r} = \frac{(x^2 + y^2)(1 - x^2 - y^2)}{r} = r(1 - r^2), \qquad \dot{\theta} = \frac{x^2 + y^2}{r^2} = 1$$

which may be solved directly. The origin is a critical point and $r = 1$ is a particular solution. When $r < 1$, $\dot{r}$ is positive; and when $r > 1$, $\dot{r}$ is negative. Also, $\dot{r}$ tends to zero as r tends to unity. We thus have the diagram below.

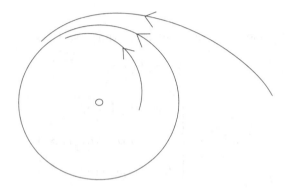

The critical point is unstable. When $r \neq 0, 1$, the phase-paths are spirals. The closed phase-path, given by $r = 1$, is called a *limit cycle*. It is isolated from any other closed phase-path, and is, in the obvious sense, 'the limit of phase-paths in its vicinity'. □

Exercise 10 Show that the second-order equation

$$\ddot{x} - \dot{x}^2 + x^2 - x = 0$$

gives rise to the plane autonomous system

$$\dot{x} = y, \qquad \dot{y} = x - x^2 + y^2$$

and also to the system

$$\dot{x} = x + y, \qquad \dot{y} = y(2x + y - 1).$$

Exercise 11 Solve the equation

$$\ddot{x} - d\dot{x} - cx = 0,$$

where c, d are non-zero real constants. Show that the equations may be represented by the system

$$\dot{x} = y, \qquad \dot{y} = cx + dy$$

and by the system

$$\dot{x} = dx + y, \qquad \dot{y} = cx.$$

Draw phase-diagrams and describe the nature of the critical points. (You should be careful to account for the different cases that arise from different values of c, d.)

Exercise 12 Solve the system

$$\dot{x} = y + x\sqrt{x^2 + y^2}, \qquad \dot{y} = -x + y\sqrt{x^2 + y^2},$$

draw its phase-diagram and classify any critical points that occur.

Exercise 13 The displacement $x(t)$ of a particle at time t satisfies

$$\ddot{x} = F(x, \dot{x})$$

where

$$F(x, \dot{x}) = \begin{cases} -x, & \text{if } |x| > \alpha \\ -\dot{x}, & \text{if } |x| < \alpha, \ |\dot{x}| > \beta \\ 0, & \text{otherwise} \end{cases}$$

for positive real constants α, β. Employing the Poincaré substitution $y = \dot{x}$, draw the phase-diagram of the resulting plane autonomous system. Describe the motion of the particle if it starts from rest at $x = \gamma$, where

(i) $0 < \gamma < \alpha$,

(ii) $\alpha < \gamma < \sqrt{\alpha^2 + \beta^2}$,

(iii) $\sqrt{\alpha^2 + \beta^2} < \gamma$.

Exercise 14 A simple pendulum, consisting of a particle of mass m connected to a fixed point by a light string of length a, and free to swing in a vertical plane (as shown below), has equation of motion

$$\ddot{x} + k \sin x = 0,$$

where k is the positive real constant g/a. Show that the Poincaré substitution $y = \dot{x}$ leads to the energy equation

$$\tfrac{1}{2}y^2 - k \cos x = c,$$

where c is another real constant.

Draw a phase-diagram to describe the motion, and identify and classify the critical points. Give a physical description of the motion corresponding to different types of phase-path.

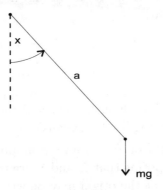

Exercise 15 In the Volterra predator–prey model, x denotes the total prey population and y denotes the total predator population. The system governing the development of the population is

$$\dot{x} = ax - cxy, \qquad \dot{y} = -by + dxy,$$

where a, b, c, d are positive real constants. Show that the equation of the phase-paths in the positive quadrant $x > 0$, $y > 0$ is

$$(a \log y - cy) + (b \log x - dx) = k,$$

where k is a real constant. Show also that the phase-paths in this quadrant are closed curves and draw a phase-diagram for the closed quadrant $x \geq 0$, $y \geq 0$.

Exercise 16 The motion of a simple pendulum, subject to non-linear damping, is described by the system

$$\frac{dx}{dt} = y, \qquad \frac{dy}{dt} = -a \sin x - by(1 + cy^2),$$

in which a, b, c are positive constants. By using properties of

$$F \equiv \tfrac{1}{2}y^2 + 2a \sin^2 \tfrac{1}{2}x$$

as a function of t, or otherwise, prove that, for any initial point $(x(0), y(0))$, there is a corresponding integer n such that

$$(x(t), y(t)) \to (n\pi, 0) \text{ as } t \to \infty.$$

Prove also that $n = 0$, if $|x(0)| \leq \pi/2$ and $|y(0)| < \sqrt{2a}$. Draw a phase-diagram, showing clearly how the character of the phase-path differs in the vicinity of $(n\pi, 0)$ for n even and for n odd. Distinguish also between the cases $b^2 < 4a$ and $b^2 \geq 4a$. (The reader will already have shown, in completing Exercise 4, that the system has no closed phase-paths.)

15.4 Linearisation

As the reader can easily surmise from the foregoing, if the nature of the critical points of the plane autonomous system

(A) $$\dot{x} = X(x, y), \qquad \dot{y} = Y(x, y)$$

is known, one is in a good position to attempt to draw a phase-diagram. One rather useful method of discussing the critical points is by approximating X and Y by linear functions in a neighbourhood of each such point and then using the linear theory. The approximation is achieved by using Taylor expansions.

By first making a translation of axes, we may suppose that the critical point under discussion is at the origin. Supposing that X and Y are twice continuously differentiable, the Taylor expansions of X, Y at the origin may be written

$$X(x, y) = ax + by + M(x, y),$$

$$Y(x, y) = cx + dy + N(x, y),$$

where
$$a = X_x(0, 0), \quad b = X_y(0, 0), \quad c = Y_x(0, 0), \quad d = Y_y(0, 0)$$

and
$$M(x, y) = O(r^2), \quad N(x, y) = O(r^2), \quad \text{as } r = \sqrt{x^2 + y^2} \to \infty.$$

Our hope is then that the *linear approximation*

(L) $$\dot{x} = ax + by, \qquad \dot{y} = cx + dy$$

will have the same type of critical point as the original system. In general, if the linear approximation has a node, saddle-point or spiral, then so does the original system. This will not be proved here, but the reader is asked to show in Exercise 19 that a centre in one system does *not* imply a centre in the other.

Example 5 Classify the critical points of the system

$$\dot{x} = x - y + 5, \qquad \dot{y} = x^2 + 6x + 8.$$

The critical points are $(-2, 3)$ and $(-4, 1)$. The translation

$$x = -2 + x_1, \qquad y = 3 + y_1$$

gives, on ignoring second-order terms

$$\dot{x}_1 = x_1 - y_1, \qquad \dot{y}_1 = 2x_1.$$

The characteristic equation

$$\lambda^2 - \lambda + 2 = 0$$

has roots

$$\lambda = \tfrac{1}{2}(1 \pm i\sqrt{7});$$

so, the critical point $(-2, 3)$ is an unstable spiral (as neither root is real but has positive real part).

The translation

$$x = -4 + x_2, \qquad y = 1 + y_2$$

gives, on ignoring second-order terms,

$$\dot{x}_2 = x_2 - y_2, \qquad \dot{y}_2 = -2x_2.$$

The characteristic equation

$$\lambda^2 - \lambda - 2 = 0$$

has roots $\lambda = -1, 2$; so, the critical point $(-4, 1)$ is a saddle-point (as the roots are real and of opposite sign). $\qquad\square$

Exercise 17 Find and classify the critical points of the following autonomous systems:

(a) $\quad \dot{x} = -2x - y - 4, \qquad \dot{y} = xy + 2x + 2y + 4,$

(b) $\quad \dot{x} = x + e^{x-y-1}, \qquad \dot{y} = 12 + 2x + 5y.$

Exercise 18 Find and classify the critical points of the system

$$\dot{x} = (a - x^2)y \quad (a \text{ constant}), \qquad \dot{y} = x - y$$

in each of the cases

(i) $a < -\tfrac{1}{4}$,

(ii) $-\tfrac{1}{4} < a < 0$,

(iii) $a > 0$.

Exercise 19 (a) Show that the system

$$\dot{x} = -y^3, \qquad \dot{y} = x$$

has a centre at the origin, but its linear approximation does not.

(b) Show that the system

$$\dot{x} = y - x(x^2 + y^2), \qquad \dot{y} = -x - y(x^2 + y^2)$$

has no closed phase-paths, but its linear approximation has a centre at the origin.

Appendix: the solution of some elementary ordinary differential equations

This appendix contains, in outline, some basic techniques which can be used in the solution of first- and second-order ordinary differential equations.

(1) $y' + p(x)y = f(x)$, where p and f are continuous

Multiply the equation through by the 'integrating factor'

$$\exp\left(\int^x p(t)\, dt\right).$$

Then the equation becomes

$$\frac{d}{dx}\left(y \cdot \exp\left(\int^x p(t)\, dt\right)\right) = f(x) \cdot \exp\left(\int^x p(t)\, dt\right)$$

and the solution is found by one integration.

Example 1 Solve $xy' + 2y = 4x^2$ for $x > 0$, where $y(1) = 2$.

The equation may be re-written

$$y' + \frac{2}{x}y = 4x.$$

With $p(x) = \dfrac{2}{x}$ (continuous as $x > 0$),

$$\exp\left(\int^x p(t)\, dt\right) \;=\; \exp\left(\int^x \frac{2}{t}\, dt\right) \;=\; e^{2\log x} \;=\; x^2.$$

So, multiplying through by this integrating factor,

$$x^2 y' + 2xy \;=\; \frac{d}{dx}\left(x^2 y\right) \;=\; 4x^3.$$

Thus,

$$x^2 y \;=\; x^4 + c \quad \text{or} \quad y \;=\; x^2 + \frac{c}{x^2},$$

where c is a constant. Using $y(1) = 2$, we deduce that $c = 1$ and hence that the solution to the problem (for $x > 0$) is

$$y \;=\; x^2 + \frac{1}{x^2}. \qquad\qquad \square$$

Exercise 1 Solve $xy' + y = e^x$ for $x > 0$, where $y(1) = 1$.

Exercise 2 Solve $y' + (\cot x)y = 2\operatorname{cosec} x$ when $0 < x < \pi$, where $y(\pi/2) = 1$.

(2) $\boldsymbol{f(x) + g(y)\dfrac{dy}{dx} \;=\; 0}$, where $\boldsymbol{f}$ and $\boldsymbol{g}$ are continuous

Such 'separable' equations are solved by re-writing the equation as

$$f(x)dx \;=\; -g(y)dy$$

and then performing one integration.

Example 2 Solve

$$\frac{dy}{dx} \;=\; \frac{3x^2 + 4x + 2}{2(y - 1)} \quad \text{for } x > -2, \text{ where } y(0) = -1.$$

'Separating' the variables gives

$$2(y - 1)dy \;=\; (3x^2 + 4x + 2)dx.$$

Integrating,

$$y^2 - 2y \;=\; x^3 + 2x^2 + 2x + c,$$

where c is a constant. Using $y(0) = -1$, we obtain $c = 3$, giving an 'implicit' solution. Solving this quadratic in y:

$$y = 1 \pm \sqrt{x^3 + 2x^2 + 2x + 4}.$$

The negative sign *must* be taken in order that $y(0) = -1$. For $x > -2$, the square root is real (exercise) and $y \neq 1$ (see original equation for the point of this remark). □

Exercise 3 Solve $\dfrac{dr}{d\theta} = \dfrac{r^2}{\theta}$ for $0 < \theta < e^{\frac{1}{2}}$, where $r(1) = 2$.

Exercise 4 Solve $y' = xy^3(1 + x^2)^{-\frac{1}{2}}$ for $|x| < \sqrt{5}/2$, where $y(0) = 1$.

(3) $\quad A(x, y) + B(x, y)\dfrac{dy}{dx} = 0$, where A, B, A_y, B_x are continuous

$\qquad$ and $A_y = B_x$ ('exact' equations)

The condition $A_y = B_x$ implies the existence of a function $\psi = \psi(x, y)$ for which

$$\psi_x = A, \qquad \psi_y = B;$$

so that the differential equation becomes

$$\psi_x dx + \psi_y dy = 0,$$

with general solution

$$\psi(x, y) = c,$$

where c is an arbitrary constant.

To determine ψ, one can work as in the following example.

Example 3 Solve $(y \cos x + 2xe^y) + (\sin x + x^2 e^y - 1)y' = 0$.

$\quad$ Putting

$$A(x, y) = y \cos x + 2xe^y \quad \text{and} \quad B(x, y) = \sin x + x^2 e^y - 1,$$

it is easy to see that

$$A_y = \cos x + 2xe^y = B_x;$$

so, the method applies. We need to find $\psi(x, y)$ such that

$$\psi_x = A = y \cos x + 2xe^y,$$

$$\psi_y = B = \sin x + x^2 e^y - 1.$$

Integrate the first of these two equations with respect to x, to give

$$\psi(x, y) \;=\; y \sin x + x^2 e^y + f(y),$$

where f is still to be determined. Differentiating this with respect to y and equating it to the right-hand side of the second of the two equations, we have

$$\sin x + x^2 e^y + f'(y) \;=\; \sin x + x^2 e^y - 1;$$

so that $f'(y) = -1$ and $f(y) = -y + \text{constant}$. Thus, the desired solution (an 'implicit' solution) is

$$y \sin x + x^2 e^y - y \;=\; c,$$

where c is an arbitrary constant.   □

The method here relies on the condition $A_y = B_x$. Sometimes this does not hold, but there exists an 'integrating factor' $\lambda = \lambda(x, y)$ for which λ_x, λ_y are continuous and

$$(\lambda A)_y \;=\; (\lambda B)_x$$

that is, which satisfies the first order partial differential equation

$$B\lambda_x - A\lambda_y + (B_x - A_y)\lambda \;=\; 0.$$

(Of course, this reduces to an ordinary differential equation if λ is a function of x alone, or of y alone.) Then our method can be applied to the original equation, re-written in the form

$$\lambda(x, y)A(x, y) + \lambda(x, y)B(x, y)\frac{dy}{dx} \;=\; 0.$$

Exercise 5 Solve $(2xy^2 + 2y) + (2x^2 y + 2x)y' = 0$.

Exercise 6 Solve $(ye^{xy} \cos 2x - 2e^{xy} \sin 2x + 2x) + (xe^{xy} \cos 2x - 3)\dfrac{dy}{dx} = 0$.

(4) $\quad \dfrac{dy}{dx} \;=\; f\left(\dfrac{y}{x}\right)$, **where f is continuous**

Put $v = \dfrac{y}{x}$, that is $y = vx$, and transform the equation to an equation in v and x by noticing that

$$\frac{dy}{dx} \;=\; v + x\frac{dv}{dx}.$$

The resulting equation is separable in v and x and the methods of (2) above apply.

Example 4 Solve

$$\frac{dy}{dx} = \frac{y^2 + 2xy}{x^2}, \qquad (x \neq 0).$$

The equation may be re-written

$$\frac{dy}{dx} = \left(\frac{y}{x}\right)^2 + 2\left(\frac{y}{x}\right).$$

The substitution $y = vx$ therefore yields

$$v + x\frac{dv}{dx} = v^2 + 2v$$

or

$$\frac{dx}{x} = \frac{dv}{v(v+1)} \qquad \text{(two steps)}.$$

Using partial fractions,

$$\frac{dx}{x} = \left(\frac{1}{v} - \frac{1}{v+1}\right) dv.$$

Integrating yields

$$\log|x| + \log|c| = \log|v| - \log|v+1|, \qquad x, c, v, v+1 \neq 0,$$

with c otherwise arbitrary. As $v/(x(v+1))$ is continuous and cannot change sign, we must have, when $x + y \neq 0$ and k is either c or $-c$, that

$$kx = \frac{v}{v+1} = \frac{y/x}{(y/x)+1} = \frac{y}{y+x}.$$

So,

$$y = \frac{kx^2}{1-kx}$$

which in fact solves the original equation when $x \neq 0$ and $x \neq k^{-1}$ (this can easily be checked). $\square$

Exercise 7 Solve $(3xy + y^2) + (x^2 + xy)\frac{dy}{dx} = 0.$

Exercise 8 Solve $\frac{dy}{dx} = \frac{x^2 + xy + y^2}{x^2}.$

(5)   **The 'linear homogeneous equation'** $y'' + p(x)y' + q(x)y = 0$,
     where p and q are continuous and one solution of the
     differential equation is known, 'the method of reduction of order'

Suppose that the known solution is $y = u(x)$, assumed never zero on its domain. We put $y = uv$ (where $v = v(x)$). Assuming v'' exists,

$$y' = u'v + uv',$$

$$y'' = u'' + 2u'v' + uv''.$$

Substituting for y, y' and y'' in the differential equation, we obtain

$$uv'' + (2u' + pu)v' + (u'' + pu' + qu)v = 0.$$

But we know that u solves that equation; so, $u'' + pu' + qu = 0$. Putting $V = v'$, we arrive at a first-order differential equation in V:

$$uV' + (2u' + pu)V = 0.$$

Where u is never zero, we may simplify the integrating factor

$$\exp\left(\int^x \left(2\frac{u'}{u} + p\right)\right) = \exp(2\log|u(x)|).\exp(\int^x p),$$

so that, using (1) above and integrating once, we have

$$V(x).(u(x))^2.\exp(\int^x p) = A,$$

where A is a constant, and hence

$$v'(x) = V(x) = \frac{A}{(u(x))^2} e^{-\int^x p(t)\,dt} \quad (A \text{ constant}).$$

One further integration gives $v(x)$ and the general solution

$$y(x) = u(x).v(x) = u(x)\left\{A\int^x \left(e^{-\int^s p(t)\,dt}\right)\frac{ds}{(u(s))^2} + B\right\}.$$

The method of variation of parameters, which is described in section 4.1, bears some resemblances to the method here.

Example 5 Given that $u(x) = x^{-1}$ is a solution of

$$2x^2 y'' + 3xy' - y = 0, \quad \text{for } x > 0,$$

find the general solution.

As the differential equation may be written

$$y'' + \frac{3}{2x} y' - \frac{1}{2x^2} y = 0, \qquad (x > 0)$$

and since, therefore, $p = \frac{3}{2} x^{-1}$,

$$v'(x) = V(x) = cx^2 e^{-\int^x (3/2t)\, dt} = cx^{\frac{1}{2}}$$

and hence

$$v(x) = \frac{2cx^{\frac{3}{2}}}{3} + k,$$

where c and k are arbitrary real constants. Therefore, the complete solution is

$$y(x) = u(x) \cdot v(x) = \frac{2cx^{\frac{1}{2}}}{3} + \frac{k}{x}, \qquad (x > 0). \qquad \square$$

Exercise 9 Solve $x^2 y'' - x(x+2)y' + (x+2)y = 0$ for $x > 0$, given that $u(x) = x$ is a solution.

Exercise 10 Solve $xy'' - y' + 4x^3 y = 0$ for $x > 0$, given that $u(x) = \sin(x^2)$ is a solution.

(6)   **The 'linear homogeneous equation with constant coefficients',**
   $ay'' + by' + cy = 0$, **where** a, b, c **are real constants and** $a \neq 0$

In order for $y = e^{\lambda x}$ (λ constant) to be a solution, λ must satisfy the *auxiliary* (or *characteristic*) *equation*

(*) $a\lambda^2 + b\lambda + c = 0$

(substituting and cancelling the positive quantity $e^{\lambda x}$). Three cases occur in searching for real solutions.

(i) If $b^2 > 4ac$ and λ_1 and λ_2 are the (real) solutions of (*), then

$$y \;=\; Ae^{\lambda_1 x} + Be^{\lambda_2 x} \qquad (A, B \text{ arbitrary real constants})$$

is a real solution of the differential equation.

(ii) If $b^2 = 4ac$ and λ is the (only and real) solution of (*), then

$$y \;=\; (Ax + B)e^{\lambda x} \qquad (A, B \text{ arbitrary real constants})$$

is a real solution of the differential equation.

(iii) If $b^2 < 4ac$ and $\lambda \pm i\mu$ are the (complex conjugate) roots of (*), then

$$y \;=\; e^{\lambda x}(A \cos \mu x + B \sin \mu x) \qquad (A, B \text{ arbitrary real constants})$$

is a real solution of the differential equation.

Example 6 (i) Find a solution of $y'' + 5y' + 6y = 0$, satisfying the 'initial conditions' $y(0) = 0$, $y'(0) = 1$.

The auxiliary equation

$$\lambda^2 + 5\lambda + 6 \;=\; 0$$

has roots $\lambda = -2, -3$. So, a solution of the differential equation is

$$y \;=\; Ae^{-2x} + Be^{-3x} \qquad (A, B \text{ arbitrary real constants}).$$

To satisfy the conditions, one needs

$$A + B \;=\; 0, \qquad -2A - 3B \;=\; 1,$$

giving $A = 1 = -B$. So, the required solution is

$$y \;=\; e^{-2x} - e^{-3x}.$$

(ii) Find a real solution of $y'' + 4y' + 4y = 0$.

The auxiliary equation this time is

$$\lambda^2 + 4\lambda + 4 \;=\; 0,$$

with (repeated) root $\lambda = -2$. So, a solution of the differential equation is

$$y \;=\; (Ax + B)e^{-2x} \qquad (A, B \text{ arbitrary real constants}).$$

(iii) Find a real solution of $y'' + y' + y = 0$.

The auxiliary equation here is

$$\lambda^2 + \lambda + 1 = 0,$$

with roots

$$\lambda = -\frac{1}{2} \pm i\frac{\sqrt{3}}{2}.$$

So, the required solution is

$$y = e^{-\frac{x}{2}}\left(A\cos\left(\frac{\sqrt{3}}{2}x\right) + B\sin\left(\frac{\sqrt{3}}{2}x\right)\right), \quad (A, B \text{ arbitrary real constants}). \quad \square$$

Exercise 11 Solve $2y'' - 3y' + y = 0$.

Exercise 12 Solve $y'' - 6y' + 9y = 0$, where $y(0) = 0$, $y'(0) = 2$.

Exercise 13 Solve $y'' - 2y' + 6y = 0$.

(7) The equation $ax^2y'' + bxy' + cy = 0$, where a, b, c are real constants and $a \neq 0$

One may reduce this equation to the case (6) of constant coefficients by means of the substitution $x = e^t$, giving

$$\frac{d}{dt} = x\frac{d}{dx}, \qquad \frac{d^2}{dt^2} = x^2\frac{d^2}{dx^2} + x\frac{d}{dx}$$

and hence

$$a\ddot{y} + (b - a)\dot{y} + cy = 0,$$

where dot denotes differentiation with respect to t. This equation may be solved by the methods of (6) above. The (composite) direct method is to use the substitution $y = x^\lambda$ directly. Notice that the general solution is

$$y = (A\log x + B)x^\lambda$$

when the auxillary equation has a double root λ.

Note The method is applicable for $x > 0$, as e^t is always positive for t real, and $x^\lambda = \exp(\lambda \log x)$ is a well-defined real number when $x > 0$.

Example 7 Solve $x^2 y'' + xy' - k^2 y = 0$, where the variable x and the constant k are strictly positive.

The substitution $y = x^\lambda$ gives

$$\lambda(\lambda - 1) + \lambda - k^2 \;=\; 0$$

and hence $\lambda = \pm k$. The general solution of the differential equation is therefore

$$y \;=\; Ax^k + Bx^{-k},$$

for $x > 0$, where A and B are constants.   □

Exercise 14 Do the same example by first making the intermediate substitution $x = e^t$.

Exercise 15 Solve $x^2 y'' - 2xy' + 2y = 0$.

Exercise 16 Solve $(1 + 2x)^2 y'' - 6(1 + 2x)y' + 16y = 0$.

(8) **The 'non-homogeneous' (or 'inhomogeneous') linear equation,**
   $$y'' + p(x)y' + q(x)y \;=\; f(x), \text{ where } p, q \text{ and } f \text{ are continuous functions}$$

Many trial and error and *ad hoc* methods are available. It will be a main aim of these notes to introduce systematic methods for some important classes of such equations. But do notice that, if $y = u_1(x)$ and $y = u_2(x)$ are 'independent' solutions of the corresponding homogeneous equation

$$y'' + p(x)y' + q(x)y \;=\; 0,$$

and if $y = v(x)$ is a solution of the given inhomogeneous equation, then

$$y \;=\; Au_1(x) + Bu_2(x) + v(x) \quad (A, B \text{ arbitrary constants})$$

is also a solution of the inhomogeneous equation. In these circumstances, $Au_1(x) + Bu_2(x)$ is called the *complementary function* and $v(x)$ a *particular integral*.

Bibliography

Because differential equations lie at the very heart of mathematics, there is a vast number of books which might be of interest to the reader, be it by way of background, collateral or further reading. The list below therefore represents only a cross-section of available recommendations. Some are cited because they contain just the right background material at just the right level, some because they hold a rightful place amongst classic mathematics texts, some because they follow up issues to which we have, for reasons of space, only paid cursory attention, and some because they will lead the reader much deeper into the various topics we have discussed.

In the following, the numbers in square brackets relate to the list of references given at the foot of this section.

Background analysis

Burkill [11] and Bartle and Sherbert [7] provide basic real analysis, whereas Apostol [3] and Rudin [43] develop the subject further. Jordan and Smith [28] and Kreyszig [31] give a variety of techniques, including the solution of some elementary differential equations (covered more briefly in the Appendix). Piaggio [37] is a classic older text, giving a number of useful techniques for solving such equations.

Background algebra

Strang [46] and Stoll [45] present the material we need on simultaneous linear equations and finite-dimensional spaces, whereas Artin [5] provides the background for consideration of infinite orthonormal sequences of vectors.

Ordinary differential equations

Birkhoff and Rota [8], Boyce and DiPrima [9], Burkill [12] and Hurewicz [25] provide, in their different ways and at different levels of mathematical sophistication, an overlap with some topics in this book and, in some respects, a complementary view. Coddington and Levinson [16] and Hirsch, Smale and Devaney [24] are more advanced; the approach of the first being more traditional, of the second more modern. Ince [26] contains a wealth of technique gathered together in the first half of the twentieth century.

Partial differential equations

Carrier and Pearson [14], Strauss [47] and Logan [32] give different treatments of some of the material in Chapters 5–7, the third being sometimes geometric where this text is analytic. Epstein [21] will interest the reader more concerned with mathematical rigour than practical application. Ockendon, Howison, Lacey and Movchan [35] and Renardy and Rogers [41] provide ample opportunities for further reading.

Fourier series

Background reading, as well as a greater selection of applications of the mathematics of Chapter 7, are provided in Brown and Churchill [10] and Seeley [44]. For information on the mathematical treatment of waves, the reader can consult Baldock and Bridgeman [6] and Coulson and Jeffrey [17].

Integral equations

At the time of writing, not many undergraduate level texts provide a good range of material on integral equations. Appropriate examples are Hildebrand [23], Jerri [27] and Pipkin [38].

Calculus of Variations

Arfken and Weber [4] cover a range of elementary material with applications, whereas Pars [36] and Courant and Hilbert [18] go far deeper into the classical theory of variational integrals. For background geometry for Chapter 11, we refer the reader to Roe [42] and (more advanced) do Carmo [19]; for mechanics, to Acheson [1], Lunn [33] and Woodhouse [48].

The Sturm–Liouville equation

Courant and Hilbert [18] and Haberman [22] give background material relating to physics.

Solutions in series

Burkill [12] is terse and theoretical where Boyce and DiPrima [9] is expansive and practical. Piaggio [37] gives advice and examples on the use of Frobenius's method. Ahlfors [2] has background on Riemann surfaces and their relationship to equations in a single complex variable.

   More general background reading on complex analysis, useful for sections 13.8 and 14.5, may be found in Priestley [39], Marsden [34] and Carrier, Crook and Pearson [13].

Fourier and Laplace transform

For background material on the Lebesgue integral approach, see Priestley [40]. For methods involving the Riemann integral, reference can be made to the first edition of Apostol [3]. The reader will find plentiful technique, with particular reference to physical problems, in Carslaw and Jaeger [15]. For the use of complex variable, see above (under *Solutions in series*).

Phase-plane analysis

Much information may be found in Jordan and Smith [29] and King, Billingham and Otto [30].

Further reading

Courant and Hilbert [18] is one of the most famous texts in mathematics and, although most suitable as a graduate reference, contains a wealth of material on many aspects of differential and integral equations.

[1] D. Acheson, *From Calculus to Chaos*, Oxford University Press, Oxford, 1997.

[2] L.V. Ahlfors, *Complex Analysis*, 3rd Edition, McGraw–Hill, New York, 1979.

[3] T.M. Apostol, *Mathematical Analysis*, 1st Edition and 2nd Edition, Addison–Wesley, Reading, Massachusetts, 1957 and 1975.

[4] G.B. Arfken and H.J. Weber, *Mathematical Methods for Physicists*, 6th Edition, Academic Press, New York, 2005.

[5] M. Artin, *Algebra*, Prentice Hall, Englewood Cliffs, 1991.

[6] G.R. Baldock and T. Bridgeman, *The Mathematical Theory of Wave Motion*, Ellis Horwood, Chichester, 1981.

[7] R.G. Bartle and D.R. Sherbert, *Introduction to Real Analysis*, 3rd Edition, Wiley, New York, 1999.

[8] G. Birkhoff and G.-C. Rota, *Ordinary Differential Equations*, 4th Edition, Wiley, New York, 1989.

[9] W.E. Boyce and R.C. DiPrima, *Elementary Differential Equations*, 8th Edition, Wiley, New York, 2005.

[10] J.W. Brown and R.V. Churchill, *Fourier Series and Boundary Value Problems*, 6th Edition, McGraw–Hill, New York, 2001.

[11] J.C. Burkill, *A First Course in Mathematical Analysis*, Cambridge University Press, Cambridge, 1978.

[12] J.C. Burkill, *The Theory of Ordinary Differential Equations*, Oliver and Boyd, Edinburgh, 1956.

[13] G.F. Carrier, M. Crook and C.E. Pearson, *Functions of a Complex Variable*, McGraw–Hill, New York, 1966.

[14] C.F. Carrier and C.E. Pearson, *Partial Differential Equations*, 2nd Edition, Academic Press, Boston, 1988.

[15] H.S. Carslaw and J.C. Jaeger, *Operational Methods in Applied Mathematics*, Dover, New York, 1963.

[16] E.A. Coddington and M. Levinson, *Theory of Ordinary Differential Equations*, McGraw–Hill, New York, 1955.

[17] C.A. Coulson and A. Jeffrey, *Waves*, 2nd Edition, Longman, London, 1977.

[18] R. Courant and D. Hilbert, *Methods of Mathematical Physics*, Volumes I and II, Interscience, New York, 1953 and 1962.

[19] M.P. do Carmo, *Differential Geometry of Curves and Surfaces*, Prentice Hall, Englewood Cliffs, 1976.

[20] H. Dym and H.P. McKean, *Fourier Series and Integrals*, Academic Press, New York, 1992.

[21] B. Epstein, *Partial Differential Equations*, McGraw–Hill, New York, 1962.

[22] R. Haberman, *Elementary Applied Partial Differential Equations*, 4th Edition, Prentice Hall, Englewood Cliffs, 2003.

[23] F.B. Hildebrand, *Methods of Applied Mathematics*, 2nd Edition, Prentice Hall, Englewood Cliffs, 1965, reprinted Dover, New York, 1992.

[24] M.W. Hirsch, S. Smale and R.L. Devaney, *Differential Equations, Dynamical Systems, and an Introduction to Chaos*, 2nd Edition, Academic Press, New York, 2004.

[25] W. Hurewicz, *Lectures on Ordinary Differential Equations*, Dover, New York, 1990.

[26] E.L. Ince, *Ordinary Differential Equations*, Dover, New York, 1956.

[27] A. Jerri, *Introduction to Integral Equations and Applications*, 2nd Edition, Wiley, New York, 1999.

[28] D.W. Jordan and P. Smith, *Mathematical Techniques*, 3rd Edition, Oxford University Press, Oxford, 2002.

[29] D.W. Jordan and P. Smith, *Nonlinear Ordinary Differential Equations*, 3rd Edition, Oxford University Press, Oxford, 1999.

[30] A.C. King, J. Billingham and S.R. Otto, *Differential Equations*, Cambridge University Press, Cambridge, 2003.

[31] E. Kreyszig, *Advanced Engineering Mathematics*, 9th Edition, Wiley, New York, 2005.

[32] J.D. Logan, *Introduction to Nonlinear Partial Differential Equations*, Wiley, New York, 1994.

[33] M. Lunn, *A First Course in Mechanics*, Oxford University Press, Oxford, 1991.

[34] J.E. Marsden, *Basic Complex Analysis*, Freeman, New York, 1973.

[35] J. Ockendon, S. Howison, A. Lacey and A. Movchan, *Applied Partial Differential Equations*, Revised Edition, Oxford University Press, Oxford, 2003.

[36] L.A. Pars, *Introduction to the Calculus of Variations*, Heinemann/Wiley, London, 1962.

[37] H.T.H. Piaggio, *Differential Equations*, Bell, Revised Edition, London, 1952.

[38] A.C. Pipkin, *A Course on Integral Equations*, Springer–Verlag, New York, 1991.

[39] H.A. Priestley, *Introduction to Complex Analysis*, 2nd Edition, Oxford University Press, Oxford, 2003.

[40] H.A. Priestley, *Introduction to Integration*, Oxford University Press, Oxford, 1997

[41] M. Renardy and R.C. Rogers, *An Introduction to Partial Differential Equations*, Springer–Verlag, New York, 1993.

[42] J. Roe, *Elementary Geometry*, Oxford University Press, Oxford, 1993.

[43] W. Rudin, *Principles of Mathematical Analysis*, 3rd Edition, McGraw–Hill, New York, 1976.

[44] R. Seeley, *An Introduction to Fourier Series and Integrals*, Benjamin, New York, 1966.

[45] R.R. Stoll, *Linear Algebra and Matrix Theory*, McGraw–Hill, New York, 1952.

[46] G. Strang, *Introduction to Linear Algebra*, 2nd Edition, Wellesley–Cambridge, Wellesley, Massachusetts, 1998.

[47] W.A. Strauss, *Partial Differential Equations: an Introduction*, Wiley, New York, 1992.

[48] N.M.J. Woodhouse, *Introduction to Analytical Dynamics*, Oxford University Press, Oxford, 1987.

Index